Spectroscopic Properties of Inorganic and Organometallic Compounds

Techniques, Materials and Applications

Volume 45

Specialist Periodical Reports Online

Application for Free Access to Electronic Chapters

Customers purchasing a print volume are now entitled to free site-wide access to the electronic version of that title. (For the definition of a site, please consult **www.rsc.org/subagree**)

To apply for free access, please contact **ejournals@rsc.org** with proof of purchase (a scanned copy is acceptable) containing details of:

- **Title of book**
- **Volume number**
- **ISBN**
- **Address of your organisation**

By providing these details you will be agreeing to the terms of the Electronic Information Licence Agreement at

www.rsc.org/subagree

ROYAL SOCIETY OF **CHEMISTRY**

A Specialist Periodical Report

Spectroscopic Properties of Inorganic and Organometallic Compounds

Techniques, Materials and Applications
Volume 45

A review of the recent literature

Editors
J. Yarwood, *Sheffield Hallam University, UK*
R. Douthwaite, *University of York, UK*
S. B. Duckett, *University of York, UK*

Authors
Jose Antonio Carrero, *University of Bilabo, Bilabo, Spain*
Gorka Arana, *University of Bilabo, Bilabo, Spain*
Nicolas Bélanger-Desmarais, *Université de Montréal, Montréal, Canada*
Kanishka Biswas, *Jawaharlal Nehru Centre for Advanced Scientific Research, Bangalore, India*
Xavier Bourrat, *BRGM and Université d'Orléan, Orleans, France*
Kristina Chakarova, *Bulgarian Academy of Sciences, Sofia, Bulgaria*
Sayandev Chatterjee, *University of Cincinnati, Cincinnati OH, USA*
William B. Connick, *University of Cincinnati, Cincinnati OH, USA*
Keith Silvio Decurtins, *Universität Bern, Bern, Switzerland*
B. Dillon, *University of Durham, Durham, UK*
Anastasia B. S. Elliott, *University of Otago, Dunedin, New Zealand*
Peter Flewitt, *University of Bristol, Bristol, UK*
Patrick Franz, *Universität Bern, Bern, Switzerland*
Keith C. Gordon, *University of Otago, Dunedin, New Zealand*
Regis Guegan, *ISTO and Université d'Orléans*
Konstantin Hadjiivanov, *Bulgarian Academy of Sciences, Sofia, Bulgaria*
Christian W. Huck, *Center for Chemistry and Biomedicine, Innsbruck, Austria*
Elena Ivanova, *Bulgarian Academy of Sciences, Sofia, Bulgaria*
Igor V. Koptyug, *International Tomography Center and Novosibirsk State University, Novosibirsk, Russia*
Abdeltif Lahfid, *BRGM and Université d'Orléan, Orleans, France*
Dong Liu, *University of Bristol, Bristol, UK*
Juan Manuel Madariaga, *University of Bilabo, Bilabo, Spain*
Nicolas Maubec, *BRGM and Université d'Orléan, Orleans, France*
Mihail Mihaylov, *Bulgarian Academy of Sciences, Sofia, Bulgaria*

Dimitar Panayotov, *Bulgarian Academy of Sciences, Sofia, Bulgaria*
Stéphanie Poirier, *Université de Montréal, Montréal, Canada*
Christian Reber, *Université de Montréal, Montréal, Canada*
Camille Sonneville, *Université de Montréal, Montréal, Canada*
Holly van der Salm, *University of Otago, Dunedin, New Zealand*
Guillaume Wille, *BRGM and Université d'Orléan, Orleans, France*

ROYAL SOCIETY OF **CHEMISTRY**

THE QUEEN'S AWARDS FOR ENTERPRISE: INTERNATIONAL TRADE 2013

Print ISBN: 978-1-84973-919-1
PDF eISBN: 978-1-78262-148-5
ISSN: 0584-8555

A catalogue record for this book is available from the British Library

Published by The Royal Society of Chemistry,
Thomas Graham House, Science Park, Milton Road,
Cambridge CB4 0WF, UK

Registered Charity Number 207890

For further information see our web site at www.rsc.org

Preface

Jack Yarwood,[a] Richard Douthwaite[b] and Simon Duckett[c]
DOI: 10.1039/9781782621485-FP005

In keeping with the changes introduced in 2009 to this volume of the series, the current edition encompasses predominantly critical or tutorial reviews from a wide spectrum of inorganic materials science and spectroscopy. The reader should be able to find unfamiliar material described in an accessible style, providing the critical context and important recent work in the field.

In this volume the articles encompass the application of spectroscopic methods from fundamental characterisation of novel complexes and compounds to 'real world' detection of environmental pollution. Continuing the trend of recent additions, combinations of spectroscopic and structural methods are increasingly common providing insight not possible using individual techniques sequentially.

Collectively the articles represent a wide cross section of activity in the application of spectroscopy to inorganic chemistry, however there are clearly common themes to the spectroscopy applied to a diverse range of materials. We hope the reader will at least learn something new or find an alternative perspective to a familiar topic and perhaps extract a little inspiration.

The editors are of course indebted to all our authors for agreeing to contribute and submitting their articles in time for publication. The editing for the majority of the articles was uneventful and of course we take responsibility for any errors and apologise in advance. We also thank Dr Merlin Fox and Mrs Alice Toby-Brant of the RSC for sending us timely reminders and keeping us on track during the commissioning and editing processes.

[a]*Materials Research Institute, City Campus, Pond Street, Sheffield S1 1WB. E-mail: J.Yarwood@shu.ac.uk*
[b]*Department of Chemistry, University of York, York YO10 5DD, UK. E-mail: richard.douthwaite@york.ac.uk*
[c]*Department of Chemistry, University of York, York YO10 5DD, UK. E-mail: simon.duckett@york.ac.uk*

CONTENTS

Cover

The cover shows the molecular structure of a molybdenum dioxocomplex with thioester functionalised amido ligands superimposed on a spectral background. Adapted from an image created by Nadia C. Mösch-Zanetti and reproduced with permission.

Magnetic resonance imaging methods in heterogeneous catalysis

Igor V. Koptyug[a,b]
DOI: 10.1039/9781782621485-00001

Applications of spatially resolved magnetic resonance in heterogeneous catalysis and related fields are considered. The chapter starts with a simple description of the basic principles of MRI and the discussion of the specific features which make MRI a powerful and versatile toolkit capable of providing useful and diverse information about catalysts, reactors and processes within them in a non-invasive manner. Next, practical aspects of constructing an MRI-compatible reactor are presented along with the methods for, and examples of, the structural MRI studies of packed beds, model reactors and related geometries. The basic principles of mass transport studies with NMR and MRI are considered next, and the literature examples of MRI studies of mass transport in model systems are briefly outlined. The rest of the chapter is devoted to the analysis of the studies of model catalytic reactors under operating conditions, and includes MRI studies of distribution and mass transport of fluids, spatially resolved spectroscopic studies of conversion, MRI thermometry of operating catalytic reactors and microreactors, and the use of the emerging techniques for nuclear spin hyperpolarization to boost the sensitivity of NMR and MRI in catalytic applications.

1 Introduction

Nuclear magnetic resonance imaging, abridged to "MRI" to stress its harmless nature, has become one of the most powerful instruments in modern medical diagnostics. In fact, MRI has revolutionized modern medicine by enabling physicians to literally see the state of internal organs in a human body and various processes taking place within it. This ability, coupled with the non-invasive nature of the technique, made it possible to abandon the "black box" approach, in which the diagnosis is often based on superficial observations and a limited number of symptoms which are often similar in many diseased states. The success of medial MRI might seem surprising given that the technique has modest spatial resolution and a number of significant limitations as compared to other modern imaging techniques, *e.g.* computer assisted (X-ray) tomography (CAT), positron emission tomography (PET), *etc.* However, the key feature of MRI is that it is best characterized as a versatile toolkit,[1] in contrast to many other techniques which are powerful but specialized tools. In addition to morphological studies, the medical MRI toolkit contains tools for angiography, thermometry, spectroscopy, elastography, functional MRI, and a lot more. The foundation for this tremendous diversity is the versatile nature of image contrast in MRI. The latter is sensitive to a wide range of properties of an

[a]*International Tomography Center, SB RAS, 3A Institutskaya St., Novosibirsk 630090, Russia*
[b]*Novosibirsk State University, 2 Pirogova St., Novosibirsk 630090, Russia. E-mail: koptyug@tomo.nsc.ru*

object under study and processes taking place within it. Furthermore, image contrast in MRI can be deliberately tailored to the needs of a particular study. As a result, MRI is able to provide a lot more than just structural information. It is this ability which makes MRI so powerful a technique.

Unlike modern medicine, chemical and process engineering practice still largely relies on the "black box" approach, trying to figure out what is happening inside an operating reactor on the basis of a number of external measurements (pressure drop, temperature, chemical composition of the feed and reactor output, transient response curves, *etc.*). A number of modern imaging techniques are being developed to overcome the existing limitations, but so far our ability to see the inner works of one of the most sophisticated bioreactors – human body – by far exceeds our ability to visualize processes inside, *e.g.*, a packed bed reactor. In fact, biomedical MRI has reached the stage when metabolic processes taking place in the cells of various tissues and organs can be interrogated in live animals,[2] and this possibility is currently being extended to in-human studies.

With all this progress in biomedical MRI applications, it might seem surprising that MRI has not become a routine technique to "diagnose" the behavior of various chemical reactors. One of the obstacles on this way is the feature that makes MRI such a powerful technique – the diversity of image contrast mechanisms, *i.e.*, the sensitivity of the detected signal to a wide range of object properties. As a result, many MRI strategies developed in medical MRI to a state of perfection perform unsatisfactorily when applied to non-biological objects. At present, non-biomedical applications of MRI are still an art rather than routine studies despite the fact that the interest in such studies is clearly on the rise, including applications to problems related to catalysis.

2 The MRI technique

NMR in general and MRI in particular explore the interaction of nuclear spins with static and oscillating magnetic fields. In NMR spectroscopy, one acquires NMR spectra which characterize local magnetic environments of nuclei possessing a non-zero spin (*e.g.*, ^{1}H, ^{13}C, ^{19}F, ^{31}P) and thereby provide information on chemical composition of a sample, its molecular structure and dynamic transformations, *etc.* In an NMR spectrum, the position of a particular resonance (the resonance frequency ω) is proportional to the applied static magnetic field B_0 and the magnetogyric ratio γ of the magnetic nucleus under consideration ($\omega \propto \gamma B_0$). This implies that if, for instance, a container with water is placed in a spatially uniform magnetic field (Fig. 1a), a single NMR resonance will be observed in the NMR spectrum (Fig. 1b). Indeed, in NMR spectroscopy, high magnetic field homogeneity is essential to obtain high spectral resolution. In contrast, in MRI the magnetic field is intentionally made to depend on a spatial coordinate, often in a linear fashion (Fig. 1c). In this example, different volumes of water reside in different static magnetic fields and therefore produce NMR resonances at different

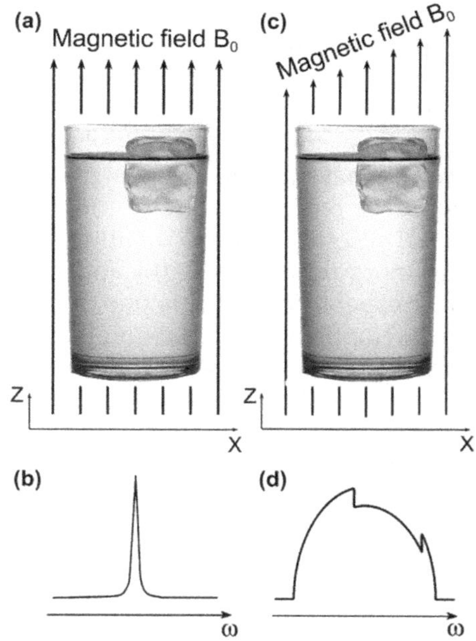

Fig. 1 A glass of water with an ice cube in a homogeneous magnetic field without (a) and with a superimposed magnetic field gradient along the x coordinate axis (c), and the expected ^{1}H NMR spectra (b,d). Usually, only liquid contributes to the detected signal. The 'spectrum' in (d) is essentially an integral projection of the sample on the x axis.

resonance frequencies. The resulting "spectrum" is essentially a one-dimensional integral projection of the sample on the direction of the gradient (Fig. 1d). If three orthogonal gradients are available, they can be combined to produce a resulting gradient with an arbitrary orientation in space, rendering an MRI measurement free from any orientational preferences and making possible the acquisition of 2D and 3D images. Gradients are often applied as pulses and are incorporated in an NMR pulse sequence along with radiofrequency pulses and delays. The borderline between MRI and non-imaging NMR applications is rather vague. Indeed, many modern magnetic resonance (MR) experiments combine spatial and spectral coordinates and are best described as experiments in a multidimensional (sub)space of spatial, spectral and temporal coordinates rather than MR spectroscopy or imaging.

NMR/MRI experiments require that a sample under study contains a large number of magnetic nuclei. Their nuclear spins serve as intrinsic probes or tracers which can convey the information about their spatial position, local environments and mobility of the groups of atoms and molecules they belong to, to the NMR detector which is usually placed externally with respect to the object under study. Therefore, an MRI experiment is non-invasive and non-destructive, does not require the introduction of any other tracers or probes into the object under study, and allows one to monitor various dynamic processes *in situ*, without a need to periodically interrupt them to make a measurement.

Modern NMR spectroscopy is performed with virtually all nuclei possessing a non-zero spin. In principle, an image can also be obtained using any nucleus with a non-zero spin. In practice, however, the majority of the MRI studies are performed using ^{1}H NMR signal detection, with a much smaller number of studies reported where other nuclei were addressed. The reason is that the sensitivity (signal-to-noise ratio, SNR) available in an experiment represents one of the major limitations for achieving better spatial resolution, implementing faster imaging strategies and developing more informative MRI applications. At the same time, the multinuclear capability of MRI is one of the highly promising directions for future developments, and the number of such applications is growing. The sensitivity achievable with various nuclei, and thus the practical feasibility of using these nuclei in an MRI experiment, are largely governed by the concentration of the nuclear isotopes in question, their magnetogyric ratios and their nuclear spin relaxation times in the objects under study. For instance, since ^{12}C and ^{16}O isotopes are spinless, NMR studies of these very important atoms require the use of ^{13}C and ^{17}O isotopes characterized by a much lower natural abundance. This greatly complicates the MRI experiments with such nuclei, because low natural abundance leads to very low sensitivity and/or requires the use of expensive isotope-enriched samples. Nuclei with very low magnetogyric ratios are usually characterized by low sensitivity, too, and therefore are seldom used in MRI.

Nuclear spin relaxation times also determine the achievable sensitivity. Furthermore, relaxation times govern spin dynamics in MRI experiments, determine image acquisition times and are important image contrast parameters. Transverse (or spin-spin) relaxation proceeds with a characteristic relaxation time T_2 and governs the decay of an observed signal after a single radiofrequency (rf) pulse in a uniform magnetic field. In practice, the apparent signal decay rate is even faster and proceeds with a characteristic time $T_2^* < T_2$ (or $T_2^* \ll T_2$). This extra decay is caused by the differences in local magnetic fields experienced by different spins which imply different precession frequencies and gradual dephasing of the ensemble of spins in a non-uniform magnetic field. However, this extra decay caused by the distribution of local fields in the sample can be at least partially reversed by means of spin echo techniques, and thus is fundamentally different from the irreversible loss of signal characterized by T_2. After any perturbation, the system tends to return to thermal equilibrium as spins realign with respect to the static field B_0. This longitudinal (or spin-lattice) relaxation process is often exponential with a characteristic recovery time denoted as T_1.

Many imaging techniques literally scan an object to produce its image. In contrast, modern NMR and MRI techniques (as well as many other techniques such as X-ray crystallography, FT IR, *etc.*) use Fourier transform to obtain spectroscopic or spatial data. In MRI, data acquisition takes place in the space of wave vector **k** (*k*-space)[3–7] defined as $k_i = \gamma G_i t/2\pi$ (i = x,y,z), where G_i is the magnitude of the magnetic field gradient along coordinate i and t is the gradient duration. The original data set detected in an experiment comprises a multidimensional array of NMR signal intensity values sampled for various values of **k**, achieved by varying the magnitude, the duration or the number of the applied gradients. Multidimensional Fourier transform of this data set is used to recover the true image of the same dimensionality.

As 'scanning' is done in *k*-space and not in the space of Cartesian coordinates, usually the entire data set in *k*-space has to be acquired before an image can be reconstructed. One-dimensional (1D) imaging requires an acquisition of a single line of sampled signal values in *k*-space and in principle can be performed very rapidly, within a millisecond or so. For samples which are nearly axially symmetric, acquisition of 1D axial projections can be a very efficient way to study fast dynamic processes. Higher image dimensionality usually increases the minimum imaging time substantially. The reason is that before the next line of samples in *k*-space can be acquired, one often has to wait (3–5)T_1 to allow the spin system to return to thermal equilibrium. As an example, for ^{1}H nuclei of water with $T_1 = 2.7$ s, a 3D image with 128×128×128 volume elements (voxels) needs 128×128 lines in *k*-space to be detected, which translates into an image acquisition time of more than 36 hours ($128 \times 128 \times 3 \times T_1$). At the same time, for a sample with $T_1 = 10$ ms the same experiment can be performed in 8 min. In practice, however, the sensitivity is often an issue, and *N* accumulations of each line in *k*-space are performed to improve SNR in the final image by a factor of $N^{0.5}$. This makes image acquisition time N times longer and makes this approach of sensitivity improvement impractical for large values of *N*.

Several approaches are used to reduce imaging time in MRI. One possibility is to reduce image dimensionality. In many cases, it suffices to acquire images of a number of carefully chosen 2D slices or even 1D bars of an object. The slices to be imaged can be non-invasively selected using appropriate combinations of gradients and frequency-selective rf pulses. As a result, the observed NMR signal is produced only by the spins residing within the selected slice, eliminating the need to define the slice of interest in any destructive fashion. As most of the image acquisition time is wasted while waiting for the spin subsystem within the imaged slice to return to equilibrium, slice-selective excitation in combination with a carefully chosen interleaved acquisition of *k*-space lines for different slices allows one to acquire not just one but several 2D slices within essentially the same imaging time. Another efficient strategy widely used in MRI is the implementation of various rapid imaging techniques. One of the approaches is based on the accumulation of multiple *k*-space lines for each pulse sequence repetition. For instance, single-shot 2D imaging allows one to acquire the entire 2D image in fractions of a second, provided that the sensitivity is sufficient. Another possibility is to repeat the pulse sequence without waiting for the complete recovery of the spin system. Unfortunately, rapid imaging techniques developed and successfully used in medical MRI, work properly for bulk liquids but often fail for samples with short T_2 times such as liquids in porous media, gases and solids. At the same time, such materials are characterized by shorter T_1 times as well, sometimes allowing faster imaging using standard imaging sequences. In certain cases, T_1 times of liquids can be reduced artificially by dissolving paramagnetic species, as it is done in medical MRI when paramagnetic contrast agents are administered.

For any imaging technique, an important feature is its spatial resolution. In contrast to many other techniques, however, the spatial resolution in MRI is sample-dependent. Images with the best resolution achieved to date have voxel sizes of the order of $5\times5\times5\ \mu m^3$, with some studies reporting voxels as small as $3.7\times3.3\times3.3\ \mu m^3$ or a volume of 40 femtoliters which contains 3×10^{12} proton spins of water.[8] However, the resolution of the MRI technique in many cases is limited by SNR, *i.e.*, by the sample rather than the instrument. Indeed, if the voxel size is reduced, the NMR signal intensity associated with each voxel diminishes leading to the reduced SNR in the image. Therefore, the very high spatial resolution could be obtained only using samples with large water content and many hours of imaging time.[8] In most practical applications the resolution will be (much) coarser. The resolution doesn't have to be isotropic, however, and in many cases the slice thickness can be a few mm while the in-plane resolution is from a few to a few hundred microns. The sensitivity limit on the attainable resolution means that the best applications for the MRI technique are those where the ultimate spatial resolution is not required. Higher SNR can be achieved using signal averaging, development of better hardware and more efficient pulse sequences, implementation of more sensitive signal detection methods and polarization of nuclear spins to increase the available nuclear spin magnetization. Nevertheless, one micron appears to be the ultimate limit for conventional imaging schemes which will be difficult to overcome even if a dramatic progress in sensitivity improvement can be achieved in the future. Indeed, there are other factors which limit the attainable resolution to roughly the same value of a few microns, which include diffusive displacements of molecules (especially for gases) and available magnetic field gradients (for materials characterized by short T_2 and T_2* relaxation times).

It follows that the attainable spatial resolution may not be the strongest feature of MRI as compared to other imaging techniques. As mentioned above, the essence of MRI is the nature and flexibility of image contrast which extends well beyond the purely structural information. For instance, one might think that an n-fold increase/decrease in concentration of the detected nuclear spins should lead to an n-fold increase/decrease in the observed signal intensity. However, in many practical cases it doesn't. This is because the detected NMR signal depends not only on spin density but also on many other parameters as well. Furthermore, many essential changes in an object under study are not accompanied by a pronounced concentration change, and thus would be impossible to observe in the maps which reflect concentration only. For instance, upon freezing of water, proton spin density variation is relatively small. Nevertheless, NMR signal intensities of water and ice can differ by orders of magnitude, allowing one to detect the phase transition easily. The reason is that the NMR signal intensity can depend dramatically on molecular mobility, and this dependence provides the necessary contrast mechanism to detect the phase transition. The true power of MRI is that the range of the properties and processes affecting image contrast is virtually unlimited and includes spin density, molecular mobility, diffusive and convective transport, chemical transformations, viscosity, temperature, paramagnetic additives, interactions with surfaces, electric currents, magnetic susceptibility, degree of cross-linking and ageing of polymeric materials, just to name a few. This means that, at least in principle, all these properties and processes can be studied by MRI.

This also means that the MRI measurements are *indirect*. Indeed, the only thing NMR/MRI can measure *directly* is an NMR signal and its characteristics such as magnitude, phase, resonance frequency and line width (and shape). An important feature of the MRI technique is the flexible and adjustable nature of image contrast. As a result, those sample properties which are of interest to a chemist or a chemical engineer can be made to influence the detected NMR signal. The problem is that many of sample properties usually affect the NMR signal simultaneously. Therefore, the real challenge is to single out the effect caused by a particular quantity of interest. In many cases, this is feasible but requires a lot of experience and care in the design of experiments and proper interpretation of their results.

Sometimes, the influence of object properties on the NMR signal is direct, as in the case of diffusion or flow of fluids in a magnetic field gradient which directly affect signal intensity or phase, respectively. In many cases, however, this influence is imposed *via* certain mediators. For instance, in the example with water and ice considered above, the mobility of molecules affects nuclear spin relaxation times (T_1, T_2), which in their turn affect the NMR signal. Relaxation times of nuclear spins are directly related to molecular mobility and therefore can provide useful information on the frequency spectrum of molecular motions, their activation energies and correlation times, and distances at the molecular scale. Apart from relaxation times, there are other characteristics of a spin system which can act as mediators, for instance chemical shift (or local magnetic fields in general), spin polarization, spin–spin interactions, *etc.* All this makes it possible to tailor an MRI experiment to the needs of a particular study and to extract useful and diverse information about objects and processes. It is worth mentioning that while appropriately weighted images usually give qualitative information only, one can also obtain quantitative maps of various parameters by evaluating the parameter of interest for each image voxel. This allows one to obtain maps of local velocities, diffusivities, temperatures, chemical composition and so on.

3 Practical aspects and structure visualization

NMR instruments with 200–500 MHz ^{1}H NMR frequency equipped with (micro)imaging accessories provide high sensitivity and high spatial and spectroscopic resolution. In addition, they are normally equipped with a vertical bore superconducting magnet, which is most suitable for the studies of certain reactor types such as trickle bed and fluidized bed reactors. The geometry of the magnet and the rf probe limits the maximum reactor diameter to a few cm which has to fit into the rf coil/gradient coils assembly. While the reactor and catalyst bed lengths are essentially unlimited, the axial dimension of the sensitive region is also limited to a few cm. These size restrictions combined with the limitations on the materials that can be present in the magnet/probe assembly during an NMR/MRI experiment (see below) require the construction of an MRI-compatible model reactor. NMR/MRI instruments equipped with horizontal bore magnets often have wider bores, but cannot accommodate long vertical reactors and thus require different reactor designs. Sensitivity and/or spatial resolution in the detected images decreases significantly for rf probes with larger volumes, therefore some compromise is necessary.

In high resolution NMR of liquids, higher magnetic fields (600–1000 MHz ^{1}H NMR frequency) provide even higher spectral resolution and SNR. However, for heterogeneous objects, higher applied magnetic fields lead to larger local magnetic field gradients within the sample and may cause various image artifacts and significant loss in SNR. Systems with low magnetic fields (<200 MHz) can accommodate larger reactors, but such experiments usually suffer from reduced sensitivity and thus lower absolute spatial and temporal resolution. Nevertheless, one of the recent trends in the applications of NMR and MRI in materials science and chemical engineering is the design and construction of instruments that use low and ultra-low magnetic fields.[9–12] In addition to lowering the costs, this allows one to remove some of the restrictions associated with the use of high-field systems, and in particular to take an instrument out of the rigorously controlled environment of a scientific NMR/MRI laboratory and bring it to a chemistry lab or even to an industrial site. Furthermore, some have an open design and thus eliminate the need to place an object under study inside a restricted space of a probe. The necessary prerequisite for many such applications is a significant boost in detection sensitivity of NMR/MRI, which happens to be another hot topic in modern magnetic resonance. While the utility of low-field MRI is still to be demonstrated for catalytic applications, an increased SNR is also very useful for high-field studies, and recent examples of such applications are discussed below.

In addition to size restrictions, many conventional reactor designs are not suitable for MRI studies for a number of other reasons. It is therefore necessary to construct a model reactor compatible with an MRI instrument. For MRI studies of a reactor (*e.g.*, packed catalyst bed) structure, various transport processes and those catalytic reactions that can be carried out at ambient temperature and pressure, a piece of glass or plastic tubing will work as the main construction element of the reactor, even though plastic materials may contribute to the detected image (*e.g.*, ^{1}H MRI). However, the most interesting and also challenging applications require substantially elevated pressures and temperatures within an operating reactor, especially when dealing with highly exothermic reactions. This poses certain problems, as the expensive MRI hardware has to be protected from high temperatures as well as from accidental damage associated with high pressures inside the vessel residing inside the magnet/probe assembly. At the same time the most obvious choice of construction materials, *viz.* metals, is generally not an option. Electrically conducting containers, including non-magnetic metals, cannot be used for constructing MRI-compatible reactors as they are impermeable to rf electromagnetic fields that are used in NMR/MRI for signal transmission and reception. Bulk metallic parts should be avoided also because pulsed magnetic field gradients used in MRI induce eddy currents in electrically conducting materials. For studies at ambient pressures or slightly above, thick-walled temperature-resistant glass or quartz can be used. For instance, in the design reported in ref. 13 exothermic reactions were safely performed at temperatures around or slightly above 100 °C (catalyst temperatures up to 250 °C during exothermic hydrogenation reactions) with the use of an evacuated double-walled glass Dewar. Essentially, the reactor comprised three concentric glass tubes. The catalyst was placed inside the inner tube (10 mm i.d.). The annular space between the inner and the middle tubes was used to heat the catalyst with a stream of air preheated to 80–100 °C. The entire reactor was inserted into the rf coil/gradient coils assembly located in the vertical bore of the NMR magnet (300 MHz, Avance DRX 300 NMR spectrometer, Bruker). The gradient coils of the instrument can be damaged if heated above 60 °C. To prevent this, the middle and the outer tube of the reactor and the annular space between them formed a vacuumed Dewar. Alternatively, an active heating/cooling with a liquid can be implemented. For instance, a fluorinated liquid hydrocarbon will allow one to avoid a strong background ^{1}H NMR signal.[14] However, the passive heat protection of the design reported in ref. 13 avoids heat removal from the reaction zone. The inner tube accommodated the catalyst and also served as a duct for supplying gaseous reactants, while liquid reactant, if required, was supplied to the catalyst through an additional capillary. In the experiments performed with a single catalyst pellet, the latter was suspended on a thermocouple implanted into the cylindrical pellet though its side. In the experiments with a catalyst bed, the latter rested on a stainless steel mesh. Pulsed magnetic field gradients used in MRI induce eddy currents in electrically conducting materials and can severely distort or even destroy the image. Therefore, the mesh was located well outside the sensitive region of the rf probe. The thermocouple was only 0.2 mm in diameter, but nevertheless it destroyed the signal in its immediate vicinity, and in addition worked as a pick-up antenna which fed extra noise and interferences from outside the magnet directly into the rf coil. The reactor design[13] ensured that the temperature at the outer wall of the reactor never exceeded 50 °C. The entire reactor was *ca.* 25 mm in diameter and thus could be inserted in a 25 mm i.d. birdcage rf insert (Bruker), but the probe body had to be redesigned to have a 25 mm bore through the entire probe. The reactor in ref. 14 was used at temperatures up to 125 °C and pressures of 10 bar but under non-reactive conditions. Recently, MRI experiments were reported that used a reactor capable of withstanding temperatures of *ca.* 300 °C and pressures of *ca.* 30 atm, with the reactor units made of Si_3N_4.[15] This brings the MRI studies one step closer to practically relevant conditions of reactor operation.

Knowing the internal structure of a reactor (*e.g.*, a packed catalyst bed) is important as it can have important influence on the heat and mass transport processes within the operating reactor and the overall reactor efficiency. MRI can be used to visualize the internal structure of a model reactor in a non-invasive manner, even for optically opaque materials as long as they are NMR/MRI-compatible (see above). Structure visualization of a multi-phase system can be done in several different ways. MRI is best performed using an NMR signal of a liquid phase. Therefore, the most straightforward way of visualizing a structure of, *e.g.*, a bed of solid elements is to completely flood it with a suitable liquid (water, organics) and acquire an image of the liquid phase.[6,14,16–22] The voids of the bed will thus be visible in the image directly as they will give a strong signal, whereas solid phase will be imaged indirectly as areas that give no observable signal (Fig. 2a). For porous solids, the image resolution (tens to hundreds of microns) is usually much coarser than the sizes of individual pores. This means that image pixels corresponding to liquid in the inner regions of porous solids would give a signal with intensity lower than the bulk liquid because of the porosity value being less than unity, and thus can be distinguished from both liquid-filled voids (maximum signal intensity) and bulk solid phase (no signal) in beds comprised of porous particles. In addition, the signal of an intrapore liquid is often additionally reduced because of its reduced nuclear spin relaxation times. This latter fact can be employed to manipulate image contrast because relaxation effects are sensitive to the parameters of an imaging pulse sequence used, providing a researcher with additional means of revealing the structure of an object (reactor) under study.

Another possibility to image a structure consisting of porous solid elements and voids is to saturate the porous material with a suitable

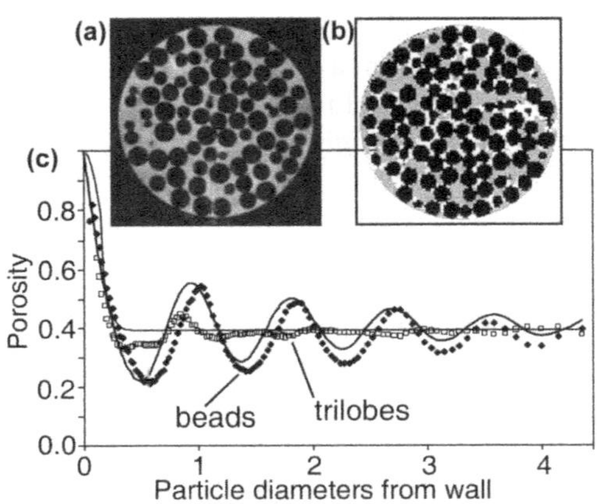

Fig. 2 (a) 2D image of a bed of glass beads 2 mm in diameter saturated with a liquid (solid phase is shown in black, liquid phase in grey). (b) An image detected during the two-phase gas-liquid flow in the bed (solid phase is shown in black, liquid phase in white, and gas phase in grey). (c) Radial distributions of porosity for beds of spherical glass beads and Al_2O_3 trilobes (diameter 1.3 mm, length 4 mm) extracted from the images of the liquid-flooded beds. Solid lines show the results of model calculations. Reproduced from ref. 17 with permission of Wiley-VCH Verlag GmbH & Co. KGaA, Weinheim.

liquid. In this case, the liquid-containing phase can be imaged directly (high signal intensity in an image) while liquid-free voids will appear in the image as areas void of any signal. As bulk liquid is not present in such an experiment, the SNR will be lower and acquisition times longer as compared to the experiment with the liquid-flooded reactor. A variant of these two strategies is to use a gas instead of a liquid. If the gas used is not adsorbed in the pores of a solid phase, the image contrast will be similar to that obtained for a liquid-flooded reactor, but SNR will be significantly lower because of a much lower spin density of gases compared to liquids. On the other hand, if gas adsorbs strongly in the pores of a solid phase, the latter may appear in the image as hyperintense areas as compared to gas-filled voids.

Yet another possibility is to image the solid phase directly. Usually, imaging of solid materials is much more difficult as compared to liquids because of very short transverse relaxation times and significant broadening of the signals in NMR spectra of solid materials. Therefore, imaging of solids is often performed with the use of specialized MRI hardware and pulse sequences.[4,23–28] However, because line broadening in solid samples is often inhomogeneous, certain solids can be addressed using simple imaging strategies such as the spin echo method.[29–37] In particular, the ^{27}Al nucleus can be used for direct imaging of aluminium-containing materials[29–31,33,34,38] such as γ-Al_2O_3, glass, refractory cement/concrete, *etc.* Considering that γ-Al_2O_3 is often used as catalyst support, this provides some interesting possibilities to use ^{27}Al MRI, including the ^{27}Al NMR/MRI thermometry considered in more detail in the corresponding section below. At the same time, direct imaging of solids is usually time-consuming because of the unfavorably low signal intensity in most studies.

Once the packed bed structure is imaged, the data can be processed to extract essential information such as distribution of pore (void) sizes, connectivity, radial porosity distribution (Fig. 2c), *etc.*[14,17–22] In addition, the known structure of a packed bed can be used as a direct input in model calculations of transport and reaction within it.[21,22,39] Knowing the partitioning of the structure into solid phase and interparticle voids is also essential when studying processes in three-phase systems such as gas-liquid-solid reactions. In most cases, the gas is not (or cannot be) observed in the images of a liquid-containing reactor. However, if the geometry of the void space is already known, then that part of void space which does not contain liquid (*i.e.*, does not give a strong signal associated with a liquid) obviously carries the gas phase (Fig. 2b).[6,7,14] Sometimes, both the liquid and the gas phases in the bed voids can be visualized directly, for instance using ^{1}H MRI of a liquid (*e.g.*, water) and ^{19}F MRI of a gas (*e.g.*, SF_6).

In certain cases, reactors may have a dynamic 'structure'. An example is a bed of solid particles fluidized with a gas flowing through it. At present, MRI is too slow to map the instantaneous positions of the solid phase in a fluidized bed of solid particles. At the same time, the technique is suitable for visualizing the time-averaged structure of a fluidized bed. In most cases, liquid-containing solid particles (oil-containing seeds,

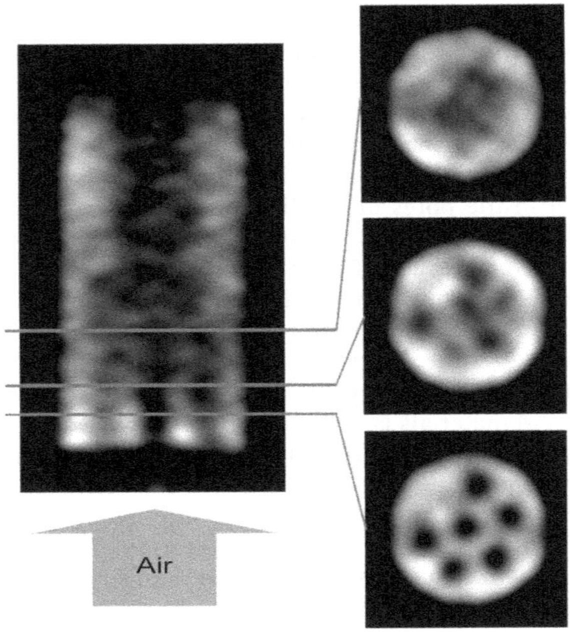

Fig. 3 ^{27}Al MR images of a fluidized bed of alumina powder in a vertical cylinder. Air is supplied through the orifices at the bottom (one orifice got plugged). One vertical (left) and three horizontal 2D slices (right) of the full 3D data set are shown. Lighter shades correspond to higher signal intensities, *i.e.*, to higher time-averaged concentrations of the solid phase.

porous grains saturated with non-volatile liquids, *etc.*) are used for this purpose to get the highest possible NMR signal.[40–42] At the same time, γ-Al_2O_3 powder used as the solid phase can be imaged directly using ^{27}Al MRI (Fig. 3). Usually, the images show the density of the fluidized phase averaged over the image acquisition time, with the signal intensity in the images proportional to the average concentration of the solid phase. Such images can reveal spatial structure of gas jets and can provide time-averaged flow velocities. Dynamic events such as bubbling can be addressed using 1D imaging or fast 2D imaging schemes.[40]

4 Mass transport in channels, voids, and porous materials

NMR has a long and successful history of application for the studies of molecular diffusion of liquids and solutes. Pulsed field gradient (PFG) NMR experiment is based on the application of a pair of gradient pulses, which lead to the NMR signal reduction for stochastically moving molecules.[3,5,43] Analysis of the results of such an experiment can yield effective (but possibly time scale dependent) diffusivity value or even the complete diffusion tensor. PFG NMR is extensively used to study diffusion in various systems including porous solid materials of relevance to catalytsis.[42–53] Such studies are often performed without the imaging capabilities (*i.e.*, without spatial resolution).

Mass transport involving liquids, and to a lesser extent – gases and granular solids, can be also addressed with MRI using two general approaches.[43,54] The first approach is based on creating an initial gradient of a tracer substance and then using MRI to repeatedly map the evolution of tracer concentration in space and time. If a tracer gives a detectable NMR signal which changes as the local tracer concentration is changed, it can be imaged directly. For instance, transport of fluids in porous materials during drying and sorption processes is sometimes used as a model of transport in porous media and can be studied by detecting the images of the fluid in the porous space of a sample and analyzing their changes in time.[54] Certain tracers that do not give an observable signal can provide negative contrast, for instance when diffusion of a deuterated liquid into a sample saturated with its protonated analog is monitored using ^{1}H MRI (H_2O/D_2O exchange). Transport of some NMR-silent tracers can nevertheless produce observable changes in the images as well. For instance, paramagnetic ions reduce nuclear spin relaxation times of a solvent, and thus their transport can be tracked by detecting the relaxation-weighted images of the solvent. Clearly, the mass transport distances addressed by all such experiments cannot be smaller than the spatial resolution of the detected images.

There is an entirely different possibility provided by NMR/MRI which can be used to study transport even in the absence of any tracer gradients. Furthermore, it essentially does not require an introduction of any tracers in a system under study. It is based on tracking the displacement of molecules of the fluid phase which is already present in the object under study, *i.e.*, uses nuclear spins of the fluid molecules as 'tracers' and manipulates those spins to detect fluid transport. Depending on the character of transport, the length scale and the information required, the MRI toolkit provides several different instruments to address the problem. The spatial tagging method essentially draws "tags" on the image of a sample under study, and after a certain delay visualizes their transformation caused by fluid transport.[7,43,55–57] The tags are non-invasive; they are produced by manipulating nuclear spin magnetization of the fluid molecules and appear as contrast stripes in the images. The molecules carry this altered magnetization as they are transported by the fluid flow, making the original tags to transform accordingly. The method is usually applicable to transport in relatively large channels. The time-of-flight is another method[43,55,57] based on disturbance of the NMR signal in a particular region of the sample by the rf pulses, but involves observation of subsequent changes in the signal intensities of the pixels of an MR image caused by the displacements of the perturbed and unperturbed spins in the sample, providing qualitative or semi-quantitative information on transport.

Quantification of velocities and mapping the velocity fields of flowing fluids are often carried out using the phase method.[3,7,43,55–59] The corresponding pulse sequence is designed to make the phase of the acquired signal proportional to the fluid flow velocity. As the phase of an NMR signal depends on many things, usually two measurements are required to extract the contribution to the signal phase caused by flow.

Measurements with and without flow can be used if the latter can be stopped; alternatively, measurements with and without flow-encoding magnetic field gradients can be implemented. Once the phase change associated with flow is extracted, it can be used to quantify flow velocity. Strictly speaking, the method measures the displacement of fluid during the time interval between two gradient pulses. Therefore, if velocity is not constant in time, the obtained effective velocity will depend on the time interval of the measurement. In addition, if different regions of the fluid travel at different velocities, a single value of effective velocity will be measured that in general does not correspond to the true average velocity. When the phase method is combined with MRI, it will yield the spatial velocity map showing effective velocities for each image pixel (voxel). The technique is applicable to the studies of flow in macroscopic channels and in some porous materials such as packed beds if the spatial resolution is sufficient to clearly resolve the transport channels. If the characteristic pore sizes are much smaller than the spatial resolution achieved, only effective velocities for a large number of pores covered by image pixel (voxel) can be evaluated. Transport of liquids and gases can be addressed, with gases showing lower SNR because of a much lower density. At the same time, the successful attempts to perform gas flow imaging show that large diffusive displacements of molecules do not prevent one from addressing convective displacements with MRI. This is because chaotic fluid displacements lead to signal attenuation but do not affect its phase. Therefore, even in the presence of relatively large diffusive displacements, smaller convective displacements can be evaluated from the signal phase if SNR is sufficient. Signal reduction caused by chaotic motion also makes unsteady, transient and turbulent flows much more difficult to address. Nevertheless, implementation of faster and more robust pulse sequences makes it possible to successfully address turbulent and unsteady flows. Velocities from a few μm/s up to ~10 m/s can be addressed using phase method.

In certain cases, the information on the distribution of flow velocities or diffusive or dispersive displacements in a sample (a histogram of velocities or displacements) may be more relevant than the spatial assignment of average transport velocities (flow velocity map). The tool for this is based on the measurement of the average propagator of motion, $P(R,\Delta)$, which gives the relative quantity of the fluid that traveled over distance R in time Δ.[3,7,43,55,58,59] The displacements R are along the direction of the applied gradients, and Δ is the separation of the gradient pair, so both the direction and the time scale can be varied, if necessary. The propagator $P(R,\Delta)$ reflects all displacements of the fluid independent of their nature. If diffusive displacements dominate and are isotropic, $P(R,\Delta)$ will have a Gaussian shape, will be centered at $R=0$ and will get broader for larger time scales (larger Δ). If convection dominates, $P(R,\Delta)$ will be shifted away from the origin and will reflect distribution of flow velocities and flow dispersion effects in the direction of the applied gradient. Many such studies of diffusion, flow, filtration and other transport processes are performed without spatial resolution, *i.e.*, for the entire sample or some part of it. The corresponding experiments are often considered as imaging in the space of molecular displacements ('q-space imaging') in contrast to the conventional MRI performed in the space of position coordinates ('k-space imaging'). Furthermore, measurements of the average propagator may be combined with an MRI experiment. This will provide spatially localized distributions of displacements for each pixel (voxel) of the image; however, such experiment will be significantly more time-consuming. There are more sophisticated tools in the MRI toolkit; they allow one to measure acceleration rather than velocity, to correlate diffusive or convective displacements at two different time instants, *etc.*[58,60–65] All these methods, however, usually require a prolonged signal accumulation and may still suffer from low SNR and low spatial resolution.

Applications of MRI to transport studies in model reactors and other geometries were extensively reviewed in the literature.[7,42,43,56,66–68] In particular, distribution of liquid phase in a packed bed during gas-liquid two-phase flow in the absence of any reaction was mapped at ambient conditions[7,42,43,56] or at elevated temperatures and pressures (see Fig. 2b).[14,17] Through the detection of liquid phase distribution it was demonstrated that local liquid pulsing events can promote transition from trickle to pulsing flow regime.[7,42] Detection of the images of the liquid phase was also used to address bubble-train and film flow in the channels of ceramic monoliths.[7,69] Phase method was applied to obtain velocity maps for single-phase flow of liquids or gases in packed beds,[7,18–20,70–72] porous ceramic materials,[73] monoliths,[70,74–77] model string-type reactor (Fig. 4a),[78] diesel particulate filters.[79,80] In addition, for the two-phase gas-liquid flow in a packed bed of glass beads, the velocity maps of both the gas and the liquid phase in the bed were acquired[42,67,81] (Fig. 4b). The flow imaging times amounted to tens of minutes for both liquid (water, ^{1}H MRI,) and gas (SF_6, ^{19}F MRI), and the

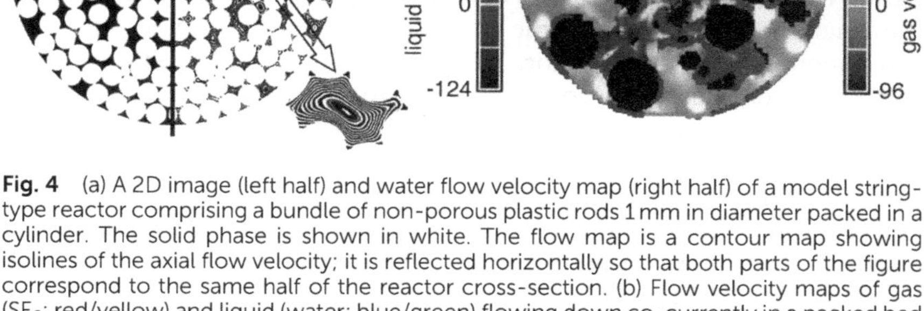

Fig. 4 (a) A 2D image (left half) and water flow velocity map (right half) of a model string-type reactor comprising a bundle of non-porous plastic rods 1 mm in diameter packed in a cylinder. The solid phase is shown in white. The flow map is a contour map showing isolines of the axial flow velocity; it is reflected horizontally so that both parts of the figure correspond to the same half of the reactor cross-section. (b) Flow velocity maps of gas (SF_6; red/yellow) and liquid (water; blue/green) flowing down co-currently in a packed bed of 5 mm glass beads. Color is available in the electronic version of this figure. Fig. 4 (b) is reproduced from ref. 81 with permission of Elsevier BV.

in-plane spatial resolution in water images/maps (177 μm) was significantly better than for the gas phase (708 μm). The average propagator method was applied to obtain the distribution of velocities in numerous studies of flow and dispersion of liquids in porous media, including packed beds,[58,60,70–73,77,82–87] and was also extended to flow and filtration of gases in various geometries[60,68,70,74–76,88–90] and gravity-driven transport of fine alumina powder through a packed bed of alumina beads.[60,91]

One of the recent trends is an attempt to significantly reduce flow imaging times in order to approach snapshot MRI velocimetry of unsteady-state processes. Comparison of the performance of different imaging sequences in this context is summarized in ref. 67. In single-phase flow, significant interest focuses on unsteady-state flows and the transition to turbulence. For instance, sequences based on echo-planar imaging (EPI) method were used to address turbulent flow and laminar-turbulent flow transition for single-phase liquid flow in a pipe,[67,92] with the maps of all three velocity components detected in 60 ms.[67] The fastest approach appears to be the one based on spiral imaging method which uses data sampling along a spiral trajectory in the *k*-space. In particular, spiral imaging was used to address pipe flow at Re = 4500, providing 91 images acquired per second with an in-plane spatial resolution of 625 μm × 625 μm.[93] Spiral flow imaging was also demonstrated for gas-liquid bubbly flow, and in particular to monitor a single rising bubble and to measure the vertical velocity component of the liquid around it.[67] Further reduction in flow imaging time was achieved based on the implementation of compressed sensing schemes which allow one to strongly undersample the signal in the *k*-space without the penalty of a significantly compromised data quality.[94] Sampling of only 33% of the number of data points in *k*-space provided flow maps for SF_6 in a bed of spheres of essentially the same quality and at the same spatial resolution as compared to the acquisition of the conventionally acquired full set of *k*-space data points.[67] This approach was used to acquire velocity maps in 3 ms and monitor dynamic transport events at the rate of 188 velocity maps per second for flow mapping of a single bubble rising in a liquid to study its wobbling motion and observe vortex shedding in its wake.

Many other transport processes relevant to chemical engineering and catalysis have been addressed. In particular, MRI was successfully used to monitor the process of preparation of supported catalysts by impregnation from solution of an active component precursor. The properties of the final catalyst, including the macroscopic distribution of an active component within the support pellet, depend critically on the mass transport processes in the course of impregnation, which in turn depend on numerous factors such as solution composition, impregnation conditions, interaction of solutes with each other and with the support surface, *etc.* In this context, direct imaging of the transport is possible for only a limited number of chemicals that give a strong NMR signal. For instance, ^{31}P MRI was used to visualize the transport of $H_xPO_4^{(3-x)-}$ into a γ-Al_2O_3 pellet after it was immersed in an aqueous solution of H_3PO_4.[16,37,95–98] Transport of $H_xPO_4^{(3-x)-}$ into the pellet was shown to be slow even under conditions of a fast capillary imbibition of water into the dry pellet. The images detected during impregnation were also compared with the ^{31}P images detected after terminating the impregnation at various stages and drying the pellet,[36,95] showing that the distributions of phosphate in the liquid phase during impregnation and of the adsorbed phosphate were quite similar. The same direct approach was used to study the transport of ^{13}C-labeled citric acid in an alumina pellet at different solution pH values using ^{13}C NMR signal detection.[99] Another illustrative example is the direct detection of the ^{195}Pt NMR signal to image the impregnation of a γ-Al_2O_3 pellet with an aqueous solution of H_2PtCl_6, albeit the concentration had to be high to ensure acceptable SNR.[95]

Another demonstrated approach is indirect in the sense that the distribution of a certain chemical is mapped by detecting the NMR signal of the intrapore liquid. Nuclear spin relaxation times of the latter can change significantly in the presence of certain molecular species either dissolved in the liquid filling the pores or adsorbed or precipitated on the pore walls. In such cases, the relaxation-weighted image of the solvent (*e.g.*, ^1H MRI) in the pores of the support will reflect local concentrations of a solute or adsorbate. This approach is advantageous since high concentration of the solvent often gives the highest SNR. In particular, many supported catalysts are prepared using solutions containing complexes of paramagnetic metal ions such as Co^{2+}, Ni^{2+}, Cu^{2+} or Fe^{3+} which significantly reduce nuclear spin relaxation times of liquids. Transport of Co^{2+} aqua complexes ($[Co(H_2O)_6]^{2+}$) during the incipient wetness impregnation of individual cylindrical γ-Al_2O_3 pellets was addressed with this approach.[16,37,98,100] Repeated detection of relaxation-weighted 2D images of water in the pellet during the impregnation process visualized the transport of $[Co(H_2O)_6]^{2+}$. Using an independent calibration experiment, the images were then converted into quantitative maps of Co^{2+} concentration in the pellet at different stages of impregnation. Such experiments performed under various conditions (solution pH, various concentrations of additives, *etc.*) revealed differences in the spatial distribution of Co^{2+} (egg-yolk, egg-white, egg-shell or uniform) in the support.[101] Similar studies were performed to visualize the transport and quantify distribution of Ni^{2+} [102] and Cu^{2+} [99] complexes in γ-Al_2O_3 pellets upon their impregnation with aqueous solution of corresponding salts. Redistribution of an active component during the drying stage can be also addressed. In this case, after removing water from the pellets they were saturated with cyclohexane[99,103] which is not expected to alter the distribution achieved in the drying stage but which gives suitable NMR signal sensitive to the presence of paramagnetic species on the pore walls of the support.

Furthermore, it was also established that some diamagnetic solutes and/or adsorbates can change the relaxation times of liquids in the pores of a support as well (Fig. 5).[104] This fact was used to study the dynamics of $PtCl_6^{2-}$ transport in a support pellet immersed in an aqueous solution of H_2PtCl_6 and oxalic acid, with the ^1H NMR signal of water used to detect the image. Similar studies were performed with $PdCl_4^{2-}$ in alumina pellets, and also to study the distribution of Pt and Pd in honeycomb monoliths.[105] Similarly, differences in the relaxation times of n-heptane

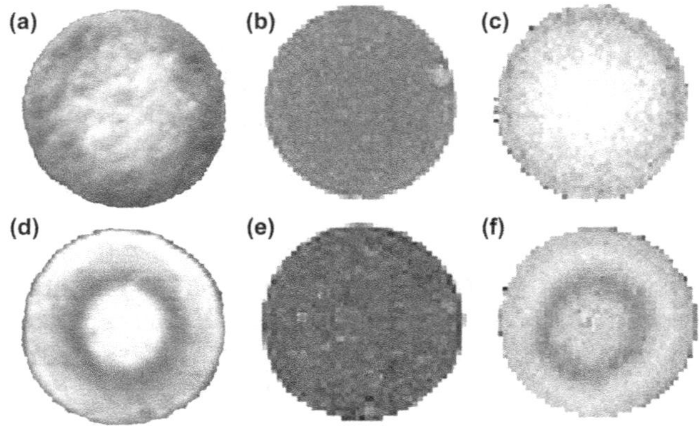

Fig. 5 Alumina pellets with an egg-shell (a–c) and egg-white (d–f) distribution of $PtCl_6^{2-}$. Digital camera images (a,d) were obtained after the pellets were cut and treated with a $SnCl_2$ solution; darker shades correspond to higher Pt contents. Proton spin density (b,e) and T_1 (c,f) maps of cyclohexane were constructed from the MR images obtained after dry pellets were saturated with cyclohexane; darker shades in (c,f) correspond to longer cyclohexane T_1 times.

in Pd/Al_2O_3 and Al_2O_3 particles were used to distinguish the catalyst and support particles in a bed comprising their random mixture.[106] The relaxation contrast approach was also used to study transport of $Mo_7O_{24}{}^{6-}$ ions in γ-Al_2O_3 pellet during its impregnation with an aqueous solution of $(NH_4)_6Mo_7O_{24}$ [37,98,99] and to map the distribution of Mo after pellet drying.

An approach similar to the one described above can be used to study distribution of carbonaceous deposits in a catalyst upon its deactivation, since formation of weakly paramagnetic deposits in a porous catalyst reduces relaxation times of the intrapore fluids and blocks pore space which leads to a reduced NMR signal intensity. These facts were used to visualize macroscopic distribution of coke in cylindrical alumina pellets after their 50–100% deactivation in ethanol dehydration to ethylene by saturating the pellets with water before imaging.[107] Similar studies were performed with HY zeolite pellets partially coked in n-heptane cracking, with 2,3-dimethylpentane used as a probe molecule for MRI.[108,109] In all these studies, distribution of coke in the catalyst pellets was found to be non-uniform. Direct imaging of coke using its ^1H NMR signal was also successfully demonstrated in the study that addressed HZSM-5 catalyst pellets prepared with binder and coked in methanol-to-gasoline conversion to 20.5 wt% coke content.[110] Such study requires the application of specialized MRI techniques because the detected signals have very short T_2 times. In the same study, an alternative indirect approach was demonstrated based on significant differences in the T_1 times of propane in the coked and uncoked catalyst. Images of coke distribution were also obtained for naphtha reforming catalyst (Pt-Re/Al_2O_3) and alumina pellets using T_1 maps of adsorbed n-heptane.[111] The experiments revealed the presence and non-uniform removal of residual coke after the catalyst regeneration at different temperatures.

5 Operando studies of model reactors

5.1 Distribution and transport of liquids and gases

As discussed in the preceding section, mass transport processes in model reactors are often studied under non-reactive conditions, for instance with the use of plastic or glass beads instead of the actual catalytic solids, and with water/air supplied instead of the actual reactants. While this approach may be useful to address some general features of mass transport processes in model packed beds, the observed picture may be very different under reactive conditions. In particular, in trickle bed reactors the distribution of the liquid phase in the void space and in the pores of catalyst particles may change drastically once, *e.g.*, an exothermic reaction is turned on. This happens because in an operating reactor the mass and heat transport processes may be essentially coupled to each other and to the catalytic conversion, exhibiting highly non-linear behavior even under steady-state conditions, and in addition may lead to complex dynamic behavior of an operating reactor. Therefore, it is not surprising that currently there is a significant interest in the development and application of various spectroscopic and imaging techniques for the *in situ* and operando studies of operating catalytic reactors.[112–115]

The dramatic difference between the distribution of a liquid phase in a fixed catalyst bed with a cocurrent gas-liquid flow under reactive and non-reactive conditions was demonstrated by MRI using heterogeneous catalytic hydrogenation of 1-octene as a model process (Fig. 6).[116] The reaction could be turned off by switching from H_2 to N_2 in the gas feed while keeping the flow of liquid 1-octene unchanged. The 2D images of the liquid phase distribution were detected and processed to yield the 2D maps of the spatial distribution of the external and internal liquid holdups for various regimes of the reactor operation. The results revealed that under reactive conditions the distribution of the liquid phase in the catalyst bed was highly non-uniform. In contrast, in the absence of reaction, the bed was largely filled with the liquid phase. Even though the reaction conditions were far from those used in practice, the difference in behavior with a non-reacting system was quite dramatic.

One of the interesting and useful models of an operating system is a single-catalyst-pellet reactor. While it may not exhibit the behavioral complexity of an operating packed bed, it is still a viable model for many dynamic processes that may take place on the length scale of a single catalyst pellet and thus may provide an insight into the appearance of critical phenomena and their expansion to a larger scale. In the very first MRI studies of gas-liquid-solid catalytic processes,[16,28,32,60,70,98,117–120] heterogeneous catalytic hydrogenation of α-methylstyrene (AMS) was chosen as a suitable model process. The experiments imaged the spatial distribution of liquid phase in the catalyst pellet during the reaction, without discriminating the contributions of AMS and the hydrogenation product cumene. To mimic various experimental conditions, AMS was supplied to a cylindrical $Pt/\gamma\text{-}Al_2O_3$ pellet either as a liquid (through a capillary to the top surface of the pellet), or as a vapor (by bubbling the supplied H_2 through liquid AMS), or both. The results clearly

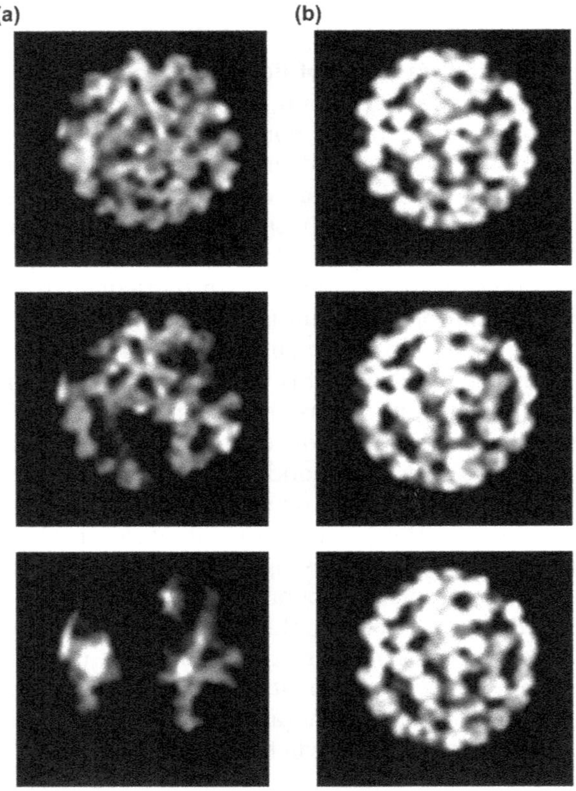

Fig. 6 ^{1}H MR images showing the distribution of the liquid phase in the catalyst bed under reactive conditions (a) and without reaction (b). Liquid 1-octene was supplied at different flow rates along with H_2 (a) or N_2 (b) to the bed of 1% Pd/γ-Al_2O_3 catalyst beads 3.2 mm in diameter.

demonstrated that even if AMS was supplied only as a vapor, some amount of liquid could be seen in the pores of the catalyst owing to AMS and/or cumene vapor adsorption or condensation. When liquid AMS supply to the top edge of the pellet *via* a capillary was turned on, the upper part of the pellet was observed to contain substantially more liquid than its lower part. Essentially, the upper part provides a substantial reservoir of the liquid reactant available for evaporation, while the lower dry part is able to efficiently perform gas-phase hydrogenation of AMS vapor, with short distances between the evaporation and reaction zones that AMS vapor has to travel to enter the active zone. The catalyst pellet temperature increased significantly once the liquid AMS supply was turned on, confirming the efficient progress of the reaction. At higher flow rates of liquid AMS, the pellet was filled with the liquid phase progressively until it was completely flooded, which leads to a significant reduction of the reaction rate as it becomes limited by the diffusion of H_2 through the liquid phase to the active catalytic centers. These results confirmed the existence of the high- and low-temperature steady states for the operating catalyst. In the experiments performed without AMS vapor supply,[60,70,117,118,120] different initial pellet conditions and variation of the liquid AMS flow rate were used to monitor the distribution of the liquid phase in the pellet on the upper and lower branches of the temperature hysteresis curve as well as during the transition periods.

The 2D images in the early studies discussed above were acquired in *ca.* 260 s using the conventional spin echo pulse sequence. Subsequent experiments were carried out with grains of the 1 mass% Pd/γ-Al_2O_3 catalyst containing 0.1 mass% Mn[13,32,35,120–122] since paramagnetic ions reduce the T_1 time of the pore liquid and proportionally reduce the image acquisition time. This made it possible to acquire two-dimensional images in 34 s and thus to visualize a number of interesting dynamic processes in the operating single-pellet reactor, including the dynamics of liquid phase redistribution within the pellet during single and multiple catalyst ignition events, oscillations of the liquid content and a reciprocating motion of the liquid front in the catalyst pellet (Fig. 7).[13,32,121] The results obtained demonstrate the ability of MRI to visualize complex dynamic behavior of an operating single-pellet catalytic reactor resulting from a non-linear coupling of the transport processes involved with the exothermic hydrogenation of an unsaturated compound. Mathematical modeling of the experimental results[123] allowed the authors to confirm that the oscillations in temperature and liquid content distribution

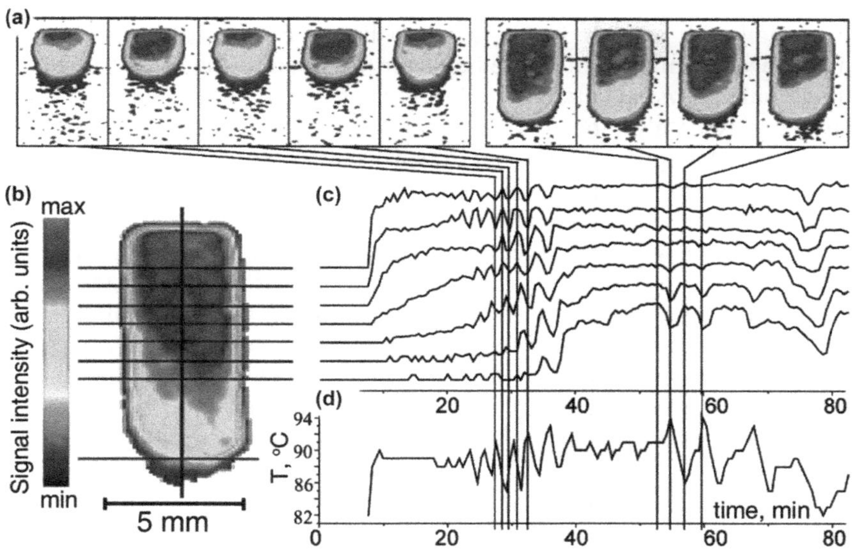

Fig. 7 ^{1}H MRI of an operating single-pellet catalytic reactor during α-methylstyrene hydrogenation. (a) Distribution of the liquid phase within the imaged axial slice of the 1% Pd/γ-Al_2O_3 pellet 4.5 mm in diameter and 12 mm long. (b) One of the images shown together with the signal intensity scale. The intersections of the set of equidistant horizontal lines with the vertical axial line define the spatial locations for the data shown in (c). The bottom horizontal line indicates the level at which the thermocouple is implanted into the pellet. (c) Signal intensity (liquid content) *vs.* time for the spatial locations shown in (b). Vertical lines indicate when the specific images shown in (a) were detected. (d) Temporal behavior of the pellet temperature measured at the location indicated with the bottom horizontal line in (b). Color is available in the electronic version of this figure. Reproduced from ref. 121 with permission of Elsevier BV.

observed experimentally were caused by the differences in the rates of heat and mass transfer and phase transitions. In particular, the calculated temperature variations and their period were in agreement with those observed experimentally, including the phase shift between the oscillations of temperature and liquid content.

MRI was also applied successfully to study mass transport processes in operating packed bed reactors. Such studies can address processes that require a cooperative action of a group of pellets not possible in a single catalyst pellet and thus take place on larger length scales. The advantage of MRI is that studies on these larger length scales can provide the spatially resolved information about liquid distribution in the reactor while still providing access to the distributions within each individual catalyst particle. The first studies addressed hydrogenation of olefins in a relatively regular bed of 1% Pd/γ-Al_2O_3 catalyst beads *ca.* 4 mm in diameter.[13,32,35,120] Liquid distribution was thus visualized both within the entire bed and in each individual bead, revealing a number of dynamic phenomena. On the length scale of the bed, distribution of the liquid phase was much less uniform under reactive conditions than in the absence of the reaction, and in addition was altered by interrupting and then restarting the flow of liquid reactant. On the length scale of individual catalyst pellets, the possibility of an ignition of an individual catalyst pellet was observed, which could remain essentially dry for prolonged periods of time even if surrounded by liquid-filled neighbors. In partially dry catalyst beads, a sharp liquid front and its oscillating motion were reminiscent of the behavior observed in the MRI studies on individual catalyst pellets. The redistribution of the liquid phase into the partially dry catalyst beads from the liquid-containing neighbors was visualized. This potentially makes such local structures very efficient in the hydrogenation reaction, as a substantial area of dry catalyst coexists with significant nearby pools of the liquid reactant.

Beds of smaller catalyst beads were used in the hydrogenation of various substrates (AMS, 1-octene, 1-heptene).[13,32,35,120] The distribution of the liquid phase in the individual catalyst beads was still available for the 2–3 mm beads, but not for the 1 mm ones. In the latter case, the steady-state distributions of the liquid phase on the length scale of the bed were not changing in time and revealed the presence of dry areas which were larger for lower liquid flow rates. These distributions depended on whether the catalyst bed was dry or prewetted at the time when the liquid supply was turned on. Under certain conditions, rapid drying indicative of the runaway of the trickle bed was observed.

Another approach for significantly reducing imaging times often used in medical MRI is the implementation of rapid imaging techniques that are able to detect a large number of data points in the *k*-space after each application of the pulse sequence. Unfortunately, most of the rapid pulse sequences developed for medical MRI are not applicable to the studies of liquids in porous materials due to the very different relaxation properties of such liquids compared to soft tissues of an organism. Very short T_2* times of the liquid phase in porous solids often require the use of spin-echo – based sequences. Therefore, more recent studies[16,36,37,97,98,124] introduced a fast multiecho imaging sequence that could be used to detect 3D images of the bed in 20–30 s or 2D images of one or several slices of the bed in 2–3 s. This approach was used to study in detail the dynamics of the liquid phase redistribution in regular layers of spherical catalyst grains. The results confirmed the existence of partially wet catalyst grains over long periods of time, replenishment of dry or partially wet grains with a liquid from adjacent grains, and the fluctuating liquid front propagation in partially wet grains. Evidently, the observation of such effects is impossible in a model system without an actual reaction.

Periodic modulation of the liquid reactants feed can lead to increased reactor productivity,[125] and MRI can be used to reveal the processes which lead to higher conversions.[124,126] In particular, hydrogenation of 1-octene was used to demonstrate the applicability of MRI in such studies.[124] The distribution of the liquid phase within the catalyst bed during the periodic feed of 1-octene and a continuous feed of H_2 was visualized in the bed comprised of 1% Pd/γ-Al_2O_3 catalyst beads with diameters 1 mm, 2–3 mm, or 4.2 mm. The effects of the liquid flow rate and the period of the on-off cycle of liquid 1-octene supply were studied, and the significant influence of reactant feed modulation on the liquid holdup and catalyst wetting efficiency were visualized directly for the first time through detection of 2D MR images. For continuous supply of 1-octene, the liquid phase distribution was relatively uniform and stationary. In contrast, the bed was largely dry, the distribution of the liquid phase was changing in time and the amount of liquid in the bed was oscillating when 1-octene supply was periodically modulated. Longer periods of liquid feed modulation led to larger liquid contents in the bed during the liquid pulse but also to a substantial or even complete drying of the bed between the pulses. The internal wetting efficiency of the catalyst was found to be lower for the periodic modulation of 1-octene feed as compared to the continuous one. Independent measurements of catalyst temperature and 1-octene conversion demonstrated significant intensification of the hydrogenation process in the periodic regime as compared to the continuous supply of 1-octene. Based on the MRI results, the intensification of the reaction for the periodic feed regime was explained by the reduction of the transport resistance for the H_2 gas and faster hydrogenation of 1-octene in the gas phase as compared to liquid phase hydrogenation.

Another example of an MRI study of an exothermic heterogeneous catalytic reaction is the decomposition of hydrogen peroxide (H_2O_2). MRI studies using a single Fe/Al_2O_3 catalyst pellet[127] revealed chaotic motion of the intrapellet liquid caused by gas and heat evolution during the reaction. Only relatively slow changes in the liquid phase distribution could be monitored as image acquisition time was *ca.* 4 min. In another study,[28,119] a composite pellet consisted of two parts, with one part representing the Ag/γ-Al_2O_3 catalyst with a moderate activity in the reaction while the other one made of γ-Al_2O_3 didn't contain any silver. The pellet was initially filled with water and then placed in the solution of H_2O_2 (0.03–3 M). The NMR signals were detected separately for the two parts of the pellet using spatially selective excitation. It was established that under the experimental conditions, hydrogen peroxide did not diffuse into the catalytically active part of the pellet and the decomposition of H_2O_2 occurred only in a thin surface layer of the pellet. Velocities of convective transport of the liquid phase around the catalyst pellet induced by gas evolution during the reaction were mapped as well.[119] A more recent study addressed reaction-induced transport of the liquid phase around the catalyst pellet during decomposition of H_2O_2 over a Cu- or Pt-doped Al_2O_3 pellet.[128] The temporal evolution of effective diffusion (dispersion) coefficients in the vertical and horizontal directions was addressed using pulsed-gradient stimulated echo (PGSTE) and double-PGSTE experiments. Comparing the results obtained with two techniques, the authors concluded that on the time scale of 10–90 ms stochastic transport played a minor role as compared to random stationary flow. The Pt-based catalyst exhibited a more efficient liquid transport and a shorter time to reaction completion. The T_2 time of the detected ^{1}H NMR signal was shown to change as the reaction proceeded and thus could be also used to qualitatively monitor its progress.[129]

In the context of wastewater bioremediation research, reduction of Cr(VI) was studied in a MRI-compatible reactor.[130] Matrix-immobilized biofilm produced by Serratia sp. bacteria was first used to reduce Pd(II) to produce biofilm-immobilized Pd nanoparticles and then used for reduction of Cr(VI) with formate as the electron donor. The fact that Cr(III) is paramagnetic was employed to monitor the reduction of Cr(VI) to Cr(III) with MRI and to evaluate local concentrations of Cr(III)(aq) in the reactor.

MRI is also increasingly used to study formation, spatial distribution and removal of water in model PEM (proton exchange, or polymer electrolyte, membrane) fuel cells under operating conditions.[131–133]

5.2 Spatially resolved NMR spectroscopy

An important advantage of magnetic resonance is the fact that in essence it is a spectroscopic technique. In the experiments performed using the imaging modality, the spectroscopic information is usually ignored and/or unavailable, *i.e.*, an image often reflects the distribution of the liquid phase irrespective of its composition. Nevertheless, it is possible to recover spectroscopic information from a properly designed spatially resolved MR experiment. This allows one, at least in principle, to obtain spatially resolved quantitative information on conversion and selectivity during catalytic processes. Several general approaches are available for such studies. In chemical shift selective imaging (CSSI), an image is constructed using one particular NMR signal in an NMR spectrum. This approach may be relatively fast, but it loses the rest of the spatial and spectroscopic information. In addition, it is not always possible to selectively excite and/or detect a single NMR line if it overlaps with other spectral components. Another approach, termed chemical shift imaging (CSI), retains all the spectroscopic information and yields a separate NMR spectrum for each pixel (voxel) of an image. Since the spectroscopic information is acquired by detecting a free induction decay (FID) or an echo signal in the absence of any applied gradients, spatial information has to be addressed using phase encoding of the signal, which can make the experiment quite lengthy. If spectroscopic information is desired only for a small number of selected volume elements, volume-selective NMR spectroscopy (VOSY) is a better choice because it reduces the acquisition time as compared to the CSI approach.

The CSI method was applied to study the gas-liquid-solid hydrogenation of AMS to cumene in an operating trickle bed reactor packed with 1% Pd/γ-Al_2O_3 catalyst beads 1 mm in diameter.[13,16,32,98,121] While in modern liquid state high-resolution NMR the signals can have line widths below 1 Hz, they are often much broader for liquids in porous materials. This is caused by the large inhomogeneity of the local magnetic fields induced by the magnetic susceptibility differences between different phases (solid, liquid, and gas), and also by the reduced nuclear spin relaxation times of intrapore liquid. For AMS in the catalyst bed, the ^1H NMR spectra detected at 300 MHz exhibited resonances *ca.* 300 Hz wide at 130 °C. Nevertheless, the chemical shift differences between the peaks of AMS and cumene are large enough for the corresponding signals of the two compounds in the pores of alumina to be distinguished. CSI was used to obtain spectroscopic information for each pixel of a 2D image of a 2 mm thick axial slice within the catalyst bed. Essentially, a ^1H NMR spectrum was made available for each image pixel. The expected increase in conversion of AMS to cumene down the catalyst bed was observed, and in addition the AMS/cumene ratio was observed to vary in the radial direction (Fig. 8).

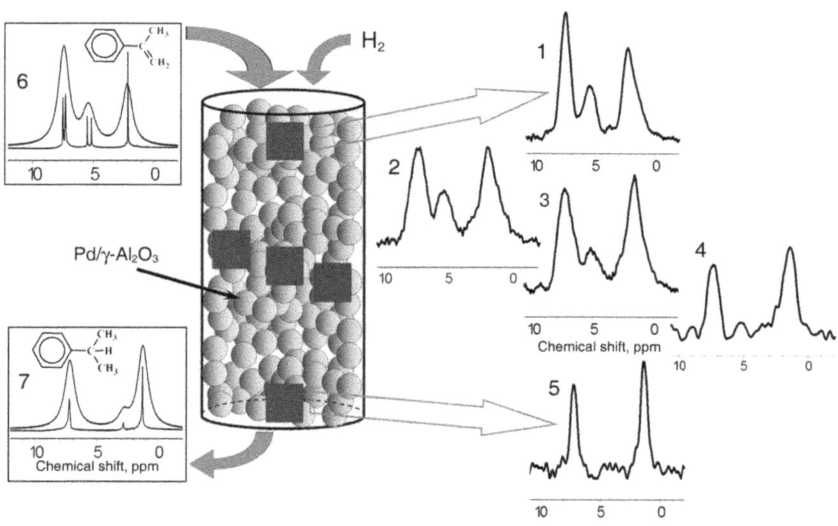

Fig. 8 ^1H NMR spectra detected with spatial resolution during AMS hydrogenation in a bed of 1 mm Pt/γ-Al_2O_3 catalyst beads. Each spectrum corresponds to a $170\times330\times2000\ \mu m^3$ voxel. Spectra 1,3,5 correspond to the same radial position within the operating reactor and are detected in its top, middle and bottom parts, respectively, while spectra 2-4 correspond to voxels at the same vertical position in the operating reactor but shifted horizontally by 1.3 mm relative to each other. In the spectra 6 and 7, the traces with narrow lines were detected for liquid AMS (6) and cumene (7), while the traces with broad lines were obtained by mathematically broadening the experimentally observed lines to 300 Hz.

One possibility to overcome the problem associated with severe line broadening is to perform an MR experiment using nuclei other than ^1H. In particular, the chemical shift range for ^{13}C NMR is at least an order of magnitude larger than that for ^1H NMR. Unfortunately, the price to pay when switching from ^1H signal detection to other nuclei is an inevitable loss in SNR. This is particularly true for ^{13}C, in part due to a four-fold lower magnetogyric ratio, but mostly because the natural abundance of the ^{13}C isotope is only *ca.* 1.1 %, while the main isotope of carbon (^{12}C) is NMR-silent. As a result, the sensitivity of ^{13}C NMR at natural abundance of the ^{13}C isotope is lower by a factor of ~5870 as compared to ^1H NMR. Such dramatic loss in sensitivity can be avoided by using ^{13}C-enriched compounds in the experiments. However, the amounts that one would need to feed a trickle bed reactor for a reasonable period of time would be unacceptably expensive.

In many NMR experiments, polarization is transferred between coupled nuclear spins by means of appropriate NMR pulse sequences. Polarization of spins in a magnetic field is proportional to the magnetogyric ratio of a nucleus, and is therefore *ca.* 4 times larger for ^1H as compared to ^{13}C spins. This can be used for a matching enhancement of ^{13}C NMR by transferring polarization from ^1H to ^{13}C nuclei. Combination of a 1D MRI experiment and the ^{13}C DEPT (distortionless enhancement by polarization transfer) spectroscopy pulse sequence was used in the studies of 1-octene transformations in the presence of H_2 in a gas-liquid fixed bed model reactor operated under conditions of co-current gas-liquid downflow.[7,81,134] The catalyst bed comprised 1 wt% Pd/Al_2O_3 catalyst trilobes. In this 2D spatial-spectral experiment, ^{13}C NMR spectra were obtained at each axial position along the reactor with a spatial resolution of 3 mm, with the NMR signal intensities integrated in the radial direction. The data acquisition time was *ca.* 17 min. From the analysis of the spatially resolved spectra it could be established that 1-octene readily isomerized into 2-octene, and n-octane was formed upon double bond hydrogenation. At higher flow rates of H_2, 3- and 4-octenes were produced in significant quantities. The concentrations of 1-octene, 2-octene and octane were shown to reach their steady-state values at different axial positions after reactor start-up using the same experimental approach. In a recent study, the novel reactor design mentioned earlier[15] was used to study oligomerization of ethylene at 100–200 °C.

MRI was also employed to study liquid-solid heterogeneous catalytic reactions. One of the early examples is the liquid-phase esterification of methanol and acetic acid into methyl acetate and water performed at room temperature in a packed bed reactor containing small beads of an acidic ion exchange resin.[7,135] To obtain spatially resolved information on chemical conversion, either CSI with one spatial coordinate along the bed axis or VOSY with 30 selected volume elements were applied. Only the NMR signal of the interparticle liquid phase contributed to the acquired data. The conversion of reactants to products was observed to increase along the bed and to decrease with increasing liquid flow rate. In the transverse direction, variations of conversion as large as ~20% could be observed. In the later studies, liquid flow velocities in the bed were both measured under non-reactive conditions using MRI and calculated using the lattice-Boltzmann approach, and then the transport data were correlated with the data on conversion provided by the MRI experiments.[21]

The spectroscopic approach used in the studies of esterification reaction[7,21,135] was based on the detection of the ^{1}H NMR signals of the OH groups of water, acetic acid and methanol which give a single exchange-broadened and shifted NMR line due to the fast proton exchange. In a later study, an approach based on the spectroscopically resolved CSI was demonstrated for an enzymatically catalyzed esterification of propionic acid and 1-butanol into butyl propionate and water carried out on an individual calcium alginate bead containing an immobilized enzyme.[136] The reaction within the alginate bead was monitored by detecting the spectra spatially resolved along one spatial coordinate. The increase in the signals of water and butyl propionate and decrease in the intensity of the butanol signal with time were clearly observed.

The competitive liquid phase etherification of 2-methyl-2-butene (MB) into tert-amyl methyl ether (TAME) and its hydration to tert-amyl alcohol (TAOH) in a fixed bed of H^+ ion-exchange resin beads was addressed in ref. 7. The 3D experiments performed had two spatial and one spectral coordinates, so that even with the use of the ^{13}C DEPT signal enhancement and a moderate spatial resolution (2.5 mm$\times$3.75 mm) the reaction had to be run for 16 hours to acquire the entire spectral-spatial data set. The spectral resolution was sufficient to evaluate conversion of MB and selectivity to TAME and their variations in the axial and transverse directions.

The spatially resolved ^{13}C DEPT spectroscopy experiment was also used to quantify the composition of the interparticle liquid phase along the reactor in the continuous liquid phase esterification of acetic acid and ethanol to ethyl acetate and water over an ion-exchange resin at ambient temperature and pressure.[137] In the same study, both the intra- and interparticle molecular diffusivities were evaluated separately for different components of the reacting mixture and with a spatial resolution of 6 mm along the catalyst bed. The results demonstrated that in all imaged parts of the bed, the concentration of ethyl acetate was higher in the intraparticle liquid than in the liquid phase outside the particles, indicating that the reaction was mass transfer limited. The total acquisition time of the spatially and chemically resolved diffusion experiment was $\sim$54.5 h.

5.3 Spatially resolved NMR thermometry

Heat transport processes can have an important influence on the operation of a catalytic reactor, and therefore need to be addressed in the experimental studies. This implies that the spatial distribution of temperatures and their variation in time need to be detected. Conventional thermometry techniques are either invasive and provide temperature readings for a limited number of spatial locations (local measurements with thermocouples or other probes[138–141]) or can only probe temperature fields at or near the catalyst or reactor surface (*e.g.*, IR thermography[142–147]). In general, the signals detected in NMR are sensitive to temperature, and therefore NMR/MRI can be used to measure temperatures indirectly. This is currently used in both analytical and biomedical applications of NMR and MRI.[28,148–151] In the high-resolution NMR of liquids, sample temperature is usually determined based on the property of the chemical shifts of certain liquids to appreciably change in response to temperature variation. For instance, ethylene glycol ($HOCH_2CH_2OH$) is one of the well-known NMR "thermometers". The chemical shifts of its two NMR signals depend differently on temperature, and therefore their separation provides an accurate evaluation of sample temperature with no need for any additional calibration. NMR thermometry can be also based on other characteristics of liquids and their NMR properties that are sensitive to temperature. In particular, NMR signal intensity, nuclear spin relaxation times, and diffusivity of molecules all depend on local temperature and can be evaluated using NMR. They all were employed in various studies to implement NMR-based thermometry. Unfortunately, none of the four temperature-dependent characteristics mentioned above (chemical shift, signal intensity, relaxation times, and diffusivity) are suitable for temperature measurements of liquids in operating multiphase reactors. Indeed, as demonstrated above, the local liquid contents may change drastically in space and time in an operating reactor, which affects all those temperature-sensitive characteristics even if local temperatures remain unchanged. A straightforward solution to this problem is to enclose a liquid in a small non-metallic container with impermeable walls and place it in a reactor. This strategy was implemented in ref. 81 to evaluate local temperatures during hydrogenation of 1-octene in a bed of Pd/Al_2O_3 trilobes. Four glass bulbs 4 mm in diameter filled with ethylene glycol were placed in various parts of the bed, and volume-selective ^1H NMR spectra of ethylene glycol in these bulbs were detected. Variations of local temperatures in the operating catalyst bed were evaluated with the accuracy of *ca.* ± 2 °C. However, this approach is invasive and provides temperature readings in a very limited number of locations within the bed, while an increase in the number of such temperature probes in the bed may significantly perturb reactor behavior. Even more importantly, this approach measures the temperature of a fairly large foreign object, which can be quite different from that of the solid catalyst or the fluid phases in the reactor.

To evaluate the catalyst temperature by NMR, it would be desirable to address the solid phase directly. This requires the detection of the NMR signal of the nuclei of the solid material (catalyst, support). Among the advantages of such approach are the eliminated need to introduce any temperature sensors and the fact that the bulk solid phase remains unchanged during the reaction and thus provides an NMR signal which should be sensitive to temperature variations only. While NMR/MRI of solid materials is often associated with much lower sensitivity as compared to the studies on liquids, some solids and their nuclei can be imaged using conventional MRI approaches.[29–37] Of particular interest in this context is the ^{27}Al nucleus of Al_2O_3 which is often used as a support for heterogeneous catalysts. Indeed, both the relaxation times and the signal intensity of ^{27}Al nuclei of γ-Al_2O_3 are sensitive to temperature.[16,30,98,122] Based on this fact, ^{27}Al MRI was applied to reveal spatial and temporal temperature variations in an operating catalytic reactor.[98,152] In the experiment, a mixture of propylene and H_2 was supplied to a fixed bed of Pd/γ-Al_2O_3 catalyst. One-dimensional profiles of the ^{27}Al NMR signal intensity along the catalyst bed with a sub-mm spatial resolution were detected every 3 minutes. The temperature of the bed was varied by changing the flow rate of propylene while keeping the H_2 feed constant. Thermocouple measurements in the lower part of the bed showed that the catalyst temperature increase from room temperature to 250 °C was accompanied by the decrease in the ^{27}Al NMR signal of the alumina support. Under certain regimes, the signal intensity distribution along the catalyst bed was significantly non-uniform, which likely reflected the gradient in the catalyst temperature. Control experiments performed at various temperatures demonstrated that ^{27}Al NMR signal intensity of alumina was insensitive to the composition of the gas phase in contact with the catalyst, confirming that signal intensity can be used as a measure of the catalyst temperature.

The ^{27}Al MRI thermometry approach was recently extended from the 1D qualitative imaging of temperature distribution to a 2D quantitative temperature mapping.[153] A single cylindrical Pt/γ-Al_2O_3 pellet was used to perform catalytic oxidation of H_2 with O_2 (Fig. 9). The pellet (4.2 mm in diameter and *ca.* 1 cm long) was glued to the wall on the inside of a 10 mm glass tube reactor. The transverse images visualized the geometry of the experiment since both the glass tube and the catalyst contain ^{27}Al

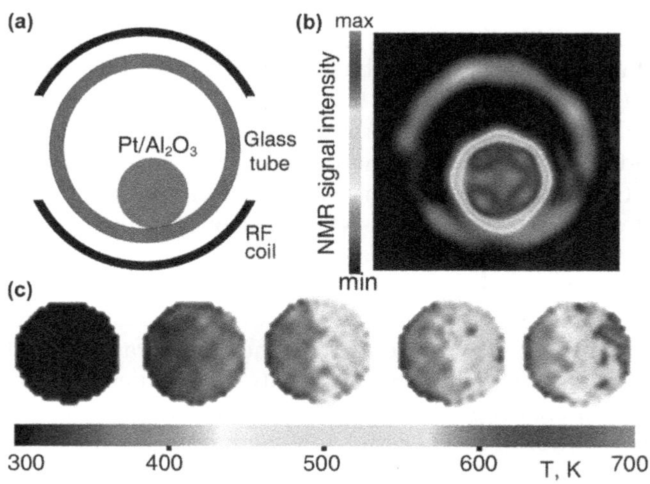

Fig. 9 (a) Geometry of Pt/Al_2O_3 pellet positioning for the MRI thermometry experiment during catalytic gas phase oxidation of H_2. (b) 2D ^{27}Al MR image of the catalyst pellet and the reactor glass tube. (c) 2D temperature maps of the catalyst pellet derived from the images of the catalyst pellet detected at different H_2 flow rates using the calibration curve that relates the ^{27}Al NMR signal intensity to the actual temperature of alumina. Color is available in the electronic version of this figure.

nuclei and thus contribute to the image. At the same time, the signal of glass was much weaker than that of the catalyst pellet, and could be neglected in the temperature mapping experiments. The ^{27}Al NMR signal intensity was uniform across the catalyst pellet in the absence of H_2. Once the H_2 flow was started, the signal decreased and exhibited a gradient in the pellet in the direction toward the contact between the pellet and the tube wall. Both effects became more pronounced as the H_2 flow rate was increased further. As the acquisition time of each 2D ^{27}Al MR image with the 0.4 mm×0.4 mm in-plane spatial resolution (without slice selection) and an acceptable SNR was fairly long (21 min), the H_2 flow rates were changed only between image acquisitions to achieve steady-state conditions.

An independent calibration experiment was used which related the measured ^{27}Al NMR signal intensity of alumina to its actual temperature which was measured by a thermocouple. This was used to convert the obtained images into quantitative 2D temperature maps. These maps revealed a pronounced temperature gradient in the transverse plane of the catalyst pellet associated with its asymmetric positioning in the reactor. Catalyst temperatures up to 700 K were achieved, demonstrating the applicability of this approach to the studies of high-temperature processes. The uncertainties in the temperature evaluation were governed by the MR signal measurement noise and were estimated as *ca.* 10–12 K at 400 K, *ca.* 30 K at 500 K and even larger at higher temperatures. This trend is explained by the overall decrease in the measured signal intensity with increasing sample temperature. It is reasonably straightforward to extend this approach to the thermometry of multiple catalyst pellets and packed catalyst beds. Since the experiment does not use any solid-state imaging methods and hardware, it can be implemented relatively easily on many instruments equipped with pulsed magnetic field gradients, including the *ex situ*, inside-out or mobile MRI devices developed recently,[11,154,155] which could be the next step to thermometry of operating industrial reactors. It is essential that this approach allows one to evaluate the true temperature of the solid catalyst pellets and packed beds in a non-invasive manner and with spatial resolution.

Despite the sensitivity problems associated with the direct detection of the signal of a solid phase, ^{27}Al NMR/MRI of alumina appears to be a useful temperature sensor. However, this approach is not problem-free as well. For some other Al-containing materials including zeolites, this approach may fail since surface adsorption processes may significantly affect the ^{27}Al NMR signal intensity. Higher temperatures lead to lower signal intensities, reducing the sensitivity and achievable spatial resolution and/or increasing measurement time. The approach also needs an independent calibration of signal intensity *versus* temperature for the measurements to be quantitative. Nevertheless, the results already obtained are quite encouraging and demonstrate that non-invasive solid state MRI thermometry is feasible in favorable cases, and may become a useful technique capable of revealing temperature gradients and formation and evolution of hot spots in operating catalytic reactors. An extension to liquid-solid and gas-liquid-solid processes should be straightforward as the presence of a liquid should not affect the ^{27}Al NMR signal in any way other than through the catalyst temperature.

5.4 MRI of microreactors and catalytic signal enhancement

One of the recent trends in chemical technology and catalysis is the use of microreactors and microfluidic devices. Their growing popularity is attributed to a number of advantages that include a better control over mass and heat transport, enhanced performance, faster and more efficient chemical processes, improved safety even when operated under extreme conditions, and easier scale-up. At the same time, from the point of view of NMR and MRI studies, microdevices present some additional challenges. In a conventional NMR/MRI experiment, the size of an rf coil would be dictated by the external size of such device, which is often much larger than the characteristic dimensions of the reaction zone of a microreactor or the channel diameters of a microfluidic chip. The low filling factor of the rf coil results in a poor SNR for liquids and makes such experiments essentially unfeasible for gases.

A useful approach particularly suitable for addressing gas-phase reactions in microreactors is based on the so-called remote detection (RD) method.[90,94,156–158] The idea behind it is to separate the encoding and the detection parts of NMR pulse sequence both in space and time. The large rf coil in which a microreactor or a microchip resides is used for encoding the spatial information into the longitudinal magnetization of the fluid in the voids or channels. The fluid flows out of the reactor continuously through a thin capillary which passes through a microcoil used for a periodic detection of the NMR signal. In addition to spatial and spectroscopic information, the time-of-flight information is obtained using a set of delays between the encoding event and the multiple detection events to recover the temporal data.

In ref. 159 RD was used to address hydrogenation of propylene in a microreactor comprising 36 parallel channels with a $50\times50\,\mu m^2$ cross-section and 5, 10 or 20 mm long, produced by etching a silicon wafer. The channel walls were coated with a Pt layer. The microreactor was placed in a 25 mm i.d. rf coil located in a 7.05 T NMR magnet. The detection microcoil was wound directly around the 360 µm o.d. outlet capillary and was 3 mm long and located *ca.* 1 cm below the encoding coil. The proximity of the detection microcoil to the magnet center ensures the reasonable homogeneity of the static magnetic field at its location provided by the shimming system of the NMR magnet. It also leads to relatively short travel times of the gas from the encoding to the detection region which reduces the relaxation-induced losses. After each repetition of the encoding part of the pulse sequence, 30 consecutive FIDs were collected, separated by a 21 ms delay. With the spatial resolution of 2D images obtained (1.9 mm along the channels and 0.5–1 mm in the transverse direction), each pixel covered 2–5 individual channels in the transverse direction. Despite the small dimensions of the microcoil (53 nl of gas in the detection volume), the NMR spectrum of the gas mixture could be detected. The line width for the static gas was *ca.* 30 Hz and increased to *ca.* 100 Hz when the gas was flowing because the gas residence time in the microcoil was *ca.* 20 ms at the chosen flow velocity. This was sufficient to resolve the chemical shifts of the reactant (propylene) and the product (propane). From the set of spatially phase-encoded spectra, a set of images could be reconstructed, with each image in the set depicting the spatial location of the chosen compound with a given time-of-flight (TOF) to the detection microcoil. Such TOF images were extracted for both propylene and propane. From these images it was concluded that gas flow velocities and propylene conversions were different in different parts of the reactor, with smaller flow rates providing larger product yields. The interpretation of the RD CSI experiment results can be rather non-trivial. The authors demonstrated, however, that a detailed understanding of the experiment combined with an appropriate choice of the pulse sequence parameters can provide the true TOF images of the reaction product. Possible ways to improve the sensitivity of the experiment were discussed, which may lead to the spatial resolution sufficient to resolve individual channels.

The limited sensitivity is a major obstacle for improving the spatial and temporal resolution in most existing MRI experiments, and is also a significant obstacle for developing novel powerful tools for the MRI toolkit. Imaging of gases and visualization of narrow channels at high spatial resolution are particularly demanding in this respect. It was estimated[159] that the use of the microcoil for signal detection in the RD experiments improved the detection sensitivity by a factor of *ca.* 800 compared to a hypothetical experiment in which the large encoding rf coil would be used for signal detection. Nevertheless, further improvement in SNR would be highly desirable not only for the studies of microreactors, but also to address larger model reactors as well. In fact, the entire field of NMR/MRI can significantly benefit from an improved SNR. It is therefore not surprising that the so-called hyperpolarization methods[2,28,160–173] are becoming extremely popular in magnetic resonance. This family of techniques is being developed in an attempt to maximize the polarization of nuclear spins in a magnetic field, thereby providing signal enhancements of up to 2×10^4 for ^{1}H NMR/MRI in a 14 T field (600 MHz NMR instrument), and even higher enhancements at lower fields and/or for other nuclei.

One of the hyperpolarization techniques, called parahydrogen-induced polarization (PHIP, or PASADENA) is particularly suitable for the development of hypersensitive NMR/MRI tools for catalysis. This is a rare example of a situation when catalysis can help magnetic resonance to solve its problems, while usually it is the other way around. Signal enhancement with PHIP[2,174–179] is based on a catalytic hydrogenation reaction in which parahydrogen (pH_2) is used instead of normal H_2. Parahydrogen is one of the two nuclear spin isomers of H_2 with the total nuclear spin of two H atoms $I=0$. It gives no NMR signal by itself, but the highly correlated state of its nuclear spins can be converted into an NMR signal enhancement. To achieve this, the equivalence of the two H atoms of the H_2 molecule has to be broken, which can be achieved when pH_2 participates in a suitable chemical reaction. PHIP is observed in many homogeneous hydrogenation reactions of suitable unsaturated precursors catalyzed by transition metal complexes. Since in PHIP the hyperpolarization of nuclear spins is produced in a catalytic reaction, this technique can be used as a highly sensitive NMR-based tool for the studies of the mechanisms and kinetics of homogeneous hydrogenations catalyzed by transition metal complexes in solution.[180–182] The high sensitivity provided by PHIP enables the detection not only of reaction products, but also of short-lived reaction intermediates that are normally present in solution at very low concentrations and thus escape NMR detection.

More recently, it has been demonstrated that PHIP effects can be also observed in heterogeneous hydrogenations of unsaturated substrates catalyzed by transition metal complexes immobilized on suitable porous solid supports[178,179,183–186] as well as by supported metal nanoparticles.[178,179,187] The latter observation was quite surprising as metal surfaces were expected to be unable to achieve the required incorporation of both H atoms of a H_2 molecule into the same product molecule because of the dissociative chemisorption of H_2 and rapid surface migration of the resulting H atoms. Nevertheless, PHIP effects were successfully observed in hydrogenation reactions catalyzed by many supported metal catalysts.[178,179,188–191] This provides an opportunity for the development of a powerful hypersensitive NMR-based technique for the characterization of the heterogeneous catalytic processes of practical importance.

Combination of PHIP with MRI was already used to study processes in a model catalytic reactor.[98,178,192] A small packed bed reactor comprised a piece of tubing packed with a heterogeneous Rh-based catalyst. The reactor was positioned inside the NMR probe and was imaged during the hydrogenation of propylene with parahydrogen. Those parts of the catalyst bed where the hydrogenation of propylene to propane was taking place were visualized in the images as the regions with hyperintense NMR signal. Furthermore, the signal enhancement provided by PHIP made it possible to detect the velocity map for the hydrocarbon gas flowing in the packed bed of the operating reactor. The experiments demonstrated that both the catalyst packing and the flow field were spatially non-uniform.

PHIP-enhanced MRI in combination with the RD technique considered above was used to address packed bed microreactors.[193] Three different continuous flow microreactors packed with Rh/SiO_2 catalyst were used in the experiments. The diameters of the packed beds were 800, 405, and 150 μm, and their lengths were 5, 14, and 15 mm, respectively. The signal enhancement provided by PHIP in the hydrogenation of propylene to propane using pH_2 was estimated as *ca.* 60, and the overall sensitivity enhancement factor given by RD (820) and PHIP was estimated as *ca.* 5×10^4. The NMR signal of the hyperpolarized gas (propane) significantly exceeded that of the thermally polarized gas (propylene), which simplifies the interpretation of the results of the RD experiment. The average gas flow rates, both in the packed bed and in the outlet capillary, were determined from the 2D data sets with one spatial (axial) and one TOF coordinate, directly from the slope of the corresponding ridges.

Comparison of the gas flow velocities and signal intensities in the packed bed and in the outflow capillary revealed a substantial adsorption of the gas on the porous catalyst support in the packed bed, which was then quantified. Furthermore, summation of all 1D axial profiles over the TOF coordinate provided the axial distribution of propane concentration assuming that the packed bed was uniform and relaxation effects negligible. Such distributions obtained for the three reactors were simulated based on the plug-flow reactor behavior and the appropriate kinetic models. The experiments were further extended to obtain 3D data sets, namely the 2D images detected for different times-of-flight (Fig. 10). The images revealed the plug-like gas flow in the two smaller reactors, while some deviation from the plug-like behavior and a larger dispersion was revealed for the larger one. The build-up of hyperpolarization along the microreactor can be clearly observed.

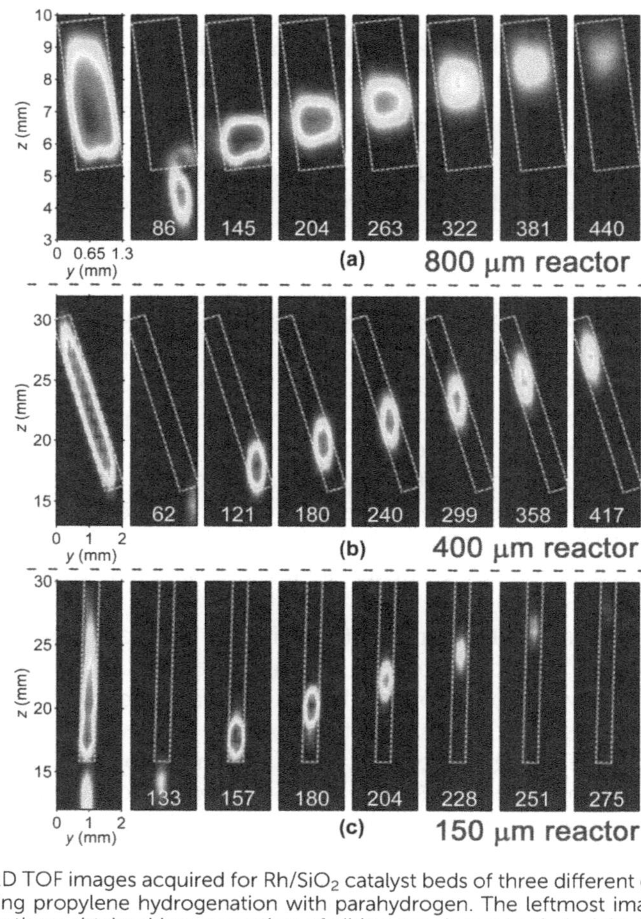

Fig. 10 RD TOF images acquired for Rh/SiO_2 catalyst beds of three different diameters at 60 °C during propylene hydrogenation with parahydrogen. The leftmost images are the time projections obtained by summation of all images shown to their right. Travel times are indicated in the panels in milliseconds. The catalyst bed regions are outlined with white dashed lines. The complete data set for each reactor was acquired in 13 min with a time resolution of 12 ms and a spatial resolution of 160–250 μm in the horizontal direction and 0.62–2.2 mm in the vertical direction. Reproduced from ref. 193 with permission of Wiley-VCH Verlag GmbH & Co. KGaA, Weinheim.

6 Summary and outlook

Applications of MR imaging in heterogeneous catalysis and related disciplines are still an art rather than everyday routine. Nevertheless, the technique has reached a stage when it can provide useful data on the processes even in operating model reactors, including information about the internal structure, mass and heat transport processes, and chemical conversion of reactants into products. Given the complexity of processes that take place in the reactors, the most powerful approach is to use MRI in combination with other experimental techniques and mathematical modeling. The data provided by MRI can be used as a direct input to limit the number of unknowns in calculations as well as to validate the results of such calculations. This approach has synergistic potential and is capable of providing significantly more information than MRI studies or modeling alone.[56]

In addition, this field of research holds promise for significant growth in the future, which will be fueled by the ongoing developments in other areas of application of NMR and MRI, in particular in the broad field of porous media MR, where the development of faster imaging techniques and techniques applicable to samples with short T_2 times is highly desirable and attracts a lot of attention at present. Another major problem is insufficient sensitivity which poses significant obstacles on the way to higher spatial and temporal resolution and to the development of novel applications. This calls for the implementation of significant signal enhancement, which can be achieved through the ongoing development of the hyperpolarization techniques. Furthermore, it is already quite clear that the utilization of the properties of the so-called long-lived spin states[194–199] can go a long way in extending the lifetime of hyperpolarization well beyond the T_1 limit and thus can significantly broaden the scope of new and exciting developments. Hypepolarization methods are already being applied to the studies of enzymatic processes *in vivo* and in bioreactors,[2] and thus can provide new impetus for applications of MRI in the studies of model biocatalytic reactors.[200–203]

References

1 I. V. Koptyug, *Appl. Magn. Reson.*, 2007, **32**, 1.

2 I. V. Koptyug, Spin Hyperpolarization in NMR to Address Enzymatic Processes In Vivo, *Mendeleev Commun.*, 2013, in press.

3 P. T. Callaghan, *Principles of Nuclear Magnetic Resonance Microscopy*, Oxford University Press, New York, 1991.

4 B. Blümich, *NMR Imaging of Materials*, Oxford University Press, New York, 2003.

5 R. Kimmich, *NMR: Tomography, Diffusometry, Relaxometry*, Springer-Verlag, Berlin, 1997.

6 I. V. Koptyug and R. Z. Sagdeev, *Russ. Chem. Rev.*, 2002, **71**, 593.

7 L. F. Gladden, M. D. Mantle and A. J. Sederman, *Adv. Catal.*, 2006, **50**, 1.

8 L. Ciobanu, D. A. Seeber and C. H. Pennington, *J. Magn. Reson.*, 2002, **158**, 178.

9 S. K. Kuster, E. Danieli, B. Blumich and F. Casanova, *Phys. Chem. Chem. Phys.*, 2011, **13**, 13172.

10 B. Blumich, F. Casanova, M. Dabrowski, E. Danieli, L. Evertz, A. Haber, M. Van Landeghem, S. Haber-Pohlmeier, A. Olaru, J. Perlo and O. Sucre, *New J. Phys.*, 2011, **13**, 015003.
11 B. Blumich, J. Perlo and F. Casanova, *Progr. NMR Spectr.*, 2008, **52**, 197.
12 F. Dalitz, M. Cudaj, M. Maiwald and G. Guthausen, *Progr. NMR Spectr.*, 2012, **60**, 52.
13 A. A. Lysova, I. V. Koptyug, A. V. Kulikov, V. A. Kirillov, R. Z. Sagdeev and V. N. Parmon, *Chem. Eng. J.*, 2007, **130**, 101.
14 N. L. Nguyen, V. van Buren, A. von Garnier, E. H. Hardy and R. Reimert, *Chem. Eng. Sci.*, 2005, **60**, 6289.
15 M. P. Renshaw, S. T. Roberts, M. Lutecki, M. D. Mantle, J. McGregor, A. J. Sederman, L. F. Gladden. Operando MRI: Fixed-bed heterogeneous catalysis at elevated temperature and pressure. Extended abstract, 11th European Congress on Catalysis - EuropaCat-XI, Lyon, France, September 1–6, 2013.
16 I. V. Koptyug, *Encyclopedia of Magnetic Resonance*, ed. R. K. Harris and R. Wasylishen, John Wiley & Sons, Ltd., 2011, emrstm1266.
17 N. L. Nguyen, R. Reimert and E. H. Hardy, *Chem. Eng. Technol.*, 2006, **29**, 820.
18 N. L. Nguyen, V. van Buren, R. Reimert and A. von Garnier, *Magn. Reson. Imaging*, 2005, **23**, 395.
19 A. Feinauer, S. A. Altobelli and E. Fukushima, *Magn. Reson. Imaging*, 1997, **15**, 479.
20 X. Ren, S. Stapf and B. Blumich, *AIChE J.*, 2005, **51**, 392.
21 E. H. L. Yuen, A. J. Sederman, F. Sani, P. Alexander and L. F. Gladden, *Chem. Eng. Sci.*, 2003, **58**, 613.
22 S. P. Sullivan, F. M. Sani, M. L. Johns and L. F. Gladden, *Chem. Eng. Sci.*, 2005, **60**, 3405.
23 P. Jezzard, J. J. Attard, T. A. Carpenter and L. D. Hall, *Progr. NMR Spectr.*, 1991, **23**, 1.
24 D. G. Cory, *Annu. Rep. NMR Spectrosc.*, 1992, **24**, 87.
25 J. B. Miller, *Progr. NMR Spectr.*, 1998, **33**, 273.
26 A. J. Fagan and D. J. Lurie, *Annu. Rep. NMR Spectrosc.*, 2005, **56**, 97.
27 D. E. Demco and B. Blumich, *Concepts Magn. Reson.*, 2000, **12**, 188.
28 I. V. Koptyug and R. Z. Sagdeev, *Russ. Chem. Rev.*, 2003, **72**, 165.
29 I. V. Koptyug, A. A. Lysova, A. V. Khomichev and R. Z. Sagdeev, *Diff. Fundam.*, 2007, **5**, 2-1.
30 I. V. Koptyug, D. R. Sagdeev, E. Gerkema, H. Van As and R. Z. Sagdeev, *J. Magn. Reson.*, 2005, **175**, 21.
31 I. V. Koptyug and A. A. Lysova, *Bruker Spin Report*, 2006, **157–158**, 22.
32 I. V. Koptyug and A. A. Lysova, *NMR Imaging in Chemical Engineering*, ed. S. Stapf and S.-I. Han, Wiley-VCH, Weinheim, 2006, 570–589.
33 M. S. Conradi, *J. Magn. Reson.*, 1991, **93**, 419.
34 I. V. Koptyug, A. V. Khomichev, A. A. Lysova and R. Z. Sagdeev, *Appl. Magn. Reson.*, 2007, **32**, 321.
35 I. V. Koptyug, A. A. Lysova, R. Z. Sagdeev, V. A. Kirillov, A. V. Kulikov and V. N. Parmon, *Catal. Today*, 2005, **105**, 464.
36 I. V. Koptyug, A. A. Lysova, R. Z. Sagdeev and V. N. Parmon, *Catal. Today*, 2007, **126**, 37.
37 I. V. Koptyug, A. A. Lysova, K. V. Kovtunov, V. V. Zhivonitko, A. V. Khomichev and R. Z. Sagdeev, *Russ. Chem. Rev.*, 2007, **76**, 583.
38 J. L. Ackerman, L. Garrido, J. R. Moore, B. Pfleiderer and Y. Wu, *Magnetic Resonance Microscopy. Methods and Applications in Materials Science, Agriculture and Biomedicine*, ed. B. Blumich and W. Kuhn, VCH, Weinheim, 1992, pp. 237–260.

39 D. A. Graf von der Schulenburg and M. L. Johns, *Chem. Eng. Sci.*, 2011, **66**, 3003.
40 S. Harms, S. Stapf and B. Blumich, *J. Magn. Reson.*, 2006, **178**, 308.
41 C. R. Muller, D. J. Holland, A. J. Sederman, M. D. Mantle, L. F. Gladden and J. F. Davidson, *Powder Technol.*, 2008, **183**, 53.
42 L. F. Gladden and J. Mitchell, *New J. Phys.*, 2011, **13**, 035001.
43 I. V. Koptyug and R. Z. Sagdeev, *Russ. Chem. Rev.*, 2002, **71**, 789.
44 J. Karger and S. Vasenkov, *Microporous Mesoporous Mater.*, 2005, **85**, 195.
45 C. Chmelik and J. Karger, *Chem. Soc. Rev.*, 2010, **39**, 4864.
46 J. Karger, *Mol. Sieves*, 2008, **7**, 85.
47 F. Stallmach and J. Karger, *Adsorption*, 1999, **5**, 117.
48 P. Kortunov, S. Vasenkov, J. Karger, M. Fe Elia, M. Perez, M. Stocker, G. K. Papadopoulos, D. Theodorou, B. Drescher, G. McElhiney, B. Bernauer, V. Krystl, M. Kocirik, A. Zikanova, H. Jirglova, C. Berger, R. Glaser, J. Weitkamp and E. W. Hansen, *Chem. Mater.*, 2005, **17**, 2466.
49 A. Pampel, M. Fernandez, D. Freude and J. Karger, *Chem. Phys. Lett.*, 2005, **407**, 53.
50 S. Vasenkov, O. Geier and J. Karger, *Eur. Phys. J. E*, 2003, **12**, S35.
51 M. Gratz, M. Wehring, P. Galvosas and F. Stallmach, *Microporous Mesoporous Mater.*, 2009, **125**, 30.
52 I. Ardelean and R. Kimmich, *Annu. Rep. NMR Spectrosc.*, 2003, **49**, 43.
53 F. Stallmach and P. Galvosas, *Annu. Rep. NMR Spectrosc.*, 2007, **61**, 51.
54 I. V. Koptyug, *Progr. NMR Spectr.*, 2012, **65**, 1.
55 E. Fukushima, *Annu. Rev. Fluid Mech.*, 1999, **31**, 95.
56 C. J. Elkins and M. T. Alley, *Exp. Fluids*, 2007, **43**, 823.
57 J. M. Pope and S. Yao, *Concepts Magn. Reson.*, 1993, **5**, 281.
58 J. D. Seymour and P. T. Callaghan, *AIChE J.*, 1997, **43**, 2096.
59 P. T. Callaghan and Y. Xia, *J. Magn. Reson.*, 1991, **91**, 326.
60 I. V. Koptyug, A. A. Lysova, A. V. Matveev, L. Yu. Ilyina, R. Z. Sagdeev and V. N. Parmon, *Magn. Reson. Imaging*, 2003, **21**, 337.
61 S.-I. Han, S. Stapf and B. Blumich, *J. Magn. Reson.*, 2000, **146**, 169.
62 S.-I. Han and B. Blumich, *Appl. Magn. Reson.*, 2000, **18**, 101.
63 B. Blumich, P. T. Callaghan, R. A. Damion, S.-I. Han, A. A. Khrapitchev, K. J. Packer and S. Stapf, *J. Magn. Reson.*, 2001, **152**, 162.
64 A. A. Khrapitchev, S. Han, S. Stapf and B. Blumich, *J. Magn. Reson.*, 2002, **159**, 36.
65 P. T. Callaghan and B. Manz, *J. Magn. Reson. A*, 1994, **106**, 260.
66 I. V. Koptyug, A. A. Lysova, A. V. Matveev, L. Yu. Ilyina R. Z. Sagdeev and V. N. Parmon, *Fluid transport in nanoporous materials*, ed. W. C. Conner and J. Fraissard, Springer, Dordrecht, 2006, 353–374.
67 L. F. Gladden and A. J. Sederman, *J. Magn. Reson.*, 2013, **229**, 2.
68 B. Newling, *Progr. NMR Spectr.*, 2008, **52**, 31.
69 A. K. Heibel, T. W. J. Scheenen, J. J. Heiszwolf, H. Van As, F. Kapteijn and J. A. Moulijn, *Chem. Eng. Sci.*, 2001, **56**, 5935.
70 I. V. Koptyug, A. A. Lysova, A. V. Matveev, V. N. Parmon and R. Z. Sagdeev, *Topics in Catalysis*, 2005, **32**, 83.
71 Y. E. Kutsovsky, L. E. Scriven, H. T. Davis and B. E. Hammer, *Phys. Fluids*, 1996, **8**, 863.
72 D. Mertens, C. Heinen, E. H. Hardy and H. W. Buggisch, *Chem. Eng. Technol.*, 2006, **29**, 854.
73 T. R. Brosten, S. L. Codd, K. V. Romanenko, S. W. Sofie and J. D. Seymour, *AIChE J.*, 2009, **55**, 2506.
74 I. V. Koptyug, S. A. Altobelli, E. Fukushima, A. V. Matveev and R. Z. Sagdeev, *J. Magn. Reson.*, 2000, **147**, 36.

75 I. V. Koptyug, L. Yu. Ilyina, A. V. Matveev, R. Z. Sagdeev, V. N. Parmon and S. A. Altobelli, *Catal. Today*, 2001, **69**, 385.
76 I. V. Koptyug, A. V. Matveev and S. A. Altobelli, *Appl. Magn. Reson.*, 2002, **22**, 187.
77 I. V. Koptyug, L. Y. Ilyina, A. V. Matveev, R. Z. Sagdeev and V. N. Parmon, *Magnetic resonance in colloid and interface science*, ed. J. Fraissard and O. Lapina, Kluwer Academic Publishers, Dordrecht, 2002, 197–208.
78 I. V. Koptyug, K. V. Kovtunov, E. Gerkema, L. Kiwi-Minsker and R. Z. Sagdeev, *Chem. Eng. Sci.*, 2007, **62**, 4459.
79 S. Tsushima, S. Hirai, Y. Yamamoto and Y. Nakasuji, *Magn. Reson. Imaging*, 2003, **21**, 430.
80 N. P. Ramskill, L. F. Gladden, A. P. E. York, A. J. Sederman, J. Mitchell and K. A. Hardstone, *Catal. Today*, 2013, **216**, 104.
81 L. F. Gladden, F. J. R. Abegao, C. P. Dunckley, D. J. Holland, M. H. Sankey and A. J. Sederman, *Catal. Today*, 2010, **155**, 157.
82 J. Park and S. J. Gibbs, *AIChE J.*, 1999, **45**, 655.
83 U. Tallarek, D. van Dusschoten, H. Van As, E. Bayer and G. Guiochon, *J. Phys. Chem. B*, 1998, **102**, 3486.
84 U. Tallarek, E. Bayer and G. Guiochon, *J. Am. Chem. Soc.*, 1998, **120**, 1494.
85 U. Tallarek, F. J. Vergeldt and H. Van As, *J. Phys. Chem. B*, 1999, **103**, 7654.
86 U. Tallarek, D. van Dusschoten, H. Van As, G. Guiochon and E. Bayer, *Angew. Chem. Int. Ed.*, 1998, **37**, 1882.
87 J. J. Tessier, K. J. Packer, J.-F. Thovert and P. M. Adler, *AIChE J.*, 1997, **43**, 1653.
88 L. G. Kaiser, J. W. Logan, T. Meersmann and A. Pines, *J. Magn. Reson.*, 2001, **149**, 144.
89 S. L. Codd and S. A. Altobelli, *J. Magn. Reson.*, 2003, **163**, 16.
90 E. Harel, J. Granwehr, J. A. Seeley and A. Pines, *Nature Mater.*, 2006, **5**, 321.
91 A. V. Matveev, L. V. Barysheva, I. V. Koptyug, V. M. Khanaev and A. S. Noskov, *Chem. Eng. Sci.*, 2006, **61**, 2394.
92 K. Kose, *Phys. Rev. A*, 1991, **44**, 2495.
93 A. B. Tayler, D. J. Holland, A. J. Sederman and L. F. Gladden, *J. Magn. Reson.*, 2011, **211**, 1.
94 V. S. Bajaj, J. Paulsen, E. Harel and A. Pines, *Science*, 2010, **330**, 1078.
95 A. A. Lysova, I. V. Koptyug, R. Z. Sagdeev, V. N. Parmon, J. A. Bergwerff and B. M. Weckhuysen, *J. Am. Chem. Soc.*, 2005, **127**, 11916.
96 J. A. Bergwerff, L. G. A. van der Water, A. A. Lysova, I. V. Koptyug, T. Visser, K. P. de Jong and B. M. Weckhuysen, *Stud. Surf. Sci. Catal.*, 2006, **162**, 175.
97 I. V. Koptyug, A. A. Lysova, V. N. Parmon and R. Z. Sagdeev, *Kinet. Catal.*, 2007, **48**, 464.
98 A. A. Lysova and I. V. Koptyug, *Chem. Soc. Rev.*, 2010, **39**, 4585.
99 A. A. Lysova, J. A. Bergwerff, L. Espinosa-Alonso, B. M. Weckhuysen and I. V. Koptyug, *Appl. Catal. A: General*, 2010, **374**, 126.
100 J. A. Bergwerff, A. A. Lysova, L. Espinosa Alonso, I. V. Koptyug and B. M. Weckhuysen, *Angew. Chem. Int. Ed.*, 2007, **46**, 7224.
101 J. A. Bergwerff, A. A. Lysova, L. Espinosa-Alonso, I. V. Koptyug and B. M. Weckhuysen, *Chem. Eur. J.*, 2008, **14**, 2363.
102 L. Espinosa-Alonso, A. A. Lysova, P. Peinder, K. P. de Jong, I. V. Koptyug and B. M. Weckhuysen, *J. Am. Chem. Soc.*, 2009, **131**, 6525.
103 I. V. Koptyug, L. Yu. Khitrina, V. N. Parmon and R. Z. Sagdeev, *Magn. Reson. Imaging*, 2001, **19**, 531.
104 L. Yu. Khitrina, I. V. Koptyug, N. A. Pakhomov, R. Z. Sagdeev and V. N. Parmon, *J. Phys. Chem. B*, 2000, **104**, 1966.

105 Z. R. Ismagilov, S. A. Yashnik, A. V. Matveev, I. V. Koptyug and J. A. Moulijn, *Catal. Today*, 2005, **105**, 484.
106 A. von Garnier, E. H. Hardy, J.-M. Schweitzer and R. Reimert, *Chem. Eng. Sci.*, 2007, **62**, 5330.
107 K. Y. Cheah, N. Chiaranussati, M. P. Hollewand and L. F. Gladden, *Appl. Catal. A: General*, 1994, **115**, 147.
108 J.-L. Bonardet, T. Domeniconi, P. N'Gokoli-Kekele, M.-A. Springuel-Huet and J. Fraissard, *Langmuir*, 1999, **15**, 5836.
109 T. Domeniconi, J.-L. Bonardet, M.-A. Springuel-Huet, J. Fraissard and J.-M. Dereppe, *Stud. Surf. Sci. Catal.*, 1997, **111**, 647.
110 N.-K. Bar, F. Bauer, D. M. Ruthven and B. J. Balcom, *J. Catal.*, 2002, **208**, 224.
111 S. Stapf, X. Ren, E. Talnishnikh and B. Blumich, *Magn. Reson. Imaging*, 2005, **23**, 383.
112 B. M. Weckhuysen, *Phys. Chem. Chem. Phys.*, 2003, **5**, 4351.
113 A. Urakawa and A. Baiker, *Topics in Catalysis*, 2009, **52**, 1312.
114 S. Mitchell, N.-L. Michels, G. Majano and J. Perez-Ramirez, *Curr. Opinion Chem. Eng.*, 2013, **2**, 304.
115 B. M. Weckhuysen, *Angew. Chem. Int. Ed.*, 2009, **48**, 4910.
116 A. A. Lysova, A. von Garnier, E. H. Hardy, R. Reimert and I. V. Koptyug, *Chem. Eng. J.*, 2011, **173**, 552.
117 I. V. Koptyug, A. V. Kulikov, A. A. Lysova, V. A. Kirillov, V. N. Parmon and R. Z. Sagdeev, *J. Am. Chem. Soc.*, 2002, **124**, 9684.
118 I. V. Koptyug, A. V. Kulikov, A. A. Lysova, V. A. Kirillov, R. Z. Sagdeev and V. N. Parmon, *Doklady Phys. Chem.*, 2002, **385**, 158.
119 I. V. Koptyug, A. V. Kulikov, A. A. Lysova, V. A. Kirillov, V. N. Parmon and R. Z. Sagdeev, *Chem. Sust. Dev.*, 2003, **11**, 109.
120 V. A. Kirillov and I. V. Koptyug, *Ind. Eng. Chem. Res.*, 2005, **44**, 9727.
121 I. V. Koptyug, A. A. Lysova, A. V. Kulikov, V. A. Kirillov, V. N. Parmon and R. Z. Sagdeev, *Appl. Catal. A: General*, 2004, **267**, 143.
122 I. V. Koptyug, A. A. Lysova, A. V. Kulikov, V. A. Kirillov, V. N. Parmon and R. Z. Sagdeev, *Magn. Reson. Imaging*, 2005, **23**, 221.
123 V. A. Kirillov, I. V. Koptyug, A. V. Kulikov, N. A. Kuzin, A. A. Lysova, A. B. Shigarov and V. N. Parmon, *Theor. Found. Chem. Eng.*, 2005, **39**, 24.
124 A. A. Lysova, I. V. Koptyug, A. V. Kulikov, V. A. Kirillov and R. Z. Sagdeev, *Topics in Catalysis*, 2009, **52**, 1371.
125 A. Ayude, J. Cechini, M. Cassanello, O. Martinez and P. Haure, *Chem. Eng. Sci.*, 2008, **63**, 4969.
126 W. Dietrich, L. Anadon, A. Sederman, L. Gladden and D. Agar, *Ind. Eng. Chem. Res.*, 2012, **51**, 1672.
127 B. Blumich, L. B. Datsevich, A. Jess, T. Oehmichen, X. Ren and S. Stapf, *Chem. Eng. J.*, 2007, **134**, 35.
128 L. Buljubasich, B. Blumich and S. Stapf, *J. Magn. Reson.*, 2011, **212**, 47.
129 L. Buljubasicha, B. Blumich and S. Stapf, *Chem. Eng. Sci.*, 2010, **65**, 1394.
130 D. A. Beauregard, P. Yong, L. E. Macaskie and M. L. Johns, *Biotechnol. Bioeng.*, 2010, **107**, 11.
131 S. Tsushima and S. Hirai, *Fuel Cells*, 2009, **9**, 506.
132 K. W. Feindel, S. H. Bergens and R. E. Wasylishen, *ChemPhysChem*, 2006, **7**, 67.
133 O. H. Han, *Progr. NMR Spectr.*, 2013, **72**, 1.
134 A. J. Sederman, M. D. Mantle, C. P. Dunckley, Z. Huang and L. F. Gladden, *Catal. Lett.*, 2005, **103**, 1.
135 E. H. L. Yuen, A. J. Sederman and L. F. Gladden, *Appl. Catal. A: General*, 2002, **232**, 29.

136 M. Kuppers, C. Heine, S. Han, S. Stapf and B. Blumich, *Appl. Magn. Reson.*, 2002, **22**, 235.

137 D. Weber, D. J. Holland and L. F. Gladden, *Appl. Catal. A: General*, 2011, **392**, 192.

138 S. B. Jaffe, *Ind. Eng. Chem., Process Des. Dev.*, 1976, **15**, 410.

139 J. Hanika, R. Lange and F. Turek, *Chem. Eng. Process.*, 1990, **28**, 23.

140 P. M. Haure, S. M. Bogdashev, M. Bunimovich, A. N. Stegasov, R. R. Hudgins and P. L. Silveston, *Chem. Eng. Sci.*, 1990, **45**, 2255.

141 O. Korup, S. Mavlyankariev, M. Geske, C. F. Goldsmith and R. Horn, *Chem. Eng. Proc.*, 2011, **50**, 998.

142 G. M. Carlomagno and G. Cardone, *Exp. Fluids*, 2010, **49**, 1187.

143 M. Simeone, L. Salemme and C. Allouis, *Int. J. Hydr. Energy*, 2008, **33**, 4798.

144 B. Li, K. Maruyama, M. Nurunnabi, K. Kunimori and K. Tomishige, *Ind. Eng. Chem. Res.*, 2005, **44**, 485.

145 B. Pinkerton and D. Luss, *Ind. Eng. Chem. Res.*, 2007, **46**, 1898.

146 B. Marwaha, S. Sundarram and D. Luss, *Chem. Eng. Sci.*, 2004, **59**, 5569.

147 J. Kopyscinski, T. J. Schildhauer, F. Vogel, S. M. A. Biollaz and A. Wokaun, *J. Catal.*, 2010, **271**, 262.

148 D. German, P. Chevallier, A. Laurent and H. Saint-Jalmes, *MAGMA*, 2001, **13**, 47.

149 W. Wlodarczyk, M. Hentschel, P. Wust, R. Noeske, N. Hosten, H. Rinneberg and R. Felix, *Phys. Med. Biol.*, 1999, **44**, 607.

150 B. Quesson, J. A. de Zwart and C. T. W. Moonen, *J. Magn. Reson. Imaging*, 2000, **12**, 525.

151 V. V. Zhivonitko, I. V. Koptyug and R. Z. Sagdeev, *J. Phys. Chem. A*, 2007, **111**, 4122.

152 I. V. Koptyug, A. V. Khomichev, A. A. Lysova and R. Z. Sagdeev, *J. Am. Chem. Soc.*, 2008, **130**, 10452.

153 A. A. Lysova, A. V. Kulikov, V. N. Parmon, R. Z. Sagdeev and I. V. Koptyug, *Chem. Commun.*, 2012, **48**, 5763.

154 V. Demas and P. J. Prado, *Concepts Magn. Reson.*, 2009, **34A**, 48.

155 *Single-Sided NMR*, ed. F. Casanova, J. Perlo and B. Blümich, Springer-Verlag, Berlin, Heidelberg, 2011.

156 A. J. Moule, M. M. Spence, S. Han, J. A. Seeley, K. L. Pierce, S. Saxena and A. Pines, *Proc. Natl. Acad. Sci. USA*, 2003, **100**, 9122.

157 C. Hilty, E. E. McDonnell, J. Granwehr, K. L. Pierce, S. Han and A. Pines, *Proc. Natl. Acad. Sci. USA*, 2005, **102**, 14960.

158 T. Z. Teisseyre, J. Urban, N. W. Halpern-Manners, S. D. Chambers, V. S. Bajaj, F. Svec and A. Pines, *Anal. Chem.*, 2011, **83**, 6004.

159 V. V. Zhivonitko, V.-V. Telkki, J. Leppaniemi, G. Scotti, S. Fransilla and I. V. Koptyug, *Lab Chip*, 2013, **13**, 1554.

160 Hyperpolarization Methods in NMR Spectroscopy. *Top. Curr. Chem.*, 2013, 338. Volume Editor: Lars Kuhn.

161 S. Mansson, E. Johansson, P. Magnusson, C.-M. Chai, G. Hansson, J. S. Petersson, F. Stahlberg and K. Golman, *Eur. Radiol.*, 2006, **16**, 57.

162 J. Kurhanewicz, D. B. Vigneron, K. Brindle, E. Y. Chekmenev, A. Comment, C. H. Cunningham, R. J. DeBerardinis, G. G. Green, M. O. Leach, S. S. Rajan, R. R. Rizi, B. D. Ross, W. S. Warren and C. R. Malloy, *Neoplasia*, 2011, **13**, 81.

163 B. M. Goodson, *J. Magn. Reson.*, 2002, **155**, 157.

164 T. G. Walker and W. Happer, *Rev. Mod. Phys.*, 1997, **69**, 629.

165 G. Navon, Y.-Q. Song, T. Room, S. Appelt, R. E. Taylor and A. Pines, *Science*, 1996, **271**, 1848.

166 G. E. Pavlovskaya, Z. I. Cleveland, K. F. Stupic, R. J. Basaraba and T. Meersmann, *Proc. Natl. Acad. Sci. USA*, 2005, **102**, 18275.

167 K. F. Stupic, Z. I. Cleveland, G. E. Pavlovskaya and T. Meersmann, *J. Magn. Reson.*, 2011, **208**, 58.
168 J. H. Ardenkjaer-Larsen, B. Fridlund, A. Gram, G. Hansson, L. Hansson, M. H. Lerche, R. Servin, M. Thaning and K. Golman, *Proc. Natl. Acad. Sci. USA*, 2003, **100**, 10158.
169 *Appl. Magn. Reson.*, 2012, **43**, issue 1–2, Special issue "The Different Magnetic Resonance Communities Join Forces for Progress in DNP".
170 *Phys. Chem. Chem. Phys.*, 2010, **12**, issue 22, Themed issue on dynamic nuclear polarization.
171 K. Golman, L. E. Olsson, O. Axelsson, S. Mnsson, M. Karlsson and J. S. Petersson, *Brit. J. Radiol.*, 2003, **76**, S118.
172 E. Terreno, D. D. Castelli, A. Viale and S. Aime, *Chem. Rev.*, 2010, **110**, 3019.
173 P. Bhattacharya, B. D. Ross and R. Bunger, *Exp. Biol. Med.*, 2009, **234**, 1395.
174 R. A. Green, R. W. Adams, S. B. Duckett, R. E. Mewis, D. C. Williamson and G. G. R. Green, *Progr. NMR Spectr.*, 2012, **67**, 1.
175 C. R. Bowers and D. P. Weitekamp, *J. Am. Chem. Soc.*, 1987, **109**, 5541.
176 J. Natterer and J. Bargon, *Progr. NMR Spectr.*, 1997, **31**, 293.
177 D. Canet, C. Aroulanda, P. Mutzenhardt, S. Aime, R. Gobetto and F. Reineri, *Concepts Magn. Reson.*, 2006, **28A**, 321.
178 K. V. Kovtunov, V. V. Zhivonitko, I. V. Skovpin, D. A. Barskiy and I. V. Koptyug, *Top. Curr. Chem.*, 2013, **338**, 123.
179 K. V. Kovtunov and I. V. Koptyug, *Magnetic Resonance Microscopy: Spatially Resolved NMR Techniques and Applications*, ed. S. Codd and J. D. Seymour, Wiley-VCH, Weinheim, 2008, 101–115.
180 S. B. Duckett and N. J. Wood, *Coord. Chem. Rev.*, 2008, **252**, 2278.
181 H. G. Niessen, D. Schleyer, S. Wiemann, J. Bargon, S. Steiner and B. Driessen-Holscher, *Magn. Reson. Chem.*, 2000, **38**, 747.
182 S. B. Duckett and R. E. Mewis, *Acc. Chem. Res.*, 2012, **45**, 1247.
183 I. V. Koptyug, K. V. Kovtunov, S. R. Burt, M. S. Anwar, C. Hilty, S. Han, A. Pines and R. Z. Sagdeev, *J. Am. Chem. Soc.*, 2007, **129**, 5580.
184 K. V. Kovtunov, V. V. Zhivonitko, A. Corma and I. V. Koptyug, *J. Phys. Chem. Lett.*, 2010, **1**, 1705.
185 I. V. Skovpin, V. V. Zhivonitko and I. V. Koptyug, *Appl. Magn. Reson.*, 2011, **41**, 393.
186 I. V. Skovpin, V. V. Zhivonitko, R. Kaptein and I. V. Koptyug, *Appl. Magn. Reson.*, 2013, **44**, 289.
187 K. V. Kovtunov, I. E. Beck, V. I. Bukhtiyarov and I. V. Koptyug, *Angew. Chem. Int. Ed.*, 2008, **47**, 1492.
188 K. V. Kovtunov, V. V. Zhivonitko, L. Kiwi-Minsker and I. V. Koptyug, *Chem. Commun.*, 2010, **46**, 5764.
189 D. A. Barskiy, K. V. Kovtunov, A. Primo, A. Corma, R. Kaptein and I. V. Koptyug, *ChemCatChem*, 2012, **4**, 2031.
190 V. V. Zhivonitko, K. V. Kovtunov, I. E. Beck, A. B. Ayupov, V. I. Bukhtiyarov and I. V. Koptyug, *J. Phys. Chem. C*, 2011, **115**, 13386.
191 K. V. Kovtunov, I. E. Beck, V. V. Zhivonitko, D. A. Barskiy, V. I. Bukhtiyarov and I. V. Koptyug, *Phys. Chem. Chem. Phys.*, 2012, **14**, 11008.
192 L.-S. Bouchard, S. R. Burt, M. S. Anwar, K. V. Kovtunov, I. V. Koptyug and A. Pines, *Science*, 2008, **319**, 442.
193 V. V. Zhivonitko, V.-V. Telkki and I. V. Koptyug, *Angew. Chem. Int. Ed.*, 2012, **51**, 8054.
194 V. V. Zhivonitko, K. V. Kovtunov, P. L. Chapovsky and I. V. Koptyug, *Angew. Chem. Int. Ed.*, 2013, **52**, 13251.

195 M. Carravetta, O. G. Johannessen and M. H. Levitt, *Phys. Rev. Lett.*, 2004, **92**, 153003-1.
196 M. B. Franzoni, L. Buljubasich, H. W. Spiess and K. Munnemann, *J. Am. Chem. Soc.*, 2012, **134**, 10393.
197 W. S. Warren, E. Jenista, R. T. Branca and X. Chen, *Science*, 2009, **323**, 1711.
198 P. Ahuja, R. Sarkar, S. Jannin, P. R. Vasos and G. Bodenhausen, *Chem. Commun.*, 2010, **46**, 8192.
199 G. Pileio, S. Bowen, C. Laustsen, M. C. D. Tayler, J. T. Hill-Cousins, L. J. Brown, R. C. D. Brown, J. H. Ardenkjaer-Larsen and M. H. Levitt, *J. Am. Chem. Soc.*, 2013, **135**, 5084.
200 I. I. Koptyug, A. A. Lysova, G. A. Kovalenko, L. V. Perminova and I. V. Koptyug, *Appl. Magn. Reson.*, 2010, **37**, 483.
201 J. S. McLean, O. N. Ona and P. D. Majors, *ISME J.*, 2008, **2**, 121.
202 A. C. Humphries, I. P. Mikheenko and L. E. Macaskie, *Biotechnol. Bioeng.*, 2006, **94**, 81.
203 K. P. Nott, F. P. Heese, M. Paterson-Beedle, L. E. Macaskie and L. D. Hall, *Can. J. Chem. Eng.*, 2005, **83**, 68.

Isotopes in the FTIR investigations of solid surfaces

Konstantin Hadjiivanov,* Mihail Mihaylov, Dimitar Panayotov, Elena Ivanova and Kristina Chakarova
DOI: 10.1039/9781782621485-00043

In this chapter the recent achievements in the use of isotopically labelled molecules for characterization of surfaces by FTIR and other vibrational spectroscopy techniques is reviewed. A brief theoretical background is provided where special attention is paid on the deviations of the experimental results from the theory. Then the application of D-, ^{13}C-, ^{15}N- and ^{18}O-labelling is consecutively considered. For deuterium we first discuss the properties of surface OD groups and then the use of deuterated molecular probes (D_2, CHD_2OH, C_2D_5OH, C_6D_{12}). When describing ^{13}C-labelled compounds the application of ^{13}CO and $^{13}C^{18}O$ is compared and then other labelled compounds ($^{13}CO_2$, $D^{13}CN$, $^{13}CH_3OH$, *etc.*) are considered. In the next section $^{15}N_2$, NO isotopologues, $(^{15}NH_2)_2CO$ and aminoacids are discussed. For ^{18}O-labelling we first consider the use of surfaces enriched to ^{18}O, then different O_2 adsorption and finally, $H_2{}^{18}O$ and $C^{18}O$. In all cases the application of isotopic labelling for clarifying the mechanisms of catalytic reactions is also considered.

1 Introduction

One of the most powerful tools for characterization of surfaces is vibrational spectroscopy.[1–3] Among the different techniques used for this purpose, definitely infrared is the main protagonist, although Raman spectroscopy, HREEL, *etc.* are also often utilized. Vibrational spectra of surface species provide valuable information on the surface adsorption sites as well as on the mechanism of surface reactions, in particular catalytic processes. For these reasons the number of studies on various surfaces involving FTIR and related techniques continuously increases.

In surface chemistry the spectra are normally registered after sample activation aimed at removing adsorbed water and eventual contaminants. The infrared spectra of activated solids provide information about the existence of particular surface species, *e.g.* hydroxyl groups, but bring only little information on the surface properties. To study in detail the surface centres, usually probe molecules are applied.[1–3] These are compounds interacting specifically with one or more types of surface centres and analysis of their spectra leads to conclusion on the surface properties of solids.

However, in some cases, it is difficult (or even impossible) to interpret the spectral results. For instance, there are spectral regions where various vibrations, overtones, combinations and even electronic transitions could be observed. When coadsorption of different adsorbates is performed, it is often impossible, based on the spectra alone, to assign unambiguously all of the resulting IR bands. In these cases, however,

Institute of General and Inorganic Chemistry, Bulgarian Academy of Sciences, Sofia 1113, Bulgaria. E-mail: kih@svr.igic.bas.bg

isotopic labelling normally gives easy and unambiguous answers. The use of isotopes is most often applied to establish whether a given vibration involves or not a particular atom. However, intelligent use of labelled compound could bring much more information and thus strongly enhances the power of the vibrational spectroscopy. For instance, using isotopic mixtures one can determine the number of ligands in polyligand species, to measure the static and dynamic interaction between adsorbed dipolar molecules and to control the intensity transfer. Isotopes normally help in establishing spectral phenomena, as Fermi-resonance, *etc.* Isotopic labelling has found a wide application in the mechanistic studies of catalytic reactions. A widely used technique for this purpose is SSITTKA (steady state isotopic transient kinetic analysis). Generally, switching the reaction gas feed between different isotopic substitutions should result in similar temporal responses of the reaction product and of the reaction intermediates while spectator species should be not sensitive or slightly sensitive to the isotopic substitution.

In this chapter we will give first a brief theoretical background of the vibrational spectra of isotopically labelled compounds. Special attention will be paid on the deviations from the theory and to which extent these deviations can hinder the spectral analysis. In practice few isotopes are widely used in the studies of surfaces by vibrational spectroscopy, *i.e.* ^{2}H (D), ^{13}C, ^{15}N and ^{18}O. The isotopic shifts with heavier elements are small and therefore not convenient for spectroscopic use. We will first consider deuterated compounds, and then the use of ^{13}C-, ^{15}N- and ^{18}O-labelled ones. Application of double-labelled molecules will be also discussed. Although the review is focused on recent studies, in order to give a complete picture, some important classic works will also be considered.

2 Brief theoretical background

According to the harmonic oscillator model, the stretching frequency, ν, of a diatomic molecule (AB) depends on the force constant (k) and the reduced mass (μ) of the two atoms:[4]

$$\nu = \frac{1}{2\pi}\sqrt{\frac{k}{\mu}} \tag{1}$$

where the reduced mass is calculated according to the equation:

$$\mu = \frac{M_A\, M_B}{M_A + M_B} \tag{2}$$

(M_A and M_B are the masses of the two atoms, respectively).

Equation (1) shows that the A–B stretching frequency will be changed after replacement of one of the atoms in the molecule by its isotope (for instance A by A′). Defining the isotopic shift factor, ***i***, as the ratio of stretching frequencies, and taking into account equation (1), one can easily obtain:

$$i = \frac{\nu_{AB}}{\nu_{A'B}} = \sqrt{\frac{\mu_{A'B}}{\mu_{AB}}}. \tag{3}$$

In equation (3) the stretching frequency can be replaced in practice by the wavenumber, $\tilde{\nu}$, because it is proportional to the frequency: the two parameters are linked by the relationship $\nu = c\tilde{\nu}$ (where c is the light speed).

Although the above analysis involves two atoms only, usually the stretching vibrations are isolated and the isotopic shifts can be well calculated according to equation (3). The calculated isotopic shifts of some stretching modes involving H, C, N and O atoms when exchanged with D, ^{13}C, ^{15}N and ^{18}O are presented in Table 1. Because of the drastic difference between the masses of protium and deuterium, the H/D exchange affects most strongly the stretching frequencies. Thus, an isotopic shift of *ca.* 1000 cm^{-1} is observed when surface OH groups are exchanged with deuterium, while the isotopic shift caused by ^{18}OH/OH substitution is about two orders of magnitude lower (see Fig. 1). The isotopic effect also increases with the atomic mass of the atom to which hydrogen is bound and is approaching the value of $\sqrt{2}$.

Table 1 Calculated isotopic shift factors of some stretching modes of practical importance. The shifts are calculated on the basis of the harmonic oscillator model.

ν_1	ν_2	$i = \nu_1/\nu_2$
H–H	H–D	1.1546
	D–D	1.4137
^{12}C–H	^{12}C–D	1.3620
	^{13}C–H	1.0030
	^{13}C–D	1.3696
^{14}N–H	^{14}N–D	1.3685
	^{15}N–H	1.0022
	^{15}N–D	1.3743
^{16}O–H	^{16}O–D	1.3736
	^{18}O–H	1.0033
	^{18}O–D	1.3822
^{12}C–^{12}C	^{12}C–^{13}C	1.0199
	^{13}C–^{13}C	1.0410
^{12}C–^{14}N	^{13}C–^{14}N	1.0214
	^{12}C–^{15}N	1.0157
	^{13}C–^{15}N	1.0382
^{12}C–^{16}O	^{13}C–^{16}O	1.0228
	^{12}C–^{18}O	1.0248
	^{13}C–^{18}O	1.0493
^{14}N–^{14}N	^{14}N–^{15}N	1.0170
	^{15}N–^{15}N	1.0350
^{14}N–^{16}O	^{15}N–^{16}O	1.0182
	^{14}N–^{18}O	1.0270
	^{15}N–^{18}O	1.0468
^{16}O–^{16}O	^{16}O–^{18}O	1.0291
	^{18}O–^{18}O	1.0608

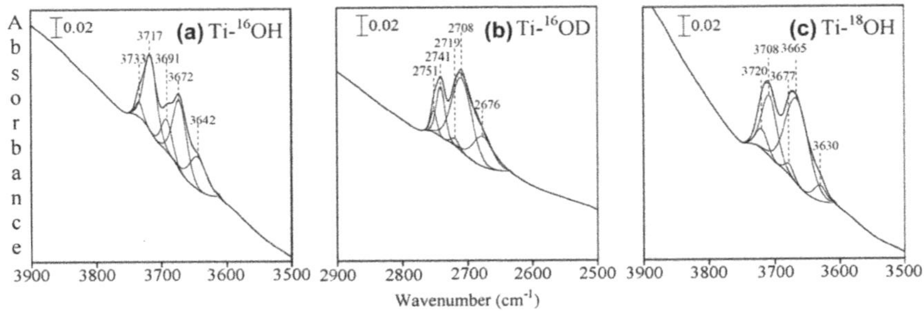

Fig. 1 Peak-fitted hydroxyl group region for (a) ^{16}O–H stretching vibrations for unlabelled TiO_2, (b) ^{16}O–D stretching vibrations for deuterated TiO_2 following D_2O exchange, (c) ^{18}O–H stretching vibrations for ^{18}O labeled TiO_2 following $H_2{}^{18}O$ exchange. Reproduced from Ref. 5.

Inspection of the data presented in Table 1 suggests that one could easily determine the nature of the atoms participating in a given vibration by just calculating the isotopic shift factor. However, some deviations between the calculated and observed values do not make this an easy task and, in some cases, even impossible. Below we shall briefly consider these deviations and to what extent they hinder analysis.

The real oscillators are not harmonic ones but are represented by the Morse potential. As a result, the experimental isotopic shift factors should be smaller than the theoretical ones. A good coincidence is reported only when the harmonic frequencies are used.[6]

The differences between the measured and calculated shifts are well pronounced with stretching vibrations involving hydrogen atoms. For instance, the experimentally observed OH/OD isotopic shift factor for OH radicals is 1.3556,[7] a value significantly lower than the theoretical value of 1.3736. Moreover, this value is lower even than the theoretical value (1.3620) for CH/CD exchange. The situation is additionally complicated by the fact that H-bonding also strongly affects the isotopic shift (for more details see below). Therefore, using ***i*** to distinguish between C–H, N–H, and O–H modes should be avoided or should be based on comparison with experimentally reported data. For instance, the CH/CD isotopic shift factor for formates on ceria has been reported to be 1.323.[8] This value is definitely lower than the theoretical one and also lower than the usually observed isotopic shift factor for isolated hydroxyls. However, for H-bonded hydroxyls, ***i*** can be even lower. The situation is complicated for two more reasons: (i) H-bonded hydroxyls are observed at relatively low frequencies and can approach the region typical of C–H stretches and (ii) very different values (up to 1.37)[9] have been reported for the νCH/νCD factor.

In fact, it is easy to distinguish O–H from M–H vibrations on the basis of the isotopic shift. Thus, the theoretical value of ***i*** for Zn–H and Ga–H stretchings is around 1.40. It was found experimentally that the H/D isotopic shift factor for ZnH groups is 1.385,[10] and for GaH groups, 1.39,[11] *i.e.* definitely higher than the value of *ca.* 1.356 observed for OH groups.

Although the isotopic shift factors are smaller with stretching vibrations involving C, N and O atoms, as compared to vibrations involving hydrogen (see Table 1), in these cases the deviations from the theory are very small and this allows distinguishing between different bonds. For instance, the experimentally obtained $CO/^{13}CO$ isotopic shift factor coincides very well with the theoretical one (1.0228) and the experimentally obtained $^{13}C–O$ stretching frequencies differ from the calculated ones by less than 1 cm^{-1}. This allows distinguishing, for instance, even between C–N and C–O modes. However, to make unambiguous conclusions one should use isotopic exchange with the two expected atoms in the bond.

The isotopic effect is also more pronounced when the masses of the two bonded atoms strongly differ. For instance, the $Cr–^{16}O/Cr–^{18}O$ and $W–^{16}O/W–^{18}O$ theoretical isotopic shift factors are 1.045 and 1.056 cm^{-1} respectively. Because the double M=O bonds are easily monitored in the overtone region around 2000 cm^{-1},[12] the expected difference between the positions of the $2\nu(Cr–^{18}O)$ and $2\nu(W–^{18}O)$ bands (providing that the $M–^{16}O$ modes coincide) is around 20 cm^{-1}. Note that, in the latter case, the factor is hardly sensitive to the nature of W isotope involved in the vibration. A similar situation is reported for Ga–H modes: the differences in the Ga–H bond due to the different natural isotopes of gallium are smaller than 1 cm^{-1}.[11]

The above considerations concern isolated stretching modes. In many cases, *i.e.* when the kinetic energy of the vibrations is localized mainly in these groups, the bending frequencies of an A–B group after isotopic exchange can also be calculated according to equation (3).[4] This is for instance the situation with vibrations involving light atoms, as hydrogen and vibrations corresponding to strong (double, triple) bonds. However, the situation with coupled vibrations is more complicated and substantial deviations from equation (3) could be observed. For a triatomic XY_2 molecule the isotopic shift of the ν_{as} modes depends on the Y–X–Y angle. If such a molecule is labelled only at one Y position, the symmetry is changed and the prediction of the band positions is complicated (although the shift is roughly the half of the predicted for a double labelled molecule). For more details the reader is directed to Ref. 4.

3 Deuterated compounds

The surface studies involving deuterated species are numerous because of two main reasons: (i) the H/D isotopic effect is very well pronounced in the vibrational spectra of the stretching modes involving hydrogen; (ii) hydrogen is present in most of the systems of practical importance. Here we will consider first surface OH and OD groups and then adsorption of various deuterated compounds.

3.1 Surface OD groups

Exchange of protium with deuterium in surface OH groups is performed for various reasons: (i) to establish the accessibility of the OH groups, (ii) to switch to a spectral region where the noise is limited and thus to obtain high quality spectra, (iii) to check for existence of spectral phenomena, (iv) to trace the origin of hydrogen in surface reaction products, *etc.*

Table 2 presents recently observed OH/OD isotopic shift factors for surface hydroxyl groups.[6,13–33] These and more data have been analyzed[6]

Table 2 OH/OD isotopic shift factors for surface hydroxyl groups. Isotopic shift factors ν(OH)/ν(OD) smaller than 1.3535 are shown by italics.

OH group	Sample	ν(OH), cm^{-1}	ν(OD), cm^{-1}	ν(OH)/ν(OD)	Ref.
SiOH	SiO_2	3748	2763	1.3565	6
SiOH	SiO_2	3746	2762	1.3563	6
SiOH	SiO_2	3740	2760	1.3551	13
SiOH	SiO_2	3738	2756	1.3563	14
SiOH	SiO_2	3670	2710	1.3542	14
SiOH	SiO_2	3530	2600	1.3577	14
SiOH	SiO_2	3520	2610	*1.3487*	14
SiOH	MCM-41	3745	2760	1.3569	15
SiOH	SiO_2-Al_2O_3	3744	2764	1.3546	16
SiOH	[Si]BEA	3525	2609	*1.3512*	6
Si(OH)Al	H–ZSM-5	3627	2676	1.3554	17
Si(OH)Al	H–ZSM-5	3619	2670	1.3554	17
Si(OH)Al	H–ZSM-5	3617	2669	1.3552	18
Si(OH)Al	H–ZSM-5	3616	2667	1.3558	19, 20
Si(OH)Al	H–ZSM-5	3606	2660	1.3556	17
Si(OH)Al	FER	3596	2653	1.3554	21
Si(OH)Al	FER	3609	2662	1.3557	21
Si(OH)Al	FER	3555	2623	1.3553	21
Si(OH)Al	HY	3546	2620	1.3534	22
Si(OH)Al	Pd/HY	3558	2623	1.3565	23
Si(OH)Al	MCM-58	3628	2679	1.3542	18
Si(OH)Al	MCM-58	3556	2628	1.3531	18
Si(OH)Al	MCM-58	3485	2576	*1.3529*	18
Si(OH)Al	MCM-68	3617	2669	1.3552	18
Si(OH)Al	2D zeolite	3594	2654	1.3542	24
AlOH	H–ZSM-5	3667	2703	1.3566	19, 20
AlOH	MCM-68	3670	2704	1.3572	18
ZnOH	Zn/H–MFI	3675	2705	1.3586	25
ZnOH	ZnO	3672	2707	1.3565	26, 27
ZnOH	ZnO	3656	2698	1.3551	26, 27
ZnOH	ZnO	3639	2683	1.3563	26, 27
ZnOH	ZnO	3620	2669	1.3563	27
ZnOH	ZnO	3564	2626	1.3572	26, 27
ZnOH	ZnO	3448	2551	*1.3516*	26, 27
TiOH	TiO_2	3733	2741	1.3619	5
TiOH	TiO_2	3717	2741	1.3561	5
TiOH	TiO_2	3691	2719	1.3575	5
Ti(OH)Ti	TiO_2	3672	2708	1.3560	5
Ti(OH)Ti	TiO_2	3642	2676	1.3610	5
ZrOH	ZrO_2	3774	2780	1.3576	28
ZrOH	ZrO_2	3733	2751	1.3570	28
ZrOH	ZrO_2	3673	2709	1.3559	28
ZrOH	MOF: UiO-66	3666	2708	1.3538	29
CeOH	CeO_2	3657	2694	1.3575	30
CeOH	CeO_2	3510	2598	*1.3510*	30
CeOH	Cu/CeO_2	3663	2702	1.3557	31
CeOH	Pt/CeO_2	3650	2685	1.3594	32
OH/NH	chitosan[a]	3435	2529	1.3582	33
OH/NH	chitosan	3290	2432	*1.3528*	33
OH/NH	chitosan	3180	2372	*1.3406*	33

[a]poly-β-(1,4)-glucosamine.

and it was concluded that in most cases the isotopic shift factor for isolated hydroxyls was around 1.356–1.358 and hardly depended on the ν(OH) value.

Based on the fundamental and overtone values of the OH and OD stretching modes, it is possible to calculate the harmonic frequencies. It was demonstrated that, for the harmonic frequencies of isolated SiOH/SiOD groups, the isotopic shift factor is very close to the calculated value, *i.e.* the difference between the measured and calculated frequencies is mainly due to the anharmonicity.[6] It was proposed that the situation is similar for other isolated hydroxyls groups.

A careful analysis of the data presented in Table 2 indicates that values of ***i*** lower than 1.3535 (shown by italics) are observed with OH groups vibrating at relatively low frequencies and believed to be H-bonded (note that with H-bonded Mg–OH groups on MgO a value of 1.308 is reported).[34] The supposition that weak H-bonding leads to a decrease in the isotopic shift factors is supported by careful investigations on the formation of H-bonds between isolated hydroxyls and weak bases during adsorption (Table 3).

These phenomena were just recently discussed in the surface chemistry literature.[18–20,24,35,36] An empiric equation for calculation of the stretching modes of surface OD groups has been proposed:[18]

$$\nu_{OD} = 0.709\ (\nu_{OH}) + 103.7 \tag{4}$$

It was also noted that H-bonded hydroxyls deviate from this relationship.

Before exposing different opinions, let us first consider some key points of the H-bonding. The drastic difference between the masses of protium and deuterium leads to measurable differences in their chemical properties that are much more pronounced than for the isotopes of other elements. The common explanation of these effects is the difference in the zero-point energy between the O–H and O–D bonds. An important consequence is that the dissociation energy of the OH and OD bonds are different, *i.e.* O–D bonds are stronger than the respective O–H bonds.

Table 3 OH/OD isotopic shift factors for surface hydroxyl groups before and after formation of H–bonds with weak bases.

Sample	OH group/complex	ν(OH) cm^{-1}	ν(OD) cm^{-1}	ν(OH)/ν(OD)	Ref.
SiO_2	SiOH	3748	2763	1.3565	6
	$SiOH\cdots CH_4$	3714	2739	1.3560	6
	$SiOH\cdots CO$	3660	2700	1.3556	6
	$SiOH\cdots CD_3CN$	3429	2552	1.3437	6
H-ZSM-5	Si(OH)Al	3616	2667	1.3558	19, 20
	$[Si(OH)Al]\cdots N_2$	3496	2584	1.3529	19
	$[Si(OH)Al]\cdots CH_4$	3494	2582	1.3532	35
	$[Si(OH)Al]\cdots CO$	3306	2460	1.3439	19, 20
2D zeolite	Si(OH)Al	3594	2654	1.3542	24
	$[Si(OH)Al]\cdots CO$	3215	2409	1.3345	24
	$[Si(OH)Al]\cdots C_2H_4$	3107	2330	1.3335	24

Also, the O–D bond is more harmonic as compared to the OH bond because the zero-point level for OD lies closer to the bottom of the Morse curve and the deviations from the harmonic oscillator model are smaller. In a recent study,[37] the OH bond distance in H_2O was found to be by 3% longer as compared to the OD bond in D_2O, while the distance of the H-bond formed is by 4% shorter than that for the deuterated system.

Very recently it was demonstrated, by adsorption of CD_3CN, that the bridging hydroxyls in H–ZSM-5 zeolite are more acidic than the corresponding O–D groups.[36] It was shown that the sensitive to coordination C–N mode appeared at 2301.2 cm^{-1} when CD_3CN was adsorbed on H–ZSM-5 and at 2300.0 cm^{-1} with deuterated form (D–ZSM-5). The same effect was established for the less acidic Al–OH groups but due to their lower acidity the spectral difference in the C–N modes was smaller. As we will show below, analysis of these results explains the anomalous isotopic shift factor of H-bonded hydroxyls.

When OH groups participate in an H-bonding, the OH stretching frequency shifts to lower wavenumbers and the shift is proportional to the strength of the bond formed. Other effects of the H-bonding are the broadening of the O–H stretching band and the increase of the extinction coefficient. It is considered that the broadening is due to coupling (sum and difference combinations) of νOH with the hydrogen bond stretching vibration. Note, that this coupling affects the band width, but not its position.

Boscoboinik *et al.*[24] assumed that the difference between the isotopic shift factors of free and H-bonded hydroxyls raised from the coupling of the OH modes with intramolecular vibrations. However, as already noted, this coupling is assumed not to affect the OH band maximum and therefore, the isotopic shift factor.

Alternative explanation was proposed by Chakarova *et al.*[19,20] Assuming that the OH and the respective OD groups possess the same acidity, the isotopic shift factor for free and H-bonded hydroxyls should be practically the same. However, due to the lower acidity of the OD groups, they are shifted to a lesser extent upon D-bonding. This can be easily seen in Fig. 2 showing the interaction of CO at 100 K with the OH/OD groups on an H–ZSM-5 sample. For convenience, the x-axis scale for panel B (OD region) is divided by the experimental isotopic shift factor for the free OH/OD groups, *i.e.* 1.3558. It is clearly seen that the principal band of the D-bonded OD groups (expected at 2438 cm^{-1}) appears at 2460 cm^{-1}. Consequently, the isotopic shift factor is 1.3439, *i.e.* definitely lower than that observed with isolated hydroxyls. The same effect occurs when an OH group is H-bonded not to a guest molecule but to oxygen from another hydroxyl group or to lattice oxygen.[22]

Inspection of Fig. 2 allows distinguishing of low frequency shoulders of the bands due to H-bonded OH (3206 cm^{-1}) and OD (2460 cm^{-1}) groups, at 3415 and 2557 cm^{-1}, respectively. However, the shoulder at 2557 cm^{-1} is less intense as compared to that at 3415 cm^{-1} and the isotopic shift factor (1.3355) is even lower than that for the principal band at 3305 cm^{-1}. Based on analysis of the spectra, it was concluded[19,20] that the shifted OH band is split into two components as a result of Fermi

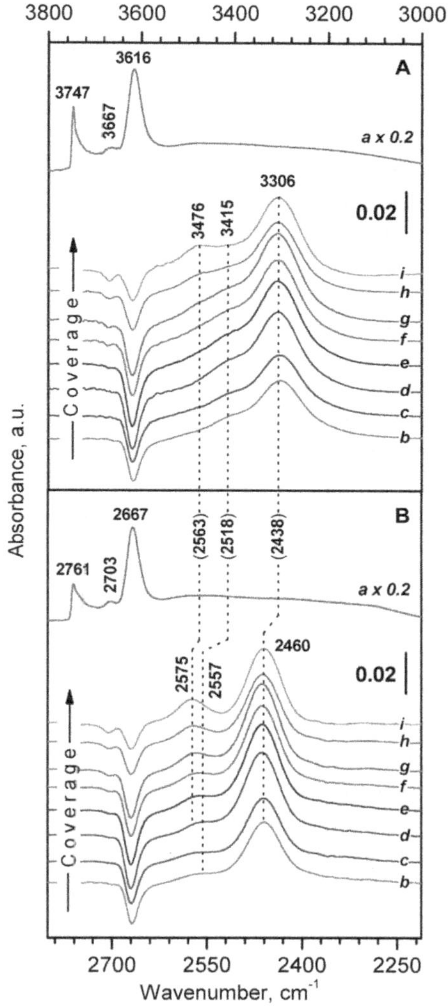

Fig. 2 FTIR spectrum of activated H–D–ZSM-5 (a) and changes in the IR spectra induced by small changes in the CO coverage after CO adsorption at 100 K (b–i). Panel A shows the OH region and panel B, the OD region. The X-axis scale for panel B is chosen as to correspond to the H–D isotopic shift factor. Reproduced from Ref. 20.

resonance with the second excited state of the δ(OH) modes which is expected slightly above 3300 cm^{-1}. The Fermi resonance is weaker with the OD bands because of the larger difference between the positions of ν(OD) and 3δ(OD) modes.

It was also found, again by the use of D-exchanged materials, that in order the discussed Fermi resonance to occur the position of the shifted OH band should be around 3300 cm^{-1}. Consequently, the effect was observed with other zeolites (H-FER)[21] and adsorbed probe molecules (C_6H_6).[38] However, when CO is adsorbed on HY, the 3640 cm^{-1} hydroxyl band shifts to 3370 cm^{-1} and no resonance effects are observed.[22] The same is valid when bases weaker than CO (*e.g.* N_2) are adsorbed on H-ZSM-5 and the shifted band is at relatively high wavenumbers.[19]

These observations are important for measuring the acidity of OH groups. In particular, they demonstrate that the appearance of two shifted bands when zeolite acidic hydroxyls interact with CO is due to spectral phenomena but not to heterogeneity of the OH groups, as supposed before.[39] In addition, it seems that the isotopic shift factor can be used to estimate the extent of H–bonding.

It should also be noted that the decrease of ***i*** upon H–bonding cannot be explained by the decrease in anharmonicity of the O–H bond, as proposed.[40] Indeed, it is well established that formation of a weak H–bond leads to increase of the harmonicity of the O–H bond oscillator.[41]

The appearance of two combination bands for SiOH groups, at *ca.* 4563 and 4530 cm^{-1}, has been debated and it was shown that only one band is observed with deuterated sample (see Fig. 3). It was proposed that this was due to the existence of two surface modes with bending character arising from the coupling of the bending Si–OH modes with Si–O–Si stretchings.[42]

It is important to distinguish between OH stretching bands of hydroxyl groups and of adsorbed water. Normally this is made on the basis of the water deformation modes around 1650–1600 cm^{-1}. However, many other surface species (nitrates, carbonates) can absorb in this region. Therefore, this criterion is often not useful. However, if a band around 1600 cm^{-1} is due to adsorbed water, it should disappear after D/H exchange. Moreover, after partial exchange, a band at *ca.* 1400 cm^{-1}, due to the deformation modes of HOD molecules, should be observed.[4]

An interesting application of deuterium was reported by Dubkov *et al.*[43] They used H_2 and D_2 to demonstrate that reactive oxygen on Fe–ZSM-5 (so-called α–oxygen) produces FeOH/FeOD groups.

As a conclusion, we can say that the exchange of surface OH groups with deuterium is a useful approach for their characterization but the

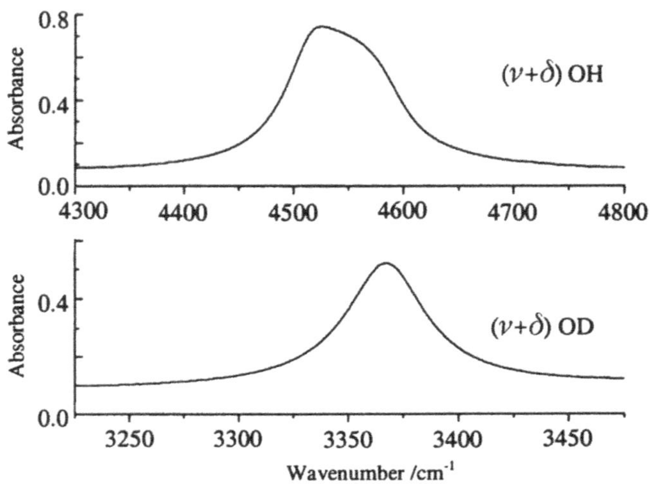

Fig. 3 Comparison of the absorption profiles of the νOH + δOH and νOD + δOD bands. The silica sample has been preheated for 20 h at 750 °C *in vacuo*. Reproduced from Ref. 42.

slight difference in the acidity of the OH and OD groups should be taken into account.

3.2 Adsorption of D-labelled molecules

The reasons for utilization of D-labelled probe molecules are various. Dihydrogen possesses several advantages as a probe: small size (allowing to penetrate into small pores); no hindrance from the gas phase; and high sensitivity of the H–H stretches to the bonding.[44] The use of D_2 instead of H_2 allows switching to less noisy spectral regions and also permits distinguishing from other effects, *e.g.* bands due to shifted OH stretching modes. In any case, the analysis of the results on D_2 adsorption should be carefully performed because the two isotopologues, H_2 and D_2, are characterized by different equilibria between the ortho- and para-forms. Here again, the measured isotopic shift factors are lower than the theoretical one (1.4137). For instance, $i = 1.3883$ and 1.3969, respectively, for H_2/D_2 adsorbed on the acidic hydroxyls and on Na^+ sites in HNaY zeolite.[44]

Panayotov and Yates[45] have studied the interaction of titania with atomic hydrogen. They observed a broad absorbance extended from *ca.* 4000 cm^{-1} to a sharp cut-off around 1000 cm^{-1}. The authors performed the same experiments with atomic deuterium and obtained essentially the same spectrum. On this basis they concluded that the broad band did not involve any vibrations with hydrogen atoms and assigned the observed feature to delocalized excited electrons in the conduction band.

Bonivardi *et al.*[9,11,46] used deuterium labelling in their studies on the surface properties of Ga_2O_3. They observed dissociative adsorption of H_2 (D_2) on reduced galia and used the O–D region to make conclusions on the nature of the OD groups formed because the O–H region was noisy.[11] They also found two kinds of Ga–H species (observed at 2003 and 1980 cm^{-1}, respectively).[46] Upon D_2 adsorption the bands were registered at 1430 and 1420 cm^{-1}, *i.e.* the isotopic shift factors were 1.4007 and 1.3944, respectively, which confirmed the assignments. Adsorption of methanol led to formation of methoxide species. In deuterated form the C–D stretching modes were reported to shift with respect to the corresponding C–H modes by the unusually high factor of 1.37.[9]

Exchange with deuterium can help in the assignment of peaks in the carbonate region. For instance, it has been found that bands at 1570 and 1305 cm^{-1}, produced after toluene adsorption on supported nickel catalysts, are shifted, after treatment with D_2 at 473 K, to 1555 and 1393 cm^{-1}, respectively.[47] On this basis the bands have been attributed to CH vibrations.

Due to simplicity in the CH stretching region, CHD_2OH is a very convenient probe molecule for establishing the bonding geometry of methoxide species produced during dissociative adsorption of methanol.[48] The authors reported a pseudo-C_{3v} local symmetry for methoxy groups on Ru(001) produced after CHD_2OH adsorption at 90 K. With coverage increase these species acquired different tilted configurations, depending on the oxygen precoverage.

Kondo *et al.*[49] studied ethoxide groups formed by dissociative adsorption of ethanol on H–MOR zeolite. It was shown that the ethoxide species once formed are not substituted during the subsequent C_2D_5OH adsorption. The observation allowed the authors to conclude that the ethoxide species are stable intermediates in the catalytic ethanol dehydration on zeolites.

It was also shown[50] that, while no C–D bonds were formed as a result of interaction between *n*-heptane and D_2 over BEA zeolite, in presence of platinum H/D exchange rapidly occurred leading to scrambling of the deuterium distribution in *n*-heptane and its isomers.

Almeida *et al.*[51] studied photooxidation of cyclohexane and deuterated cyclohexane on anatase. In particular they noted that the negative peaks of cyclohexane associated with its oxidation were shifted outside the carbonate spectral region which allowed easier interpretation of the spectra.

There are some recent SSITKA studies involving deuterium. However, they also utilize ^{13}C and will be regarded in section 4.5.

4 ^{13}C-labelled compounds

4.1 ^{13}CO

By far, the most used isotopologue of CO is ^{13}CO. The use of $C^{18}O$ (see 6.3) is more restricted and $^{13}C^{18}O$ (see 4.2) have found increasing application just recently. In surface characterization labelled CO is utilized for three main reasons: (i) to establish whether surface species contain or not a carbon atom, (ii) to check the structure of polycarbonyl species, and (iii) to control the dipole-dipole interaction between the adsorbed molecules.

The experimentally obtained $^{12}CO/^{13}CO$ isotopic shift coincides well with the theoretical value. However, it has been reported that, due to mechanical coupling between the M–C and C–O modes, the isotopic shift slightly increases with the increase in the M–C stretching vibration, respective the adsorption strength.[52] The reverse relationship has been proposed for the $CO/C^{18}O$ isotopic shift.

A classic example of the utilization of ^{13}CO is the assignment of a band at 2115 cm^{-1} on reduced ceria. Although the band position is typical of carbonyls, comparison between the results on CO and ^{13}CO adsorption demonstrated that it was not associated with carbon but was due to electronic transiton.[53] In contrast, ^{13}CO was used to establish that all bands observed after CO adsorption on Pt/CeO_2 were due to surface carbonyls and there was no essential contribution of the Ce^{3+} electronic transition (see Fig. 4).[54] The nature of the carbonyl bands on a Ru/Al_2O_3 catalyst was proven in a similar way.[55]

In many cases the number of CO molecules that can be simultaneously coordinated to one surface site corresponds to the number of coordinative vacancies of the cation and thus CO can be used to test the coordination state of surface sites.[56] This is particularly important for catalysis because the simultaneous coordination of two or more molecules to one site facilitates the interaction between them. The use of CO isotopic mixtures is unique for determination of polycarbonyl structures. Consider first dicarbonyl species. The two CO ligands coordinated to the

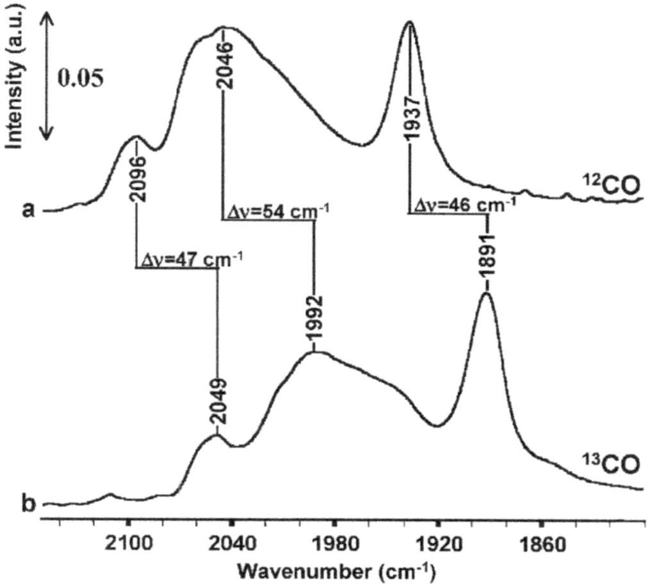

Fig. 4 FT-IR spectra recorded after carbon monoxide adsorption (0.7 kPa) at r.t. on Pt/ CeO_2 followed by evacuation at the same temperature: (a) ^{12}CO and (b) ^{13}CO. Reproduced from Ref. 54.

Table 4 Fraction (% of total) of dicarbonyl species having different numbers of CO and ^{13}CO ligands depending on the CO : ^{13}CO molar ratio.

CO : ^{13}CO ratio	$M(CO)_2$	$M(CO)(^{13}CO)$	$M(^{13}CO)_2$
1 : 1	25%	50%	25%
3 : 1	56.25%	37.5%	6.25%

same cation could interact vibrationally which results in a split of the C–O stretching modes into symmetric and antisymmetric peaks.

Two carbonyl bands changing in concert already indicate existence of geminal species. However, the dicarbonyl structure can be unambiguously proven only by isotopic labelling. If a CO + ^{13}CO mixture is adsorbed, the following species should be produced: $M(CO)_2$, $M(CO)(^{13}CO)$ and $M(^{13}CO)_2$, each of them characterized by two IR bands. On the basis of the experimentally observed ν_s and ν_{as} ^{12}C–O modes and according to the approximate force field model one can easily calculate the modes of the mixed ligand complex, $M(CO)(^{13}CO)$.[57] Another important parameter is the distribution ratio between the different species (see Table 4). For instance, after adsorption of an 1 : 1 isotopic mixture, the $M(CO)(^{13}CO)$ species should be the half of all dicarbonyls. Although the extinction coefficients of CO and ^{13}CO slightly differ,[4] the data from Table 4 can be used for estimation of the expected intensities of the different bands. By changing the molar ratio of the isotopes, one can favour the formation of different species.

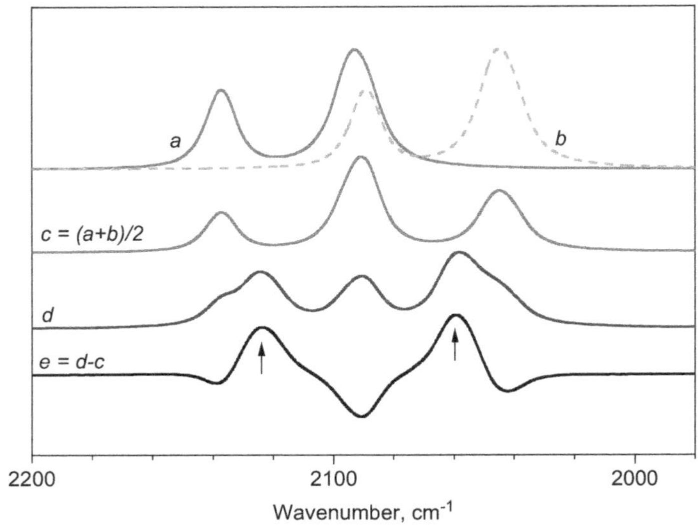

Fig. 5 Simulated FTIR spectra for: (a) $M(^{12}CO)_2$ dicarbonylic species; (b) corresponding $M(^{13}CO)_2$ species; (c) average of the spectra "a" and "b"; (d) spectrum corresponding to adsorption of a 1 : 1 $^{12}CO + ^{13}CO$ isotopic mixture; and (e) difference between spectra "d" and "c". The bands due to mixed-ligand species are arrowed.

Although the use of isotopic mixtures for determination of dicarbonyl structures seems to be unproblematic, the practical analysis is often hindered due to the superimposition of various carbonyl bands. To overcome these obstacles, our team has recently proposed a practical approach for easy understanding the spectra.[58] It is based on comparison between (i) the average of the spectra of ^{12}CO and ^{13}CO adsorbed alone (see Fig. 5, spectrum c) and (ii) the spectra registered after adsorption of isotopic mixtures (Fig. 5, spectrum d). If no polycarbonyls are formed, the registered and the averaged spectra should practically coincide. All bands due to polycarbonyls should be with lower intensity in the registered spectrum (as compared to the "averaged" one) while new bands due to mixed ligand complexes should appear. This is well seen in the difference spectra (Fig. 5, spectrum e) where the bands belonging to mixed-ligand complexes are arrowed.

Using ^{12}CO–^{13}CO isotopic mixtures, a row of dicarbonyl structures have been proven in the past years. The results are summarized in Table 5. It was also demonstrated[59] that the IR bands at 2192 and 2148 cm^{-1}, appearing after CO adsorption on Basolite C300 MOF sample, are associated with individual monocarbonyls and are not due to the ν_s and ν_{as} modes of dicarbonyls, as suggested earlier.

Let us now consider tricarbonyls. Usually their spectra are complicated because of the superimposition of too many bands. Indeed, when tricarbonyls are formed, other adsorption forms of CO, as physically adsorbed CO and CO polarized by OH groups, are usually encountered. Some dicarbonyls often co-exist with tricarbonyls. In most cases the tricarbonyls have distorted C_{3v} symmetry and, as a result, they display three IR bands. The mixed ligand species are $M(CO)_2(^{13}CO)$ and $M(CO)(^{13}CO)_2$,

Table 5 Spectral performance of surface $M(CO)_2$ and $M(CO)(^{13}CO)$. The calculated frequencies are given in brackets.

Sample	Species	$M(CO)_2$		$M(CO)(^{13}CO)$		Ref.
		ν_s, cm^{-1}	ν_{as}, cm^{-1}	$\nu(^{12}CO)$, cm^{-1}	$\nu(^{13}CO)$, cm^{-1}	
CuBEA	$Cu^+(CO)_2$	2180	2151	2170 (2169)	2111 (2112)	60
FeO_x	$Fe^{n+}(CO)_2$	2075.8	2017.6	2060.6 (2049.4)	1987.2 (1982.2)	61
FeO_x	$Fe^{n+}(CO)_2$	2138.8	2091.7	- (2124.8)	2068.8 (2058.5)	61
CoMOR	$Co^+(CO)_2$	2114	2036	2088 (2097)	2008 (2007)	62
NiBEA	$Ni^+(CO)_2$	2138	2095	2125 (2125)	2062 (2062)	63
Rh–ZSM-5	$Rh^{2+}(CO)_2$	2176	2142	2164 (2165)	2107 (2106)	64
Rh–ZSM-5	$Rh^{2+}(CO)_2NO$	2181	2153	2168 (2171)	2116 (2115)	65
Ir–ZSM-5	$Ir^+(CO)_2$	2104	2033	2088 (2088)	2002 (2004)	66
Ir–ZSM-5	$Ir^{2+}(CO)_2$	2172	2128	2157 (2158)	2090 (2093)	66
PtNaMOR	$Pt^+(CO)_2$	2135	2101	2123 (2123)	2069 (2065)	67
PtNaMOR	$Pt^{3+}(CO)_2$	2205	2167	2192 (2193)	2130 (2129)	67
Pt/ZSM5	$Pt^+(CO)_2$	2122	2092	2112 (2111)	2052 (2055)	68
Pt/ZSM5	$Pt^{3+}(CO)_2$	2212	2176	2199 (2199)	2138 (2138)	68
Pt/ZSM5	$Pt^{3+}(CO)_2$	2196	2155	2184 (2182)	masked (2120)	68

Table 6 Fraction (% of total) of tricarbonyl species having different numbers of CO and ^{13}CO ligands depending on the CO : ^{13}CO molar ratio.

CO : ^{13}CO ratio	$M(CO)_3$	$M(CO)_2(^{13}CO)$	$M(CO)(^{13}CO)_2$	$M(^{13}CO)_2$
1 : 1	12.5%	37.5%	37.5%	12.5%
3 : 1	42.2%	42.2%	14%	1.6%

but in cases of distorted C_{3v} symmetry they also differ because ^{13}CO can occupy different positions. Consequently, the total number of bands appearing after adsorption of isotopic mixtures is too large and detailed analysis is often impossible. Here, the above described comparison with "averaged" spectra could be very useful.

Another useful criterion for proving the tricarbonyl structure is the intensity of the $M(CO)_3$ bands after adsorption of isotopic mixtures (see Table 6) resulting from the statistical distribution of the ligands. Thus, using a 1 : 1 CO + ^{13}CO mixture, the intensity of bands characterizing $M(CO)_3$ species should be 12.5% of the intensity of the same bands registered after adsorption of pure CO.

In order to distinguish between the carbon- and nitrogen-containing oxo-species formed during CO and NO interaction on a $MnCe/Al_2O_3$ sample, coadsorption of ^{13}CO and NO was studied.[69] It was concluded that carbonate-like species were dominant on the surface during CO + NO interaction at $T \geq 373$ K. The ^{13}C-labelling also helped in assigning bands at 2253 and 2228 cm^{-1} to isocyanate species. ^{13}CO was used to confirm the assignment of bands appearing after CO adsorption on Cu/CeO_2 to carbonyls, carbonates and formates.[70] The same approach was used to attribute bands at 2236 and 2168 cm^{-1} to isocyanates produced as a result of CO + NO interaction in presence of H_2 on CeO_2, Au/CeO_2 and Au/CeO_2–Al_2O_3.[71]

4.2 $^{13}C^{18}O$

Consider now the utilization of $^{13}C^{18}O$. The advantage of this probe molecule is that the $^{13}C–^{18}O$ modes of the carbonyl species appear at wavenumbers that are *ca.* 100 cm^{-1} lower than for the C–O modes (*vs.* *ca.* 50 cm^{-1} for ^{13}CO). This ensures a very high resolution of the carbonyl bands. Here slight deviations from the theoretical (0.9531) and experimental (0.9536) isotopic shift factors are reported for $^{13}C^{18}O/CO$ substitution and the experimental frequencies are found at wavenumbers higher with 1–2 cm^{-1} than those expected theoretically.[58]

Mori *et al.*[72] used adsorption of CO on a Cu–ZSM-5 sample partly pre-covered by $^{13}C^{18}O$ to conclude on the existence of two types of Cu^+ adsorption sites (Fig. 6). Indeed, the sites characterized by ν(CO) at 2158 cm^{-1} (ν($^{13}C^{18}O$) at 2057 cm^{-1}) were predominantly filled with the $^{13}C^{18}O$-precovered samples and the adsorption of CO led mainly to filling of the second type of sites characterized by a carbonyl band at 2150 cm^{-1}.

Föttinger *et al.*[73] used adsorption of $^{13}C^{18}O$ to establish the mechanism of carbonate formation on Pd/Al_2O_3 catalysts. The authors noted that the use of this CO isotopologue should allow distinguishing between CO_2 formed by CO disproportionation ($2\ ^{13}C^{18}O \rightarrow {}^{13}C^{18}O_2 + {}^{13}C$) and CO_2 produced *via* WGS reaction ($^{13}C^{18}O + {}^{16}OH \rightarrow {}^{13}C^{18}O^{16}O + \frac{1}{2}H_2$). The results obtained allowed rejecting the disproportionation pathway and a mechanism of carbonate formation with the participation of surface OH groups was proposed.

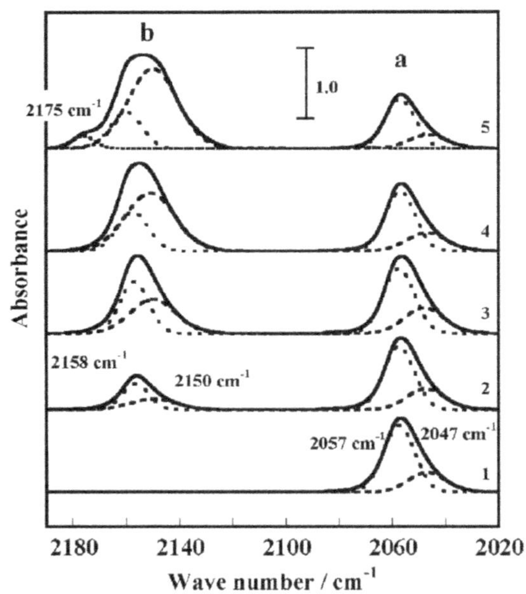

Fig. 6 IR spectra in absorption process of CO isotopes on sample CuMFI (a) in the region of $^{13}C^{18}O$ and (b) in the region of $^{12}C^{16}O$. (1) Firstly $^{13}C^{18}O$ was adsorbed, (2)–(5) followed by the adsorption of increasing pressures of $^{12}C^{16}O$, where (1)–(4) are in the irreversible adsorption process and (5) is in the reversible adsorption process. Reproduced from Ref. 72.

Isotopic mixtures of CO and $^{13}C^{18}O$ are very convenient for determination of polycarbonyl structures, especially in cases when many carbonyl bands are superimposed.[58,74] Carbonyls formed on supported ruthenium catalysts have provoked a big interest resulting in many controversial interpretations.[63,74] Adsorption of CO on different supported ruthenium samples results in appearance of three main bands: the so-called HF_1 (2156–2130 cm^{-1}), HF_2 (2090–2055) and LF (2060–1980 cm^{-1}). The HF_1 and HF_2 bands have been assigned to di- or tricarbonyls. The latter species are converted during evacuation at elevated temperatures into species characterized by HF_2 and LF bands and have often been assigned to dicarbonyls. Recently, based on the use of CO–$^{13}C^{18}O$ isotopic mixtures, it has been demonstrated that facial tricarbonyl species (bands at 2130 and 2100–2030 cm^{-1}) are converted into another kind of tricarbonyls with meridional structure (bands at 2070 and 1991 cm^{-1}).[74] Tri- and tetracarbonyl species on reduced iron sites in Fe–ZSM-5 and Fe–BEA zeolites have been identified using CO isotopic mixtures.[75] It has been demonstrated that the use of the conventional CO–^{13}CO mixtures was not utile because of superimposition. However, the better resolution of the bands registered after coadsorption of CO and $^{13}C^{18}O$ allowed detailed analysis of the spectra. The use of CO–$^{13}C^{18}O$ isotopic mixtures has recently helped also in the precise identification of the bands due to di- and tricarbonyls with different numbers of CO and $^{13}C^{18}O$ ligands in several cases: with Ni^+ sites in Ni–MOR[76] and with Cu^+ sites in Cu–ZSM-5[77] and in Cu–MCM-41.[78]

CO isotopic mixtures are often applied to measure the dynamic and static components of the shift of the carbonyl band maxima with coverage increase. Briefly, in order dynamic interaction to occur, the dipolar molecules should vibrate with the same frequency. Therefore, the dynamic interaction is suppressed when a CO molecule is surrounded by its isotopologues. Thus, the dynamic shift is simply the difference between $\nu(^{12}CO)$ in two cases: (i) when pure CO was adsorbed and (ii) after adsorption of isotopic mixtures containing a small percentage of ^{12}CO. Zecchina *et al.*[79] used a CO–^{13}CO–$^{13}C^{18}O$ (5 % $^{13}C^{18}O$) isotopic mixture to separate dynamic and static effects for the carbonyl bands registered on titania P25 nanocrystals (85% anatase + 15% rutile). They found $\Delta\nu_{dyn} = +3\ cm^{-1}$ and $\Delta\nu_{st} = -15\ cm^{-1}$ for the band at 2179 cm^{-1} and $\Delta\nu_{dyn} = +2\ cm^{-1}$ and $\Delta\nu_{st} = -8\ cm^{-1}$ for the band at 2169 cm^{-1}. For a rutile TiO_2 sample the dynamic shift was estimated by the analysis of the ^{13}CO band arising from the natural ^{13}C abundance:[80] $\Delta\nu_{dyn} = +2\ cm^{-1}$ and $\Delta\nu_{st} = -15\ cm^{-1}$ for the band at 2181 cm^{-1} and $\Delta\nu_{dyn} = +2\ cm^{-1}$ and $\Delta\nu_{st} = -5\ cm^{-1}$ for the band at 2149 cm^{-1}.

When CO is adsorbed on metal surfaces intensity transfer phenomena occur.[81–83] The bands located at higher wavenumber gain intensity at the expense of lower-frequency bands. The phenomenon is favoured by (i) higher density of the CO adsorption layer and (ii) lower difference between the maxima of the bands involved in the process. Therefore, intensity transfer from $^{13}C^{18}O$ to CO bands should be much smaller than transfer from ^{13}CO to CO bands. This has recently been demonstrated for different systems. Borovkov *et al.*,[83] studying CO adsorption on a Pt/Al_2O_3

sample, noted that the strong intensity transfer did not allow accurate determination of the position of the ^{13}CO band. However, these difficulties were avoided by the use of $CO–^{13}C^{18}O$ mixture. Based on the results the authors established that the dynamic interaction between the adsorbed CO molecules was smaller for a $PtCu/Al_2O_3$ sample as compared to Pt/Al_2O_3 and concluded that platinum sites are diluted by copper as a result of formation of alloys. Similar effects for the intensity transfer were reported with a reduced Au/SiO_2 sample.[81] A $Au^{\delta-}$–CO band at 2075 cm^{-1} was detected after CO adsorption at high coverage. The same band was also registered after adsorption of a 1 : 1 $CO–^{13}CO$ mixture, but the respective $Au^{\delta-}–^{13}CO$ band was not resolved. The latter appeared in the spectra with negligible intensity only at low CO coverages. However, even at high coverage, a $Au^{\delta-}–^{13}C^{18}O$ band was clearly detected when a $CO–^{13}C^{18}O$ mixture was utilized. Therefore, the risk of not detecting key bands due to intensity loss is much smaller when using $CO–^{13}C^{18}O$ isotopic mixtures.

Boccuzzi *et al.*[84] used $CO–^{13}CO$ isotopic mixtures to study several supported gold catalysts. They observed dynamic interaction and intensity transfer with Au/TiO_2 and Au/ZrO_2 samples and concluded that the gold CO adsorption sites were mutually interacting. On the contrary, these phenomena were not observed with a Au/CeO_2 sample which allowed the authors to state that the CO adsorption sites were isolated.

It should be underlined that $CO–^{13}C^{18}O$ mixtures are unique for some applications. Thus, it has been shown that the following scrambling reaction occurred when CO and $^{13}C^{18}O$ were contacted with AgX zeolite:[85]

$$CO + {}^{13}C^{18}O \rightarrow {}^{13}CO + C^{18}O \tag{5}$$

This is demonstrated in Fig. 7. Adsorption of a $CO–^{13}C^{18}O$ mixture resulted in appearance of two Ag^{+}–CO bands at 2176 and 2165 cm^{-1} and the respective $Ag^{+}–^{13}C^{18}O$ complexes at 2075 and 2065 cm^{-1}. With time, new bands at 2126 and 2116 cm^{-1} emerged and developed thus evidencing formation of $Ag^{+}–^{13}CO$ and $Ag^{+}–C^{18}O$. Note that if a $CO–^{13}CO$ mixture is adsorbed the scrambling should also occur but cannot be monitored because the reactants and the reaction products are identical.

4.3 $^{13}CO_2$

$^{13}CO_2$ is usually used to establish the structure of surface carbonate-like species. FTIR experiments on CO_2 and $^{13}CO_2$ adsorption on CeO_2 were combined with DFT calculations in order to precisely assign the bands appearing in the carbonate-carboxylate region.[30] It was found that the isotopic shift factors of the highest-frequency bands (around 1600 cm^{-1}) are about 1.025, slightly higher than the theoretical value (1.0228). The isotopic shift of the middle bands (1500–1200 cm^{-1}) was very close to the theoretical expectations based on diatomic harmonic oscillator. In agreement with the DFT calculated isotopic shifts, the bands below 1100 cm^{-1} were hardly affected by the isotopic substitution.

Mul *et al.*[86] studied the photocatalytic conversion of CO_2 and H_2O over Cu/TiO_2. The main product of the reaction carried out with $^{13}CO_2$ was

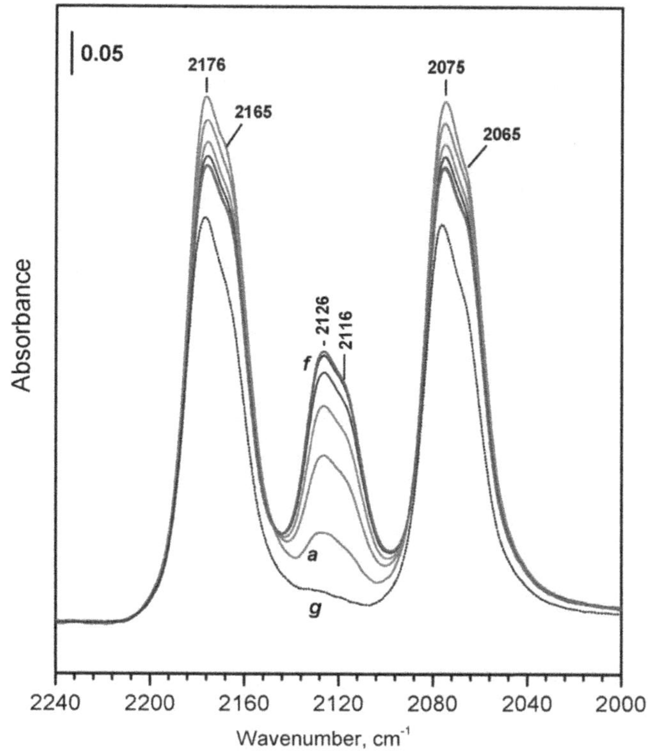

Fig. 7 FTIR spectra of a 1 : 1 CO–$^{13}C^{18}O$ isotopic mixture (500 Pa) adsorbed at 298 K on the activated AgX sample. Spectra registered after 2 (a), 12 (b), 32 (c), 62 (d), 122 (e) and 152 (f) min. Spectrum g corresponds to a CO–$^{13}C^{18}O$ isotopic mixture (500 Pa) adsorbed at 298 K (2 min of contact) on a sample reduced by CO at 523 K. The spectra are background and gas-phase corrected. Reproduced from Ref. 85.

^{12}CO. It was concluded that the reaction proceeded primary with surface carbon residues. Carbonate structures were also detected and the assignment of the respective bands confirmed by comparison of ^{12}C- and ^{13}C-species.

4.4 Other ^{13}C-labelled molecules

A band at 2342 cm^{-1} was registered in the IR spectra of ZnO prepared *via* acetate precursor. The band was assigned to occluded CO_2 and the use of ^{13}CO-labelled acetate revealed that CO_2 originated from the precursor.[87]

Baiker *et al.*[88,89] used benzyl alcohol labeled with ^{13}C at the methylene group to study its liquid-phase oxidation on Pd/Al_2O_3. The authors expected that dehydrogenation and decarbonylation of labelled benzaldehyde should produce ^{13}CO. No such product was registered with a catalyst preliminary blocked by CO, *i.e.* only part of the Pd sites were catalytically active.

An interesting study with a double-labelled compound, $D^{13}CN$, was performed by Yates *et al.*[90] It was demonstrated that in this way the dissociative and non-dissociative HCN adsorption forms could be easily distinguished. Thus, a sharp band at 3300 cm^{-1} and a broad band in the 3300–2800 cm^{-1} region (registered after HCN adsorption on alumina) were shifted by *ca.* 700 cm^{-1} after $D^{13}CN$ adsorption and assigned to C–H modes. Another band, at 2097 cm^{-1}, was isotopically shifted to 1911 cm^{-1} ($i = 1.097$). The band was assigned to the C–N stretching modes of molecularly adsorbed HCN and the high shift was attributed to the contribution of deuterium isotope effect. In contrast, bands at 2179 and 2100 cm^{-1} were shifted by factors of 1.023 and 1.019, respectively, and assigned to C–N modes, Moreover, the authors claimed that Al–CN species (2179 cm^{-1}) were more affected by isotopic substitution than the Al–NC species (2100 cm^{-1}).

4.5 SSITKA and related experiments

^{13}CO is widely used to study the mechanism of catalytic reactions. Chuang and Guzman[91] and Meunier[92] reviewed recently the use of transient infrared methods to determine the role of infrared observable species in the mechanisms of the NO + CO reaction, heterogeneous ethylene hydroformylation, water gas shift (WGS) and reverse WGS reactions.

Naito *et al.*[93] determined as inappropriate the possibility of participation of surface carbon species (produced by CO dissociation) in the hydrogenation of CO. In these experiments, ^{13}C was first accumulated on the surface *via* ^{13}CO disproportionation and then the $CO + H_2$ reaction was conducted. Reappearance of adsorbed ^{13}CO was observed but only $^{12}CH_4$ was detected in the reaction products, indicating that accumulated surface carbon was inactive for methane formation.

Different mechanisms of the WGS reaction were advocated on the basis of isotopic labelling. Kalamaras *et al.*[94] followed the reaction products on Pt/CeO_2 by DRIFT spectroscopy when switching the feed from $^{12}CO + H_2O$ to $^{13}CO + H_2O$. They observed red shift for the bands characterizing formate, carbonate, or carboxylate adsorbed species and concluded that one of them should be reaction intermediate. Similar experiments, again with Pt/CeO_2, were performed in the presence of H_2 in the feed.[95] The authors concluded that formates were reaction intermediates because they had a similar dynamic response with the CO_2 reaction product. It was also found that when the ν(CO) band for CO adsorbed on Pt was completely exchanged the CO_2 bands achieved only 50% of exchange. This result impeached the earlier proposed mechanism involving direct reaction of Pt–CO with O adatoms on ceria to produce gas phase CO_2. Based again on the use of ^{13}CO, it was concluded that the formates are only minor intermediates.[96]

Yang *et al.*[97,99] used $^{13}CO_2$ and D_2 to support the bands' assignment for bidentate formates observed as major species during methanol synthesis from CO_2 and H_2 over Cu/SiO_2 catalysts. Bands at 1580 and 1330 cm^{-1} were sensitive to both, $^{13}CO_2/CO_2$ and D/H substitutions and were assigned to $\nu_{as}(HCOO^-)$ and $\nu_s(HCOO^-)$ modes, respectively, of formate species. In contrast, a band at 1365 cm^{-1} was affected only by D/H exchange and consequently attributed to δ(C–H) modes of the same formates. Features at 2145 and 2160 cm^{-1} were associated with $D\text{-}^{13}COO$

and D-^{12}COO.[99] The isotopic response in the methanol product was substantially slower than that in the formates, indicating the possibility formates to be not rate-limiting intermediates. However, because ^{12}C-formate species were observed before and after addition of ^{13}CO to the $CO_2 + D_2$ reaction mixture, it was concluded that formates are spectator species in the synthesis of methanol.[98]

Similar conclusions were drawn for the formates and carbonates present on the surface of Au/TiO_2 during the WGS reaction.[100,101] The bands corresponding to these spectator species remained essentially unchanged for more than 120 s after switch from CO to ^{13}CO. On the contrary, the band due to CO adsorbed on gold sites completely shifted within 2 s and the Au–CO species were considered as active intermediates.

Different groups have studies the reverse WGS reaction over Pt/CeO_2 by DRIFTS.[102–105] It was demonstrated, by switching from $^{12}CO_2 + H_2$ to $^{13}CO_2 + H_2$ feed, that the CO product, surface carbonates and Pt-CO species were exchanged significantly faster than formates and consequently Pt-CO and carbonates were proposed as reaction intermediates.[103,105] However, the same effect was found in the absence of H_2 in the feed, *i.e.* when no catalytic reaction proceeded.[104] Therefore, one cannot conclusively state that Pt–CO and carbonates are reaction intermediates. Moreover, surface formates were found to exchange rapidly during wet conditions (as rapidly as Pt–CO and carbonates) which indicated that formates cannot be classified as spectator species in presence of water.

$^{13}CH_3OH$ was used in *operando* studies on methanol (photo)oxidation of different catalysts.[106,107] Similar isotopic exchange rates were established for CO_2 and surface formates in the CH_3OH oxidation on Au/CeO_2 and it was concluded that formates were intermediate species and their decomposition was the rate determining step.[106] In contrast, the exchange rate of fomates during CH_3OH photooxidation on TiO_2 was much slower as compared to this one of CO_2.[107] The results indicated that the majority of the formates on TiO_2 were spectators and only a part of them could have been involved in the reaction.

The identification of transient surface intermediates in hydroformylation reaction, $CO + C_2H_4 + H_2$, over Rh/Al_2O_3 was spectrally supported by using ^{13}CO.[108] An isotopic shift factor of 1.024 was established for the ^{12}C-bands at 1675 and 1737 cm^{-1}, attributed to propionyl intermediate and propionaldehyde, respectively.

5 ^{15}N-labelled compounds

5.1 $^{15}N_2$

Dinitrogen is a convenient probe molecule for measuring surface acidity. However, the ^{14}N–^{14}N modes appear in the region where CO_2 absorbs. Thus, small changes in the atmospheric CO_2 concentration could strongly affect the quality of the spectra. Consequently, in many cases the use of $^{15}N_2$ as a probe is determined by the fact that the ^{15}N–^{15}N stretching modes appear outside the CO_2 region.[19–21]

Recently, $^{14}N_2$–$^{15}N_2$ isotopic mixtures with different molar ratios were used to establish the formation of geminal dinitrogen species.[109]

Adsorption of $^{14}N_2$ at 100 K on CO-reduced Ni–ZSM-5 resulted in formation of two bands, at 2287 and 2270 cm^{-1}, attributed to $Ni^+(N_2)_2$ species. The geminal structure was proven by $^{14}N_2$–$^{15}N_2$ coadsorption experiments: two $Ni^+(^{14}N_2)(^{15}N_2)$ bands were registered at 2277 and 2201 cm^{-1}, in excellent agreement with the calculations based on the approximate force field model. Presently, this is the first example of geminal dinitrogen adsorption complex proven by isotopically labelled molecules.

5.2 NO isotopologues

The most used NO isotopologue is ^{14}NO. $^{15}N^{18}O$ could provide additional information but its use is restricted because of the difficulties in delivery.

Szanyi *et al.*[110] have compared the spectra of ^{14}NO and ^{15}NO adsorbed on a Cu–SSZ-13 zeolite in order to distinguish between N–O and N–N vibrations. As seen from Fig. 8, all NO bands registered below 2200 cm^{-1} are shifted after ^{15}NO adsorption by a factor of 1.018, consistent with the expected one for N–O vibrations. However, a band at 2249 cm^{-1} is shifted by a factor of 1.0331 which allows assignment of the band to N–N modes of adsorbed N_2O.

Adsorption of NO on Cu supported on dealuminated Y zeolite resulted in appearance of bands at 1638 and 1616 cm^{-1}. These bands were shifted to 1603 and 1582 cm^{-1} when ^{15}NO was adsorbed.[111] The shift coincided well with the calculated one for the NO_2 molecule and it was concluded the bands under consideration characterized NO_2 or NO_3^- species. Thielemann *et al.*[112] have studied NO adsorption and NO + O_2 coadsorption on MoOx/SBA-15 and performed parallel experiments with ^{15}NO. The authors detected (i) dinitrosyl species (^{14}NO bands at 1806 and 1702 cm^{-1} and ^{15}NO bands at 1779 and 1682 cm^{-1}) and surface nitrates (^{14}NO bands at 1618 and 1573 cm^{-1} and ^{15}NO bands at 1585 and 1534 cm^{-1}). They used the isotopic shift factor to support the proposed assignments.

Interaction of ^{15}NO with NO_2-precovered titania surface resulted in isotopic exchange of all N-containing surface species (NO^+, N_2O_3 and nitrates).[113] Because dimmeric structures were necessary for the process, formation of [$ONONO_2$] species was postulated to explain the isotopic exchange for nitrates.

In many cases ^{14}NO–^{15}NO mixtures are used to prove or reject the presence of dinitrosyl structures. Mihaylov *et al.*[114] established two kinds of dinitrosyl species on a Cr–ZSM-5 sample. The first kind, $Cr^{3+}(NO)_2$, were characterized by ν_s at 1910 cm^{-1} and ν_{as} at 1872 cm^{-1}. Mixed ligand complexes, $Cr^{3+}(NO)(^{15}NO)$, were obtained after coadsorption of NO and ^{15}NO and displayed bands at 1891 and 1760 cm^{-1}, in agreement with calculated values. The bands of the second kind of dinitrosyls, $Cr^{2+}(NO)_2$, were registered at 1902 an 1768 cm^{-1} and the mixed ligand species manifested bands at 1891 and 1749 cm^{-1}.

Very recently, a new kind of dinitrosyl copper species were reported.[115] After low temperature NO adsorption on Cu–ZSM-5 bands at 1915, 1863 and 1853 cm^{-1} were registered. The bands at 1863 and 1853 cm^{-1} were attributed to the ν_{as} modes of two kinds of $Cu^{3+}(NO)_2$ species, respectively, the symmetric modes for both complexes being at 1915 cm^{-1}. The

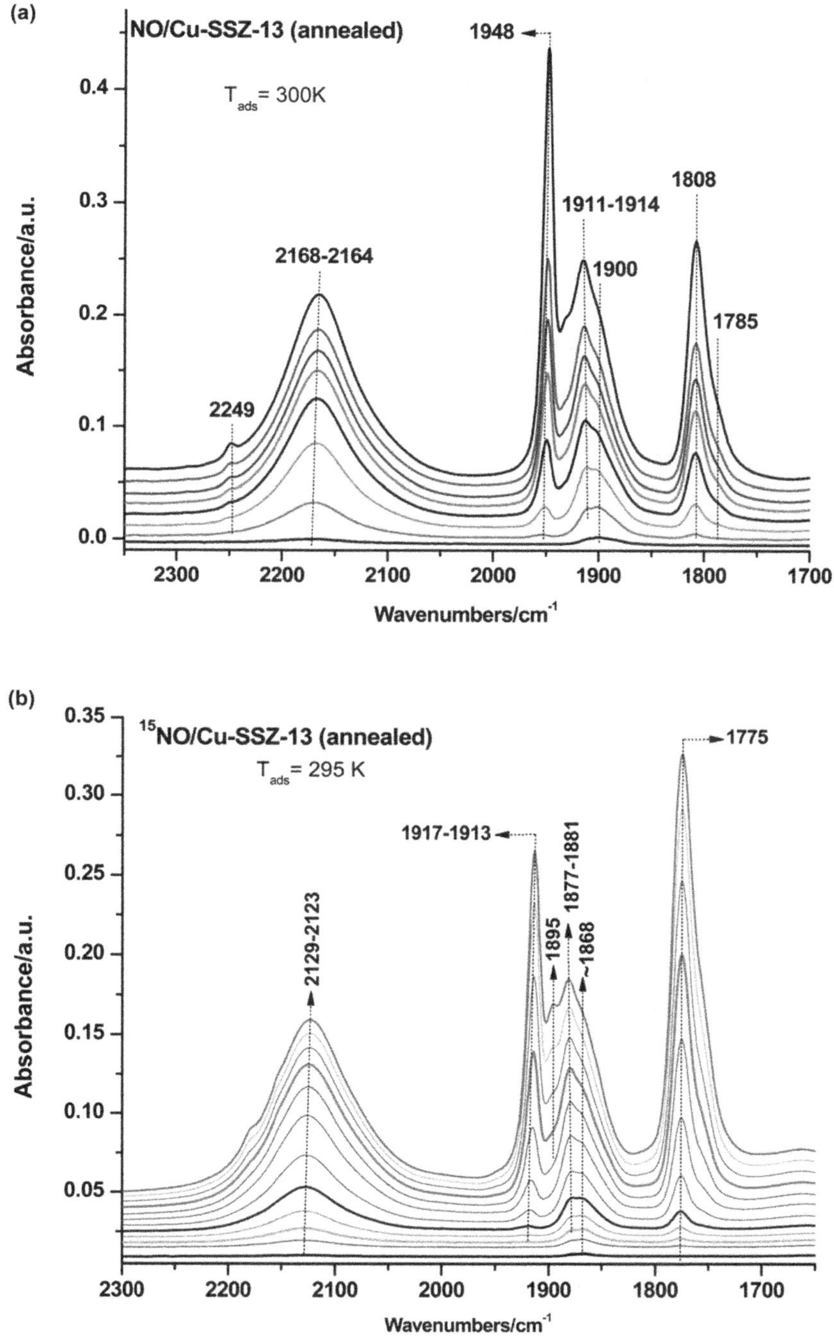

Fig. 8 Series of selected IR spectra obtained after exposure of annealed Cu–SSZ-13 samples to ^{14}NO (a) and ^{15}NO (b) at 300 K. The samples were annealed in vacuum at 773 K for 2 h prior to IR measurements. Reproduced from Ref. 110.

assignment was proven by ^{14}NO–^{15}NO coadsorption experiments: the frequencies for the mixed ligand species, $Cu^{3+}(^{14}NO)(^{15}NO)$, were detected at 1903, 1902, 1841 and 1831 cm^{-1}, in good agreement with calculated values.

On the contrary, King *et al.*[116] have used NO–^{15}NO isotopic mixtures to establish that only mononitrosyls and no dinitrosyl species were formed after NO adsorption at 95 K on pure and oxygen precovered Ir[100] surface.

5.3 Other ^{15}N-containing molecules

The photodecomposition of aminoacids on TiO_2 and Au/TiO_2 has been studied by ATR-IR.[117] The use of ^{15}N-labelled acids helped the authors to assign bands at 2133 and 2119 cm^{-1} to cyanide species. A detailed study of different ^{15}N-labelled species produced after adsorption of $(NH_2)_2CO$ and $(^{15}NH_2)_2CO$ on NaY zeolites was published by Li *et al.*[118] These authors reported that HNCO (2276 cm^{-1}) and NCO^- (2169 cm^{-1}) species were produced after interaction between adsorbed urea and NO_2. The assignment was supported by the isotopic shift of the bands by factors of 1.007–1.008, similar to those observed for gaseous HNCO.

5.4 SSITKA and related studies

Burch *et al.*[119] have performed *operando* IR spectroscopic study on the mechanism of selective catalytic reduction of NO_x over a Ag/Al_2O_3 catalyst. The explored gas feed contained NO, O_2, n-C_8H_{18}, H_2O and H_2. The spectra registered were very complicated because of the formation of different surface carbon- and nitrogen-containing species. To distinguish between the different bands the authors used replacement of NO in the feed by $^{15}N^{18}O$. As a result, several bands changed in shape and position. A band at 2145 cm^{-1}, shifted to 2112 cm^{-1}, was assigned to cyanides adsorbed on silver sites. The experimental isotopic shift factor (1.0156) coincided very well with the theoretical one for $C^{15}N$/CN substitution (1.0157). A band at 2232 cm^{-1} was shifted to 2223 cm^{-1} and attributed to isocyanates. Another band at 1305 cm^{-1} was shifted to 1272 cm^{-1} and assigned to nitrate species. Surprisingly, the authors observed very low isotopic exchange degree for the isocyanates that are believed to be SCR intermediates. However, performing the experiments with surface containing few isocyanates, the authors were able to isolate fast-reacting NCO^- species. The temporal response of these species to a $^{15}N^{18}O/^{14}NO$ switch corresponded to that obtained for dinitrogen formation.

Costa and Efstathiou[120] have studied the species formed during the $NO + O_2 + H_2$ reaction over supported Pt as well as the exchange of ^{14}N-species formed after switching the gas flow to a mixture containing ^{15}NO. The authors established that only two bands were affected by the isotopic substitution (NO^+, 2220 cm^{-1}, and bidentate nitrates, 1580 cm^{-1}) and concluded these species were possible reaction intermediates.

6 ^{18}O-labelled compounds

Oxygen has three stable isotopes: ^{16}O, ^{17}O and ^{18}O. In practice, ^{18}O-labelled compounds are used in the vibrational spectroscopy of surfaces because the isotopic shift is more significant compared to ^{17}O and because ^{18}O-labelled substances are less expensive. Two main approaches are used in these studies: (i) investigation of ^{18}O-labelled solids and (ii) studies on the adsorption of ^{18}O-containing substances.

6.1 Solids labeled by ^{18}O

Grassian *et al.*[5] used titania with $Ti^{18}OH$ surface hydroxyls (produced by treatment with $H_2{}^{18}O$, see Fig. 1) to investigate the mechanism of SO_2 adsorption. The isotopic labelling led to two important conclusions: (i) water formation during the process did not involve surface oxygen from the OH groups because the $\delta(H_2O)$ band at 1620 cm^{-1} was essentially the same as that observed with unlabelled titania (Fig. 9) and (ii) labelled oxygen went into the sulfur containing surface products, as evidenced by the different wavenumbers of the SO_x bands on unlabelled and ^{18}O-labelled samples. On the basis of these results, as well as results with deuterated sample, the authors proposed a scheme of SO_2 interaction with TiO_2 producing surface sulfites and water. In analogous way it was concluded that oxygen from the hydroxyl groups on Fe_2O_3 and Al_2O_3 participates in the formation of hydrogencarbonates when the samples interacted with CO.[121] A detailed interpretation of the IR bands of HCO_3^- species containing different numbers of ^{18}O atoms was also proposed.

A similar approach was used to determine the products of cyanogen chloride (ClCN) decomposition on γ-Al_2O_3 surface.[122] Adsorption of ClCN at 250 K on Al_2O_3 with ^{18}OH groups showed a substantial red shift in the $\nu_{as}(NCO)$ band of isocyanates (-25 cm^{-1}) with respect to the band registered with non-labelled sample (2245 cm^{-1}). This demonstrated that surface hydroxyl groups of γ-Al_2O_3 were involved in the ClCN decomposition and indicated that Ti–NCO species were reaction intermediates.

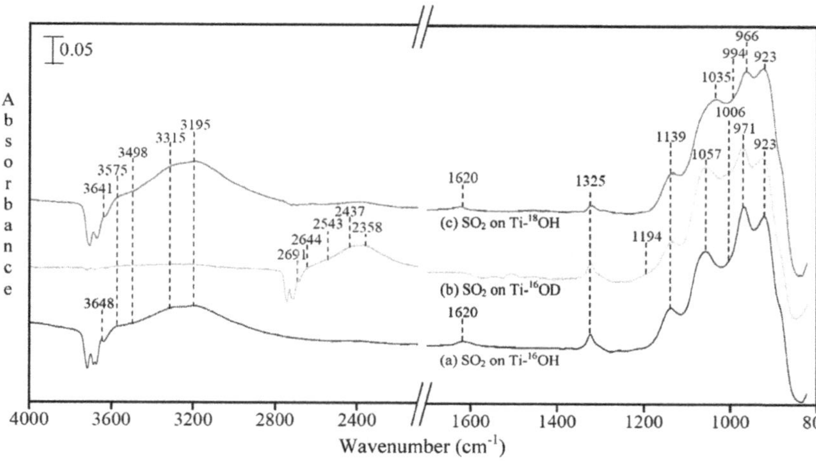

Fig. 9 Transmission FTIR spectra for TiO_2 in the presence of gas phase SO_2 at a pressure of 100 mTorr for (a) SO_2 adsorbed on TiO_2, (b) SO_2 adsorbed on deuterated-TiO_2, and (c) SO_2 adsorbed on ^{18}O-labeled TiO_2. Gas-phase absorptions have been subtracted from the spectra. Reproduced from Ref. 5.

Note that in the above cases the isotopic exchange could have also led to surface ^{18}O-species that were not involved in the formation of OH groups (*e.g.* lattice O^{2-}). Indeed, it has been demonstrated that dissociative adsorption of ammonia on TiO_2 produced OH groups vibrating at 3569 cm^{-1}.[123] These groups were observed at 3559 cm^{-1} with a sample, preliminary treated with $H_2{}^{18}O$. The results confirmed that the 3569 cm^{-1} band was due to OH and not to NH stretching modes and indicated the presence of coordinatively unsaturated $^{18}O^{2-}$ ions on the labelled sample.

Mul *et al.*[124] have studied the photocatalytic oxidation of cyclohexane over TiO_2. The catalyst surface was preliminary enriched to different extents with ^{18}O and the oxidation was carried either by $^{16}O_2$ or by $^{18}O_2$. Although using $^{18}O_2$ feed, the authors detected only ^{16}O-containing products (cyclohexanone, carboxylates) with non-labelled surface. However, when the photocatalytic reaction was preceded by isotopic exchange of the solid, ^{18}O-labelled cyclohexanone was also observed. In particular, the C=O bands for cyclohexanone (1690 and 1679 cm^{-1}) were shifted to 1657 and 1642 cm^{-1}, respectively, for the ^{18}O-containing product, in good agreement with the theoretical expectations. In addition, the authors have detected carboxylate species by bands at 1520 and 1414 cm^{-1}. Some ^{18}O-labeled carboxylates were evidenced by bands at 1510 and 1494 cm^{-1}. Because of the small shift, the latter were assigned to $C^{16}O^{18}O^-$ species. It was concluded that oxygen in cyclohexanone (as well as in the spectator species as carboxylates), originated from the catalyst surface.

6.2 $^{18}O_2$

Adsorption of dioxygen could be dissocitaive and non-dissociative. In the latter case the oxygen–oxygen bond is preserved and the ν(O–O) position depends on the O–O bond order. Possible products are O_2 (*ca.* 1550 cm^{-1}), $O_2{}^-$ (*ca.* 1140 cm^{-1}) and $O_2{}^{2-}$ (*ca.* 850 cm^{-1}). In all these cases the theoretical isotopic shift factor for $^{16}O/^{18}O$ exchange is 1.0608 and partial exchange should result in ^{16}O–^{18}O containing products. When oxygen dissociates, IR and Raman active are the M–O bonds. In particular, the double M=O bond appears in the 1050–950 cm^{-1} region and is also often monitored in the overtone region. In these cases the isotopic shift depends on the metal nature (*e.g.* $i = 1.45$ for the Cr–O bond) and no intermediate bands should be produced after partial exchange.

Treatment of a SrF_2/La_2O_3 catalyst in oxygen flow at temperatures up to 973 K led to the appearance of an IR band at 1130 cm^{-1}.[125] Subsequent introduction of $^{18}O_2$ to the system caused emergence of two new bands, at 1095 and 1064 cm^{-1}, at the expense of the band at 1130 cm^{-1}. On this basis the 1130 cm^{-1} band was assigned to the O–O stretching modes of superoxide ($O_2{}^-$) species and the bands at 1095 and 1064 cm^{-1}, to the respective O–^{18}O and ^{18}O–^{18}O modes. In addition, Raman spectra indicated that lattice oxygen was also involved in the $^{18}O/^{16}O$ isotopic exchange. A main peak at 407 cm^{-1} was detected with the non-labelled sample and shifted to 387 cm^{-1} upon isotopic exchange with $^{18}O_2$. Consequently, the 407 cm^{-1} band was assigned to La–^{16}O modes (the theoretical value for the La–^{18}O stretching mode being 383 cm^{-1}).

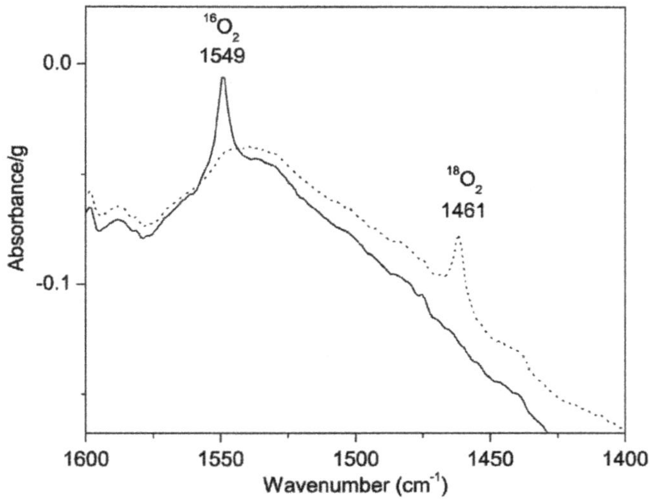

Fig. 10 $^{16}O_2$ (solid line) and $^{18}O_2$ (dotted line) IR absorption bands after adsorption of 200 mbar equilibrium pressure of oxygen on Cu/HZSM5 at 80 K. Reproduced from Ref. 127.

An IR band at 1550 cm^{-1} was detected after O_2 adsorption on TiO_2.[126] The band shifted to 1463 cm^{-1} when $^{18}O_2$ was utilized (i = 1.0594) and it was concluded that oxygen was adsorbed molecularly. Using the same approach it was established that oxygen is molecularly adsorbed on the acidic hydroxyls of Cu/H-ZSM-5:[127] the $^{16}O_2$ band was registered at 1549 cm^{-1} and the band appearing after $^{18}O_2$ adsorption, at 1481 cm^{-1} (see Fig. 10).

Based on $^{18}O/^{16}O$ isotope substitution experiments, the vibrational modes of oxygen adsorbed on CeO_2 were assigned as follows:[128] a band at 1551 cm^{-1} (obtained after O_2 adsorption at 93 K) was attributed to physisorbed O_2; bands in the 1135–1127 and 877–831 cm^{-1} regions were attributed to the O–O stretching modes of superoxide, O_2^-, and peroxide, O_2^{2-}, species, respectively. A band at 357 cm^{-1} was assigned to the cerium-oxygen vibration of the adsorbed O_2^-, whereas bands at 538 and 340 cm^{-1} were attributed to the antisymmetric and symmetric Ce–O vibrations of the adsorbed η^2 peroxo species. Similar conclusions were made as a result of a Raman study of $^{16}O_2$ and $^{18}O_2$ adsorption on CeO_2.[129] The authors assigned a band at 1140 cm^{-1} to superoxide, and bands at 862 and 830 cm^{-1}, to peroxide species. In addition, two mechanisms of exchange between superoxide and gas phase $^{18}O_2$ were proposed: exchange without O–O bond breaking, producing adsorbed $^{18}O_2^-$ species, and exchange with O–O bond breaking, leading to adsorbed $^{16}O^{18}O^-$. The peroxide species were exchanged with gas phase $^{18}O_2$ at room temperature but at a much slower rate than superoxide species did.

Woertink *et al.*[130] compared the Raman spectra of a Cu-ZSM-5 catalyst activated in flows of $^{16}O_2$ and $^{18}O_2$. Three isotope sensitive bands were found in the 900–200 cm^{-1} region: $^{16}O/^{18}O$ at 237/234, 456/448 and 870/830 cm^{-1}. No split was observed when isotopic mixture was used in the experiments which rejected any assignment to O–O vibrations.

The authors concluded on the existence of $[Cu_2O]^{2-}$ core species with δ(Cu–O–Cu) at 237 cm^{-1} and ν_s and ν_{as} Cu–O stretches at 456 and 870 cm^{-1}, respectively. The assignments were supported by analysis of overtone and combination bands as well as by DFT calculations.

Isotopic exchange with $^{18}O_2$ was used to establish whether mono-oxo (Mo=O) or di-oxo [Mo(=O)$_2$] structures exist on supported molybdenum.[12,131–133] For Mo/TiO$_2$ only a discrete shift of a Raman band at 994 to 944 cm^{-1} was observed (coinciding with the theoretical expectations of $i = 1.0513$) which allowed concluding that the supported species were monooxo ones (Fig. 11).[12,131,132] The slight red shift of the Mo=^{16}O band in the course of ^{18}O-enrichment of the surface was attributed to exchange of other oxygens in the molybdenum first coordination sphere by ^{18}O. Similar results were obtained more recently for Mo/Al$_2$O$_3$ and Mo/ZrO$_2$.[133] By analogous approach the existence of monooxo V=O structures on VO$_x$/CeO$_2$ catalysts was proven.[134] Under hydration with $H_2{}^{18}O$, the two existing V=^{16}O Raman modes (at 1015 and 1033 cm^{-1}) decreased in intensity while two new V=^{18}O modes (975 and 990 cm^{-1}) developed. The observed isotopic shifts coincided with the expected one for V=O

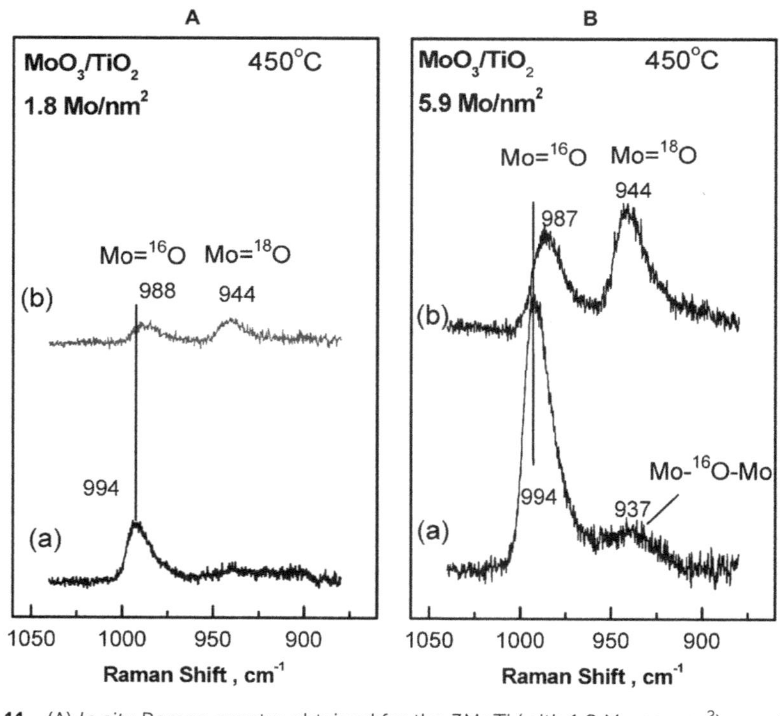

Fig. 11 (A) *In situ* Raman spectra obtained for the 3MoTi (with 1.8 Mo per nm^2) sample at 450 °C: (a) under flowing $^{16}O_2$ and (b) under flowing 2% $^{18}O_2$/He after 35 reduction (4.5% H$_2$/N$_2$)/oxidation (2% $^{18}O_2$/He) cycles. (B) *In situ* Raman spectra obtained for the 15MoTi (with 5.9 Mo per nm^2) sample at 450 °C: (a) under flowing $^{16}O_2$; (b) under flowing 2% $^{18}O_2$/He after 30 reduction (4.5% H$_2$/N$_2$)/oxidation (2% $^{18}O_2$/He) cycles. The contribution of TiO$_2$ (anatase) is subtracted from the original spectra. Recording parameters: laser wavelength, λ_0 = 488.0 nm; laser power, w = 40 mW; spectral slit width, ssw = 3.5 cm^{-1}. Reproduced from Ref. 131.

diatomic oscillator ($i = 1.045$). Noticeably, during re-oxidation of reduced VO_x/CeO_2 sample using $^{18}O_2$ the majority of oxidized VO_x species remained rich in ^{16}O. It was concluded that two sequential processes occurred: first, $^{18}O_2$ re-oxidizes reduced ceria; second, the ceria lattice oxygen (mostly still as ^{16}O) oxidizes the reduced VO_x species.

Zhang and He[135] studied the selective catalytic oxidation of ammonia over a Ag/Al_2O_3 catalyst. They found that interaction of oxygen with preadsorbed NH_3 resulted in the appearance of two bands, at 1529 and 1456 cm^{-1}, changing in concert. The bands were attributed to the N–O stretching and N–H bending modes, respectively, of –HNO species. The assignment was supported by utilizing $^{18}O_2$: in this case the 1529 cm^{-1} band was detected at 1490 cm^{-1}, in agreement with the expected shift.

Fukuoka *et al.*[136] have used $^{18}O_2$ and $D_2{}^{18}O$ to study the preferential oxidation (in presence of H_2) of CO on Pt/SiO_2. It was established that neither gas-phase $^{18}O_2$ nor $D_2{}^{18}O$ reacted with CO over Pt nanoparticles supported on FSM-16. The surface silanol groups were proposed to act as the source of oxygen in the CO oxidation.

Another example of using ^{18}O-isotope labelling is to obtain mechanistic insights into oxidative conversions of organics adsorbed on Au/TiO_2 catalyst.[137] It was found that acetic acid can be partly oxidized on Au/TiO_2 forming a gold ketenylidene ($Au_2{=}C{=}C{=}O$) intermediate (νCO at 2040 cm^{-1}). The absence of isotopic shift of the carbonyl band when $^{18}O_2$ was used as oxidant indicated that the O moiety in the Au_2CCO intermediate originated from the acetic acid, but not from the gas phase.

6.3 Other ^{18}O-labelled compounds

Ozkan and co-workers[138] have used $H_2{}^{18}O$ to study the ethanol steam reforming over zirconia- and ceria-supported cobalt catalysts. They found no indication of the formation of $CH_3C^{18}OCH_3$ which suggested that acetone was produced through the aldol condensation of two acetaldehyde molecules formed from dehydrogenation of ethanol without participation of a $H_2{}^{18}O$ molecule.

Daturi *et al.*[139] investigated coadsorption of $C^{18}O + NO$ on Ag/Al_2O_3 catalysts and established production of $C^{16}O$ indicating scrambling reaction of $C^{18}O$ with surface oxygen atoms or with oxygen atoms from NO molecules. Adsorption of $C^{18}O$ on reduced catalysts followed by NO admission at temperature where isocyanate species are stable resulted in formation of $NC^{18}O$ and $NC^{16}O$ species that were indicated by bands of comparable intensity in the spectrum (*ca.* 2230 and 2220 cm^{-1}). The results confirmed the authors' hypothesis that dissociation of both CO and NO is necessary for formation of isocyanate species in contrast to previously proposed reaction of gas molecule CO with surface $N_{ads.}$ to form surface NCO^- species. The authors proposed a mechanism of the interaction between CO and NO including initial dissociation of both molecules.

6.4 SSITKA and related studies

In several recent works $^{18}O_2$ was used to obtain information on the mechanism of CO oxidation. Wu *et al.*[140] found that the initial products of the $CO + {}^{18}O_2$ reaction on different CeO_2 nanocrystals were $C^{16}O_2$ and non-labelled bidentate, bridged and unidentate carbonates. In the course of reaction, $C^{16}O^{18}O$ started to appear and only the unidentate carbonate species were slowly exchanged with ^{18}O. The observed isotopic shifts were: 1446 to 1433 cm^{-1}, 1275 to 1268 cm^{-1}, 1040 to 1022 cm^{-1}, and 852 to 847 cm^{-1}. The results evidence that, even in the presence of gaseous $^{18}O_2$, CO reacts preferably with lattice oxygen of ceria.

Li *et al.*[141] studied reaction pathways of CO oxidation over $Au/FePO_4$ catalyst. Interaction of CO with a sample preliminary treated with $^{18}O_2$ at 473 K led only to two adsorbed $C^{16}O_2$ species (bands at 2360 and 2340 cm^{-1}), *i.e.* the active oxygen on the catalyst surface resisted oxygen isotopic exchange. Again $C^{16}O_2$ was produced when the oxidation of CO was performed with $^{18}O_2$. In the course of reaction $C^{16}O^{18}O$ (2340 and 2325 cm^{-1}) was also detected. The latter product developed in parallel with Au^0-CO species (2114 cm^{-1}) and it was concluded that metallic gold was necessary for CO oxidation by gaseous oxygen. Further support of this hypothesis was obtained from experiments with hydrogen-reduced catalysts. With this pretreatment $C^{16}O_2$ and $C^{16}O^{18}O$ appeared simultaneously from the $CO + {}^{18}O_2$ reaction beginning. Small amount of $C^{18}O_2$ was also detected and explained by isotopic scrambling between CO and $^{18}O_2$. It was concluded that two parallel reaction pathways takes place in the CO oxidation at ambient temperature: active oxygen being supplied by (i) the solid or (ii) the gaseous phase, respectively. When analogous experiments were carried out with $Au/LaPO_4$ catalyst[142] some differences were noticed. Interaction of CO with a sample preliminary treated with $^{18}O_2$ led to formation of adsorbed $C^{16}O_2$ and some amount of $C^{16}O^{18}O$, *i.e.* active ^{18}O-species were present on the catalyst surface. In agreement with this, the $CO + {}^{18}O_2$ interaction on $^{16}O_2$ pretreated sample resulted again in production of $C^{16}O_2$ and some amount of $C^{16}O^{18}O$.

It was also reported that when a fresh Au/TiO_2 catalyst was exposed to a CO stream, a band at 1242 cm^{-1} grew in the IR spectrum.[143] This band disappeared quickly when the stream was switched to O_2. At the same time, production of CO_2 was observed. If the experiments were carried out with $^{18}O_2$, the 1242 cm^{-1} band was shifted to 1236 cm^{-1}. The 1242 cm^{-1} band was assigned to a surface hydroxycarbonyl species that accumulate on the surface in the presence of CO and absence of oxygen. The suggested mechanism of CO oxidation includes rapid interaction of intermediate hydroxycarbonyls by O_2 to bicarbonate species which decompose to CO_2 and regenerate the surface hydroxyls.

Engeldinger *et al.*[144] studied the oxidative carbonylation of methanol over a CuY zeolite. When a $CH_3OH + {}^{16}O_2$ containing gas feed was switched to $CH_3OH + {}^{18}O_2$ one, only a formate band at around 1590 cm^{-1} shifted to lower wavenumbers while the band at 1643 cm^{-1} assigned to adsorbed monomethyl carbonate remained unchanged. The experiments revealed that gaseous oxygen is involved in the production of methyl formate and CO_2 but to an essentially lower extent in the formation of dimethoxymethane.

7 Conclusions and perspectives

The power of vibrational spectroscopy for characterization of solid surfaces and studying the mechanism of surface reactions is strongly enhanced by the use of isotopic labelling. In this way one can unambiguously prove the nature of atoms in a chemical bond corresponding to a given vibration. The technique allows valuable information of the structure of surface species to be obtained. It is also possible to trace the origin of atoms in surface reaction products and to obtain information on the mechanism of catalytic reactions. Although some general rules on how to use isotopically labelled molecules are generally accepted, each particular case is a challenge to the researcher how to obtain as much useful information as possible. New and interesting approaches are published each year and are often based on previous achievements. We hope that this review will help in the appropriate experiment planning. The authors believe that many presently available assignments of surface species need revision and this can be made by the use of isotopically labelled molecules combined with precise calculations, including DFT modelling.

References

1 F. Thibault-Starzyk and F. Maugé, in *Characterization of Solid Materials and Heterogeneous Catalysts: From Structure to Surface Reactivity* (M. Che and J. C. Védrine, Eds.), Wiley-VCH Verlag GmbH & Co, 2012, pp. 3–48.

2 A. A. Davydov, *Molecular Spectroscopy of Oxide Catalyst Surfaces*, John Wiley & Sons, Chchester, England, 2003.

3 O. Marie, P. Bazin and M. Daturi, *Spectrosc. Prop. Inorg. Organomet. Compd.*, 2012, **42**, 34–103.

4 S. Pinchas and N. Laulicht, *Infrared Spectra of Labelled Compounds*, Academic Press, London & New York, 1971.

5 C. E. Nanayakkara, J. Pettibone and V. H. Grassian, *Phys. Chem. Chem. Phys.*, 2012, **14**, 6957–6966.

6 K. Chakarova, N. Drenchev, M. Mihaylov, P. Nikolov and K. Hadjiivanov, *J. Phys. Chem. C*, 2013, **117**, 5242–5248.

7 B. M. Cheng, Y.-P. Lee and J.-F. Ogilvie, *Chem. Phys. Lett.*, 1988, **151**, 109–115.

8 G. Jacobs, P. M. Patterson, U. M. Graham, A. C. Crawford and B. H. Davis, *Int. J. Hydrogen Energy*, 2005, **30**, 1265–1276.

9 S. E. Collins, L. E. Briand, L. A. Gambaro, M. A. Baltanas and A. L. Bonivardi, *J. Phys. Chem. C*, 2008, **112**, 14988–15000.

10 F. Boccuzzi, E. Borello, A. Zecchina, A. Bossi and M. Camia, *J. Catal.*, 1978, **51**, 150–159.

11 M. Calatayud, S. E. Collins, M. A. Baltanars and A. L. Bonivardi, *Phys. Chem. Chem. Phys.*, 2009, **11**, 1397–1405.

12 G. Tsilomelekis and S. Boghosian, *J. Phys. Chem. C*, 2011, **115**, 2146–2154.

13 J. Hohmeyer, E. V. Kondratenko, M. Bron, J. Kröhnert, F. C. Jentoft, R. Schlögl and P. Claus, *J. Catal.*, 2010, **269**, 5–14.

14 J.-P. Gallas, J.-M. Goupil, A. Vimont, J.-C. Lavalley, B. Gil, J.-P. Gilson and O. Miserque, *Langmuir*, 2009, **15**, 5825–5834.

15 E. Gianotti, L. Marchese, M. Guidotti, N. Ravasio, R. Psaro and S. Coluccia, *Stud. Surf. Sci. Catal.*, 2005, **155**, 311–320.

16 J. N. Kondo, R. Nishitani, E. Yoda, T. Yokoi, T. Tatsumi and K. Domen, *Phys. Chem. Chem. Phys.*, 2010, **12**, 11576–11586.
17 J. Datka, B. Gil and P. Baran, *Microporous Mesoporous Mater.*, 2003, **58**, 291–294.
18 B. Gil, G. Košova and J. Cejka, *Microporous Mesoporous Mater.*, 2010, **129**, 256–266.
19 K. Chakarova and K. Hadjiivanov, *J. Phys. Chem. C*, 2011, **115**, 4806–4817.
20 K. Chakarova and K. Hadjiivanov, *Chem. Commun.*, 2011, **47**, 1878–1880.
21 K. Chakarova and K. Hadjiivanov, *Microporous Mesoporous Mater.*, 2013, **177**, 59–65.
22 K. Hadjiivanov, K. Chakarova, N. Drenchev and M. Mihaylov, *Curr. Phys. Chem.*, 2012, **2**, 151–161.
23 D. Scarano, S. Bordiga, C. Lamberti, G. Ricchiardi, S. Bertarione and G. Spoto, *Appl. Catal. A*, 2006, **307**, 3–12.
24 J. A. Boscoboinik, X. Yu, E. Emmez, B. Yang, S. Shaikhutdinov, F. D. Fischer, J. Sauer and H.-J. Freund, *J. Phys. Chem. C*, 2013, **117**, 13547–13556.
25 Yu. G. Kolyagin, V. V. Ordomsky, Y. Z. Khimyak, A. I. Rebrov, F. Fajula and I. Ivanova, *J. Catal.*, 2006, **238**, 122–133.
26 H. Noei, H. Qiu, Y. Wang, E. Löffler, C. Wöll and M. Muhler, *Phys. Chem. Chem. Phys.*, 2008, **10**, 7092–7097.
27 J. Strunk, K. Kähler, X. Xia and M. Muhler, *Surf. Sci.*, 2009, **603**, 1776–1783.
28 L. E. Davies, N. A. Bonini, S. Locatelli and E. E. Gonzo, *Lat. Am. Appl. Res.*, 2005, **35**, 23–28.
29 M. Kandiah, S. Usseglio, S. Svelle, U. Olsbye, K. P. Lillerud and M. Tilset, *J. Mater. Chem.*, 2010, **20**, 9848–9851.
30 G. N. Vayssilov, M. Mihaylov, P. Petkov, K. I. Hadjiivanov and K. Neyman, *J. Phys. Chem. C*, 2011, **115**, 23435–23454.
31 T. Tabakova, F. Boccuzzi, M. Manzoli, J. W. Sobczak, V. Idakiev and D. A. Andreeva, *Appl. Catal. A*, 2006, **298**, 127–143.
32 G. Jacobs, P. M. Patterson, U. M. Graham, D. E. Sparks and B. H. Davis, *Appl. Catal. A*, 2004, **269**, 63–73.
33 R. Valentin, B. Bonelli, E. Garrone, F. Di Renzo and F. Quignard, *Biomacromolecules*, 2007, **8**, 3646–3650.
34 J. C. Lavalley, M. Bensitel, J. P. Gallas, J. Lamotte, G. Busca and V. Lorenzelli, *J. Mol. Struct.*, 1988, **175**, 453–458.
35 K. Chakarova, N. Drenchev and K. Hadjiivanov, *J. Phys. Chem. C*, 2012, **116**, 17101–17109.
36 K. Chakarova, P. Nikolov and K. Hadjiivanov, *Catal. Commun.*, 2013, **41**, 38–40.
37 A. K. Soper and C. J. Benmore, *Phys. Rev. Lett.*, 2008, **101**, 065502.
38 K. Chakarova and K. Hadjiivanov, *Microporous Mesoporous Mater.*, 2011, **143**, 180–188.
39 S. Kotrel, J. H. Lunsford and H. Knözinger, *J. Phys. Chem. B*, 2001, **105**, 3917–3921.
40 S. P. Zhdanov, L. S. Kosheleva and T. I. Titova, *Langmuir*, 1987, **3**, 960–967.
41 C. Berthomieu and C. Sandorfy, *J. Mol. Spectrosc.*, 1965, **15**, 15–21.
42 A. Burneau and C. Carteret, *Phys. Chem. Chem. Phys.*, 2000, **2**, 3217–3226.
43 K. A. Dubkov, E. V. Starokon, E. A. Paukshtis, A. M. Volodin and G. I. Panov, *Kinet. Catal.*, 2004, **45**, 202–208.
44 E. N. Gribov, D. Cocina, G. Spoto, S. Bordiga, G. Ricchiardi and A. Zecchina, *Phys. Chem. Chem. Phys.*, 2006, **8**, 1186–1196.
45 D. A. Panayotov and J. T. Yates, Jr., *Chem. Phys. Lett.*, 2007, **436**, 204–208.
46 S. E. Collins, M. A. Baltanas and A. L. Bonivardi, *Langmuir*, 2005, **21**, 962–970.

47 D. Mukai, Y. Murai, T. Higo, S. Tochiya, T. Hashimoto, Y. Sugiura and Y. Sekine, *Appl. Catal. A*, 2013, **466**, 190–197.
48 R. B. Barros, A. R. Garcia and L. M. Ilharco, *ChemPhysChem*, 2005, **6**, 1299–1306.
49 J. N. Kondo, K. Ito, E. Yoda, F. Wakabayashi and K. Domen, *J. Phys. Chem. B*, 2005, **109**, 10969–10972.
50 J. N. Kondo, S. Yang, Q. Zhu, S. Inagaki and K. Domen, *J. Catal.*, 2007, **248**, 53–59.
51 A. R. Almeida, J. A. Moulijn and G. Mul, *J. Phys. Chem. C*, 2008, **112**, 1552–1561.
52 B. Tremblay and L. Manceron, *Inorg. Chem.*, 2008, **47**, 4531–4535.
53 C. Binet, A. Badri and J. C. Lavalley, *J. Phys. Chem.*, 1994, **98**, 6392–6398.
54 P. Bazin, O. Saur, J. C. Lavalley, M. Daturi and G. Blanchard, *Phys. Chem. Chem. Phys.*, 2005, **7**, 187–194.
55 A. S. Baird, K. M. Kross, D. Gottschalk, E. A. Hinson, N. Wood and K. A. Layman, *J. Phys. Chem. C*, 2007, **111**, 14207–14214.
56 K. Hadjiivanov and H. Knözinger, *Surf. Sci.*, 2009, **603**, 1629–1636.
57 K. Hadjiivanov, in *Ordered Poropus Solids* (V. Valtchev, S. Mintova and M. Tsapatis, Eds.), Elsevier, Amsterdam, 2008, pp. 263–281
58 M. Mihaylov, O. Lagunov, E. Ivanova and K. Hadjiivanov, *Top. Catal.*, 2011, **54**, 308–317.
59 N. Drenchev, E. Ivanova, M. Mihaylov and K. Hadjiivanov, *Phys. Chem. Chem. Phys.*, 2010, **12**, 6423–6427.
60 R. Kefirov, A. Penkova, K. Hadjiivanov, S. Dzwigaj and M. Che, *Microporous Mesoporous Mater.*, 2008, **116**, 180–187.
61 Z.-H. Lu and Q. Xua, *J. Chem. Phys.*, 2011, **134**, 034305.
62 V. Indovina, M. C. Campa and D. Pietrogiacomi, *J. Phys. Chem. C*, 2008, **112**, 5093–5101.
63 A. Penkova, S. Dzwigaj, R. Kefirov, K. Hadjiivanov and M. Che, *J. Phys. Chem. C*, 2011, **111**, 8623–8631.
64 E. Ivanova, M. Mihaylov, F. Thibault-Starzyk, M. Daturi and K. Hadjiivanov, *J. Catal.*, 2005, **236**, 168–171.
65 E. Ivanova, M. Mihaylov, H. Aleksandrov, M. Daturi, F. Thibault-Starzyk, G. Vayssilov, N. Rösch and K. Hadjiivanov, *J. Phys. Chem. C*, 2007, **111**, 10412–10418.
66 M. Mihaylov, E. Ivanova, Thibault-Starzyk, M. Daturi, L. Dimitrov and K. Hadjiivanov, *J. Phys. Chem. B*, 2006, **110**, 10383–10389.
67 M. Mihaylov, K. Chakarova, K. Hadjiivanov, O. Marie and M. Daturi, *Langmuir*, 2005, **21**, 11821–11828.
68 K. Chakarova, M. Mihaylov and K. Hadjiivanov, *Catal. Commun.*, 2005, **6**, 466–471.
69 N. Drenchev, I. Spassova, E. Ivanova, M. Khristova and K. Hadjiivanov, *Appl. Catal. B*, 2013, **138–139**, 362–372.
70 P. Bera, A. López Cámara, A. Hornés and A. Martínez-Arias, *J. Phys. Chem. C*, 2009, **113**, 10689–10695.
71 M. Kantcheva, O. Samarskaya, L. Ilieva, G. Pantaleo, A. M. Venezia and D. Andreeva, *Appl. Catal. B*, 2009, **88**, 113–126.
72 T. Mori, A. Itadani, E. Tabuchi, Y. Sogo, R. Kumashiro, M. Nagao and Y. Kuroda, *Phys. Chem. Chem. Phys.*, 2008, **10**, 1203–1212.
73 K. Föttinger, R. Schlögl and G. Rupprechter, *Chem. Commun.*, 2008, **44**, 320–322.
74 M. Mihaylov, O. Lagunov, E. Ivanova and K. Hadjiivanov, *J. Phys. Chem. C*, 2011, **115**, 13860–13867.

75 M. Mihaylov, E. Ivanova, K. Chakarova, P. Novachka and K. Hadjiivanov, *Appl. Catal. A*, 2011, **391**, 3–10.
76 H. A. Aleksandrov, V. R. Zdravkova, M. Y. Mihaylov, P. St. Petkov, G. N. Vayssilov and K. I. Hadjiivanov, *J. Phys. Chem. C*, 2012, **116**, 22823–22831.
77 N. Drenchev, P. A. Georgiev and K. Hadjiivanov, *J. Mol. Catal. A*, 2011, **341**, 7–13.
78 K. Chakarova, G. Petrova, M. Dimitrov, L. Dimitrov, G. Vayssilov, T. Tsoncheva and K. Hadjiivanov, *Appl. Catal. B*, 2011, **106**, 186–194.
79 L. Mino, G. Spoto, S. Bordiga and A. Zecchina, *J. Phys. Chem. C*, 2012, **116**, 17008–17018.
80 L. Mino, G. Spoto, S. Bordiga and A. Zecchina, *J. Phys. Chem. C*, 2013, **117**, 11186–11196.
81 K. Chakarova, M. Mihaylov, S. Ivanova, M. A. Centeno and K. Hadjiivanov, *J. Phys. Chem. C*, 2011, **115**, 21273–21282.
82 P. Hollins, *Surf. Sci. Rep.*, 1992, **16**, 51–94.
83 V. Yu. Borovkov, S. P. Kolesnikov, V. I. Koval'chuk and J. L. d'Itri, *Russ. Chem. Bull. Int. Ed.*, 2007, **56**, 863–869.
84 A. Chiorino, M. Manzoli, F. Menegazzo, M. Signoretto, F. Vindigni, F. Pinna and F. Boccuzzi, *J. Catal.*, 2009, **262**, 169–176.
85 O. Lagunov, K. Chakarova and K. Hadjiivanov, *Phys. Chem. Chem. Phys.*, 2012, **14**, 2178–2182.
86 C.-C. Yang, Y.-H. Yu, B. van der Linden, J. C. S. Wu and G. Mul, *J. Am. Chem. Soc.*, 2010, **132**, 8398–8406.
87 W. M. Hlaing Oo, M. D. McCluskey, A. D. Lalonde and M. G. Norton, *Appl. Phys. Lett.*, 2005, **86**, 073111.
88 D. Ferri and A. Baiker, *Top. Catal.*, 2009, **52**, 1323–1333.
89 D. Ferri, C. Mondelli, F. Krumeich and A. Baiker, *J. Phys. Chem. B*, 2006, **110**, 22982–22986.
90 S. Kim, D. C. Sorescu and J. T. Yates, Jr., *J. Phys. Chem. C*, 2007, **111**, 5416–5425.
91 S. S. C. Chuang and F. Guzman, *Top. Catal.*, 2009, **52**, 1448–1458.
92 F. C. Meunier, *Catal. Today*, 2010, **155**, 164–171.
93 S. Naito, S. Aida, T. Kasahara and T. Miyao, *Res. Chem. Intermed.*, 2006, **32**, 279–290.
94 C. M. Kalamaras, S. Americanou and A. M. Efstathiou, *J. Catal.*, 2011, **279**, 287–300.
95 G. Jacobs and B. H. Davis, *Appl. Catal. A*, 2007, **333**, 192–201.
96 F. C. Meunier, D. Tibiletti, A. Goguet, S. Shekhtman, C. Hardacre and R. Burch, *Catal. Today*, 2007, **126**, 143–147.
97 Y. Yang, C. A. Mims, R. S. Disselkamp, D. Mei, J.-H. Kwak, J. Szanyi, C. H. F. Peden and C. T. Campbell, *Catal. Lett.*, 2008, **135**, 201–208.
98 Y. Yang, C. A. Mims, D. H. Mei, C. H. F. Peden and C. T. Campbell, *J. Catal.*, 2013, **298**, 10–17.
99 Y. Yang, C. A. Mims, R. S. Disselkamp, C. H. F. Peden and C. T. Campbell, *Top Catal.*, 2009, **52**, 1440–1447.
100 J. Wang, V. F. Kispersky, W. N. Delgass and F. H. Ribeiro, *J. Catal.*, 2012, **289**, 171–178.
101 M. Shekhar, J. Wang, W.-S. Lee, M. C. Akatay, E. A. Stach, W. N. Delgass and F. H. Ribeiro, *J. Catal.*, 2012, **293**, 94–102.
102 D. Tibiletti, A. Goguet, D. Reid, F. C. Meunier and R. Burch, *Catal. Today*, 2006, **113**, 94–101.
103 F. C. Meunier, D. Tibiletti, A. Goguet, D. Reid and R. Burch, *Appl. Catal. A*, 2005, **289**, 104–112.

104 G. Jacobs and B. H. Davis, *Appl. Catal. A*, 2005, **284**, 31–38.
105 F. C. Meunier, A. Goguet, S. Shekhtman, D. Rooney and H. Daly, *Appl. Catal. A*, 2008, **340**, 196–202.
106 P. Bazin, S. Thomas, O. Marie and M. Daturi, *Catal. Today*, 2012, **182**, 3–11.
107 M. El-Roz, P. Bazin, M. Daturi and F. Thibault-Starzyk, *ACS Catal.*, 2013, **3**, 2790–2798.
108 N. Sivasankar and H. Frei, *J. Phys. Chem. C*, 2011, **115**, 7545–7553.
109 V. Zdravkova, M. Mihaylov and K. Hadjiivanov, *J. Phys. Chem. C*, 2012, **116**, 12706–12711.
110 J. Szanyi, J. H. Kwak, H. Zhu and C. H. F. Peden, *Phys. Chem. Chem. Phys.*, 2013, **159**, 2368–2380.
111 T. Komatsu, T. Nagai and T. Yashima, *Res. Chem. Intermed.*, 2006, **32**, 291–304.
112 J. P. Thielemann, J. Kröhnert and C. Hess, *J. Phys. Chem. C*, 2010, **114**, 17092–17098.
113 R. V. Mikhaylov, A. A. Lisachenko, B. N. Shelimov, V. B. Kazansky, G. Martra and S. Coluccia, *J. Phys. Chem. C*, 2013, **117**, 10345–10352.
114 M. Mihaylov, A. Penkova, K. Hadjiivanov and M. Daturi, *J. Mol. Catal. A*, 2006, **249**, 40–46.
115 M. Tortorelli, K. Chakarova, L. Lisi and K. Hadjiivanov, *J. Catal.*, 2013, **309**, 376–385.
116 S. Khatua, Z.-P. Liu and D. A. King, *Surf. Sci.*, 2005, **584**, 214–224.
117 I. Dolamic and T. Bürgi, *J. Phys. Chem. C*, 2011, **115**, 2228–2234.
118 G. Li, C. A. Jones, V. H. Grassian and S. C. Larsen, *J. Catal.*, 2005, **234**, 401–413.
119 S. Chansai, R. Burch, C. Hardacre, J. Breen and F. Meunier, *J. Catal.*, 2011, **281**, 98–105.
120 C. N. Costa and A. M. Efstathiou, *J. Phys. Chem. C*, 2007, **111**, 3010–3020.
121 J. Baltrusaitis, J. H. Jensen and V. H. Grassian, *J. Phys. Chem. B*, 2006, **110**, 12005–12016.
122 S. Kim, D. C. Sorescu and J. T. Yates, Jr., *J. Phys. Chem. C*, 2007, **111**, 18226–18235.
123 S. Civiš, M. Ferus, M. Zukalová, P. Kubát and L. Kavan, *J. Phys. Chem. C*, 2012, **116**, 11200–11205.
124 A. R. Almeida, J. A. Moulijn and G. Mul, *J. Phys. Chem. C*, 2011, **115**, 1330–1338.
125 L. Wang, X. Yi, W. Weng, C. Zhang, X. Xu and H. Wan, *Catal. Lett.*, 2007, **118**, 238–243.
126 I. X. Green and J. T. Yates, Jr., *J. Phys. Chem. C*, 2010, **114**, 11924–11930.
127 A. B. Ene, M. Bauer, T. Archipov and E. Roduner, *Phys. Chem. Chem. Phys.*, 2010, **12**, 6520–6531.
128 V. V. Pushkarev, V. I. Kovalchuk and J. L. d'Itri, *J. Phys. Chem. B*, 2004, **108**, 5341–5348.
129 Z. Wu, M. Li, J. Howe, H. M. Meyer and S. H. Overbury, *Langmuir*, 2010, **26**, 16595–16606.
130 J. S. Woertink, P. J. Smeets, M. H. Groothaert, M. A. Vance, B. F. Sels, R. A. Schoonheydt and E. I. Solomon, *Proc. National Acad. Sci. USA*, 2009, **106**, 18908–18913.
131 G. Tsilomelekis and S. Boghosian, *Phys. Chem. Chem. Phys.*, 2012, **14**, 2216–2228.
132 G. Tsilomelekis and S. Boghosian, *Catal. Sci. & Technol.*, 2013, **3**, 1869–1888.
133 G. Tsilomelekis and S. Boghosian, *Catal. Today*, 2010, **158**, 146–155.
134 Z. Wu, A. J. Rondinone, I. N. Ivanov and S. H. Overbury, *J. Phys. Chem. C*, 2011, **115**, 25368–25378.

135 L. Zhang and H. He, *J. Catal.*, 2009, **268**, 18–25.
136 A. Fukuoka, J. Kimura, T. Oshio, Y. Sakamoto and M. Ichikawa, *J. Am. Chem. Soc.*, 2007, **129**, 10120–10125.
137 I. X. Green, W. Tang, M. Neurock and J. T. Yates, Jr., *Faraday Discuss.*, 2013, **162**, 247–265.
138 H. Song, X. Bao, C. Hadad and U. Ozkan, *Catal. Lett.*, 2011, **141**, 43–54.
139 N. Bion, J. Saussey, M. Haneda and M. Daturi, *J. Catal.*, 2003, **217**, 47–58.
140 Z. Wu, M. Li and S. H. Overbury, *J. Catal.*, 2012, **285**, 61–73.
141 M. Li, Z. Wu, Z. Ma, V. Schwartz, D. R. Mullins, S. Dai and S. H. Overbury, *J. Catal.*, 2009, **266**, 98–105.
142 M. Li, Z. Wu and S. H. Overbury, *J. Catal.*, 2011, **278**, 133–142.
143 J. D. Henao, T. Caputo, J. H. Yang, M. C. Kung and H. H. Kung, *J. Phys. Chem. B*, 2006, **110**, 8689–8700.
144 J. Engeldinger, M. Richter and U. Bentrup, *Phys. Chem. Chem. Phys.*, 2012, **14**, 2183–2191.

Raman-in-SEM studies of inorganic materials

Guillaume Wille,*[a,c] Xavier Bourrat,[a,c] Nicolas Maubec,[a,c] Regis Guegan[b,c] and Abdeltif Lahfid[a,c]
DOI: 10.1039/9781782621485-00079

SEM-EDS and micro-Raman spectroscopy have been combined for material characterization in several recent studies. Switching from one to the other is frequently considered as a problem that cannot be solved using specific solutions. Although both techniques have followed a parallel but very different evolution since their introduction in the early 1930s, the concept of Raman-in-SEM first began in the 1980s and the first commercial systems were marketed in the early 2000s. The two main systems and techniques that have been developed and marketed by three manufacturers are presented and described in this chapter. An evaluation of their advantages and limitations is proposed.

A metrological study is then proposed for one of these systems, based on the 'on-axis' technique using a curved mirror placed under the SEM pole piece. This study allows a discussion of the performance and limitations of Raman spectroscopy when performed in a SEM.

A comprehensive review of published work is provided, although papers are rare in the open literature. The technique is essentially used for controls, expert assessments and high technology applications.

Advanced techniques that allow the use of Raman-in-SEM spectroscopy are discussed in detail using application examples taken from different fields in geosciences, materials chemistry or from expert assessments.

The conclusions of this study show that Raman-in-SEM spectroscopy is to date the first step in the combination of two well-known and mature techniques enabling the synergy between them to be maximised. Raman-in-SEM spectroscopy is relatively easy to set up and effectively complements the capabilities and efficiency of analytical SEM for material characterization. What are the most likely development perspectives that may be considered for this analytical coupling? Today, commercial systems are limited to only point-level micro-Raman analysis at the micrometre scale. In the near future, developments in both hardware and software will probably allow analysts to acquire Raman maps, or to employ multi-technique analyses based on a combination of data from SEM, EDS Raman, *etc.* New hardware developments may enhance the spatial resolution of both SEM and Raman spectroscopy.

1 Introduction

A short history of SEM and Raman spectroscopy is necessary, since technology is the driving force in this area. Scanning electron microscopy began approximately at the same time as Raman spectroscopy. However, they followed very different paths before being combined together in the coupled tool that is the subject of this chapter, namely Raman-in-SEM spectroscopy.

[a]*BRGM, 3 Avenue Claude Guillemin, BP 36009, 45060 ORLEANS Cedex 2, France. E-mail: g.wille@brgm.fr*
[b]*ISTO, 1A Rue de la Férollerie, 45071 ORLEANS Cedex 2, France*
[c]*UMR CNRS, Université d'Orléans, BRGM*

In 1922, the French physicist Leon Brillouin[1] and in the following year the Austrian physicist Adolf Smekal[2] introduced the theoretical principle of inelastic light scattering. Several researchers then started working on this phenomenon, such as L. Mandelstam and G. Landsberg (USSR) and J. Cabannes (France). The inelastic scattering of light was then demonstrated experimentally in 1928 by C.V. Raman and K.S. Krishnan at Calcutta University[3] and was given the name Raman Effect. However, despite numerous studies on its theoretical aspects (such as the work of the Czechoslovak physicist George Placzek between 1929 and 1934[4]), 'Raman spectroscopy' only really began to be one of the major analysis techniques during the 1960s–1970s, at the time of the emergence and development of laser sources.

Electron microscopy first began with the electronic lens in 1931 (E. Ruska) then the first Transmission Electron Microscopes (TEM) appeared with the first prototype in 1933 and, finally, the Siemens Super-Microscope, the first commercial TEM in 1939. The Scanning Electron Microscope (SEM) was developed by M Von Ardenne in 1938 and V. Zworykin in 1942. The first commercial SEM, the Cambridge Stereoscan, was commercialized in 1965. At the same time, R. Castaing developed the first WDS (Wavelength Dispersive X-ray Spectrometer) detector coupled to the TEM. The first Electron Probe MicroAnalyser (EPMA), the CAMECA MS85, was commercialized in 1958. R. Castaing also introduced the theory of X-ray emission and quantitative analysis by X-ray spectrometry.[5] EDS (Energy Dispersive Spectrometer) detectors were developed in 1964 and 1969 for the SEM.

SEM and Raman micro-spectroscopy appear to be essential for the micro-characterization of the morphology and the elemental and structural chemical properties of materials and minerals. Each technique provides complementary information that can be obtained at the same micrometric scale. The data provided by each technique on its own is frequently insufficient in itself and has to be combined with data from the other technique. Moreover, compared to optical microscopy, SEM imaging offers a wide range of well adapted contrasts for distinguishing regions of interest through their chemistry, morphology, crystallography, *etc.* Their coupling in a unique tool provides a promising analytical combination for such studies.

In 1986–1988, M. Truchet and M. Delhaye[6,7] proposed a description of an optical system for simultaneous micro-Raman spectroscopy, electron microscopy imaging and WDS elemental analysis in a transmission electron microscope or electron microprobe (TEM, EPMA) (Fig. 1). In several publications,[8,9] M. Truchet, M. Delhaye and other authors discussed the technical 'state of the art' and the potential interest in biology.

One of the main advantages of EPMA for coupling with micro-Raman is that EPMA is generally equipped with an optical microscope (Cassegrain-type optics) with a micrometric depth of field, placed in the objective lens of the electron optics. This optical system is required to adjust the sample surface to the exact focus position of WDS spectrometers. Cassegrain optics are a combination of a primary concave mirror and a secondary convex mirror. The concave mirror is pierced at its centre with a hole that

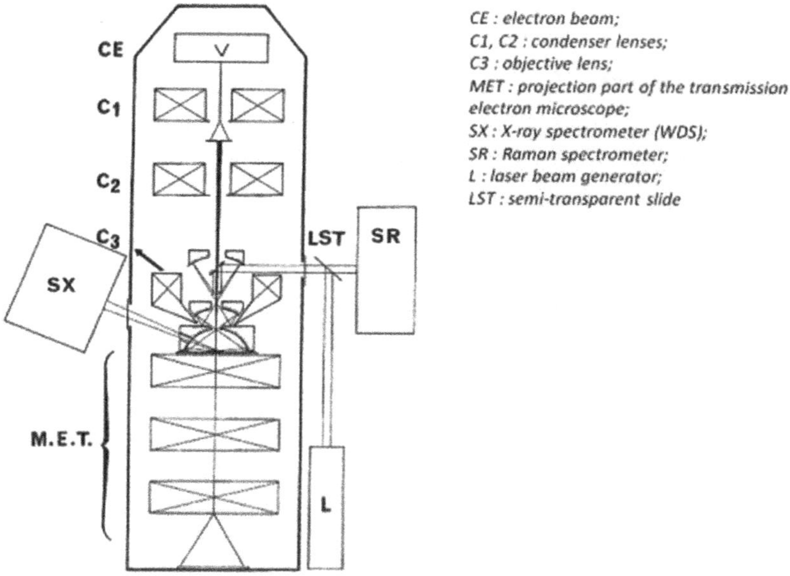

Fig. 1 Schematic section of the EPMA column equipped with the Raman microprobe device invented by Truchet and Delhaye (after[6]).

allows light to reach a camera or a light detector (for example, a spectrometer). For these authors, these optics could be modified by adding a special device equipped with a semi-transparent plate for sending a laser beam onto the sample surface and transmitting scattered light to the spectrometer. A major obstacle to such an arrangement was related to the very low numerical aperture of Cassegrain objectives used in transmission EPMA, thereby limiting the application field of such Raman spectroscopes to high scattering samples.

Why couple SEM and micro-Raman spectroscopy?

Interconnected data can be obtained from a combination of analytical methods. As detailed in the 2nd paragraph of section 1, a combination of SEM, EDS and Raman spectroscopy data is frequently required in materials sciences for numerous reasons, some of which are listed below:

- SEM with various detectors is a unique tool for imaging samples for morphological, chemical and structural description.
- EDS provides elemental composition.
- Raman spectroscopy provides structural information on minerals, such as polymorphism, the influence of trace elements on crystal structure, the presence of chemical species such as hydroxides, carbonates, organic compounds, fluid inclusions, *etc.*

The expert assessment acquired by BRGM (French Geological Survey) on particles collected in a water distribution network is a good illustration. The particles were analysed by SEM and EDS. Their morphology is not that simple, since some needles were found among massive particles (Fig. 2). The composition of the particles, as determined by EDS,

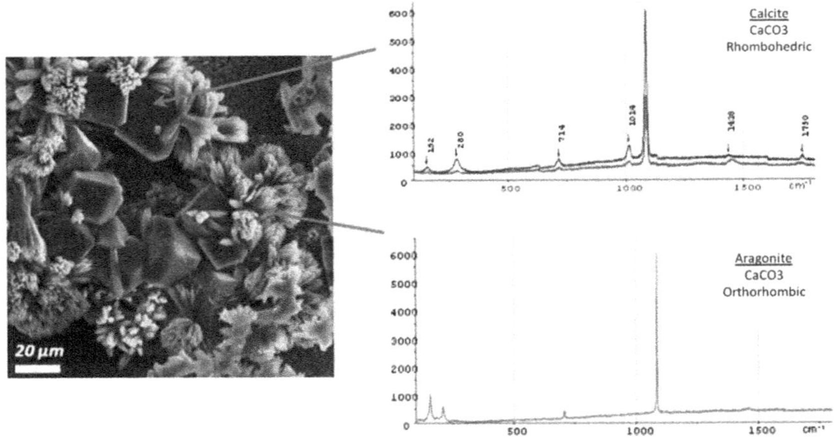

Fig. 2 Combination of SEM-EDS and Raman spectroscopy data for the identification of mineral grains.

shows that they are all made of calcium carbonate. However, polymorph identification was not possible and the presence of the needles could not be rationalized. Using Raman-in-SEM (spectra in Fig. 2), we were able to point the laser beam inside the SEM onto a massive particle or a group of needles. In this case, Raman spectroscopy provided structural information, which enabled the particles to be identified as calcite and the needles as aragonite. Calcite is a stable polymorph of calcium carbonate under normal ambient conditions, whereas aragonite is a high temperature, high pressure form of calcium carbonate, or instead originates from biomineralisation. The presence of bacterial activity is suspected in this case. The resolution of an optical microscope does not enable the spectrum of pure aragonite to be recorded. Petrologists for example have learned to use polymorphs to determine the thermodynamic conditions of geodynamic processes (temperature, pressure, *etc.*). The example of Al_2SiO_5 polymorphs may be cited:[10] sillimanite, andalusite and kyanite, for which Raman spectroscopy is required. The example of TiO_2, which crystallizes as brookite, rutile or anatase, is also often cited. Establishing the difference sometimes requires Raman spectroscopy. Raman-in-SEM also extends the limits of spatial resolution as well in all microanalytical fields.

2 Background and principles of the techniques

Raman-in-SEM spectroscopy makes sense if it is performed in the most efficient analytical devices. Consequently, it is worthwhile beginning this review of basic concepts by describing the different type of detectors that may be associated with Raman-in-SEM spectroscopy.

Scanning Electron Microscope

Thus, SEM equipped with various accessories and detectors and with elemental and structural microanalysis detectors is a powerful tool for material characterization, providing information on morphology,

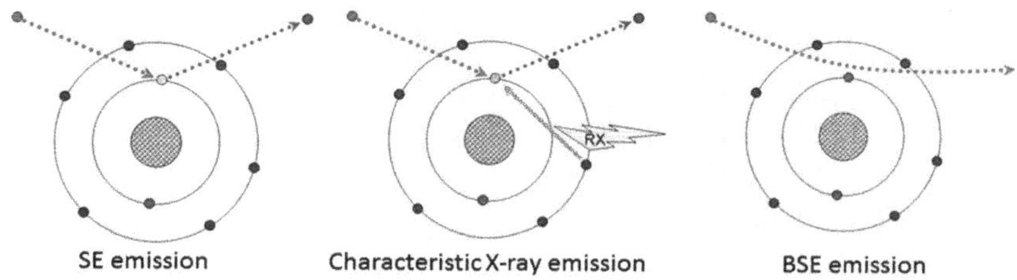

Fig. 3 Main electron/matter interactions used in SEM.

chemistry, texture, crystallography, *etc.* This technique can be applied to numerous materials in various fields such as metallurgy, ceramics, glass making, geology and life sciences. SEM is now widely used in academic and industrial laboratories, both for research and product controls. SEM images are obtained from electron-matter interactions. The main interests of SEM imaging are nanometric resolution and large depth of field imaging.

Imaging and microanalysis capability is based on several interactions from the incident electron beam and the sample (Fig. 3). Secondary electron (SE) images are generated by ionization of atoms in the sample: when the sample is bombarded with a beam of incident electrons, electrons can be ejected from atoms in the sample. By convention, the energy of these electrons is considered to be lower than 50 eV.[11] The Everhart-Thornley Detector[12] is the most frequently used SE detector in SEMs. Other SE detectors have been designed (see for example [13–16]) for low vacuum and environmental working conditions. Contrasts in SE images are strongly related to surface topography. As discussed in the technical section, coating the sample is a limitation for Raman. Working in low vacuum conditions is a real advantage for Raman-in-SEM. To get an image in a SEM, the surface of the sample is scanned point by point and line by line. The detected signal is synchronized with the (X,Y) position of the beam on the specimen in the microscope and the resulting image is therefore a distribution map of the intensity of the signal being emitted from the scanned area of the specimen. The narrower the electron beam, the higher the resolution.

A second type of image can be obtained using another interaction of the incident beam with the sample. Backscattered Electron (BSE) imaging mode results from the elastic scattering of electrons by atoms in the sample. BSE image contrasts are known to reflect variations in chemical composition (phases, zonation, *etc.*), but they can also be affected by crystallographic orientation on well-polished samples (crystalline contrast). Several types of BSE detector are available, including multi-quadrant semiconductor detectors, scintillator detectors, *etc.*

Other imaging modes can also be implemented in the SEM. Cathodoluminescence (CL) is a photonic emission in the visible and near-visible light range under the electron beam. This technique is popular for zircon or quartz or more sophisticated combined techniques for semiconductor nanostructures.[17–19] It is widely used for characterizing growth, zonation and crystallography of samples in geology,

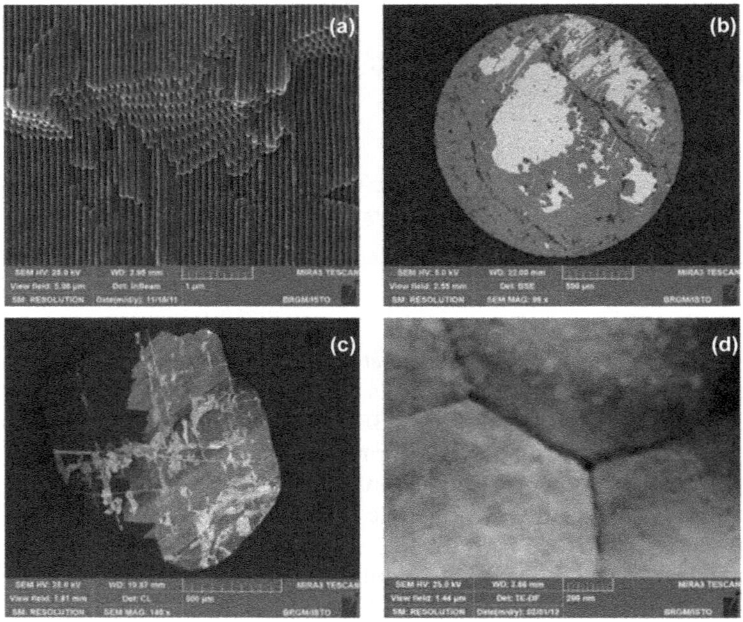

Fig. 4 Examples of SEM images: a – SE image of nanoporous alumina b – BSE image of a levitated oxidized iron droplet[20,21] c – CL image of cassiterite (SnO_2) (St Renan -France) d - STEM/Dark Field image of pearl.[22]

materials sciences and the semiconductor industry. Scanning Transmitted Electron Microscopy (STEM) employs a specific type of detector recently implemented in SEM for observing thin sections of materials in parallel with Transmission Electron Microscopy (TEM). Examples of different SEM imaging modes are given in Fig. 4.

One of the main interests of SEM is the capacity to combine observations (images) with microanalysis at a micrometric scale (or below). Characteristic X-ray emission resulting from the relaxation of atoms ionized by electron beam bombardment can be detected for elemental qualitative and quantitative microanalysis. Two techniques with different characteristics (spectral resolution, detection limits, *etc.*) are used to detect this emission: Energy Dispersive Spectroscopy (EDS) and Wavelength Dispersive Spectroscopy (WDS) (Fig. 5). Most SEMs are equipped with EDS detectors. WDS can also be installed on SEMs, but the use of WDS for X-ray detection in Electron Probe Microanalysis (EPMA) is well known.

Another microanalysis technique is also used in SEM. Electron diffraction can be observed and characterised on well-polished and correctly tilted (inclined) samples. It offers crystallographic phase and orientation mapping: EBSD (Electron BackScattered Diffraction) (Fig. 6). This technique, which began as an "experimental curiosity" during the 1970s, has turned out to be a powerful crystallographic technique with the beginning of on-line computer-assisted indexing of the EBSD patterns.[23,24] It is now a well-known and mature technique with numerous applications in various areas (see for example[24–26]). It is sometimes

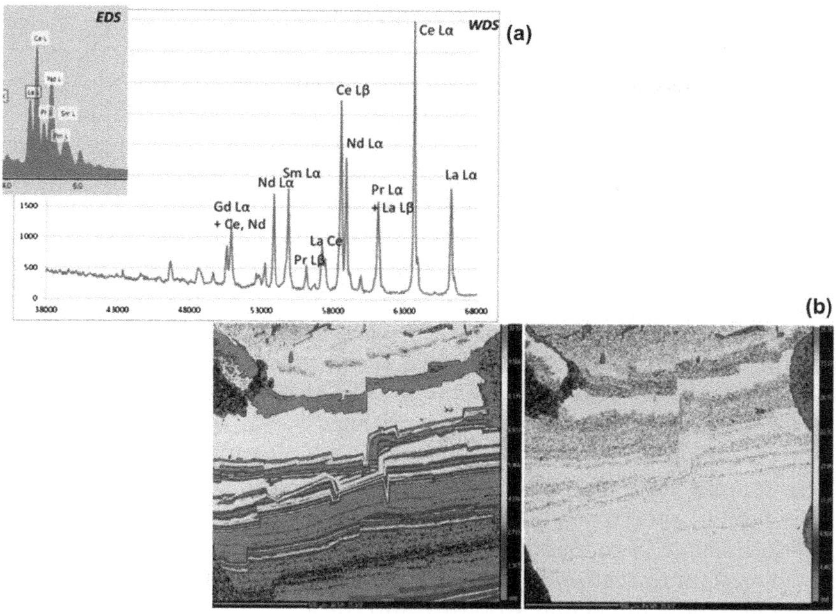

Fig. 5 Examples of X-ray spectrometry data: a – Comparison of EDS and WDS (LiF crystal) spectrum of a monazite b – Al (left) and Fe (right). WDS mapping on a zoned andradite.

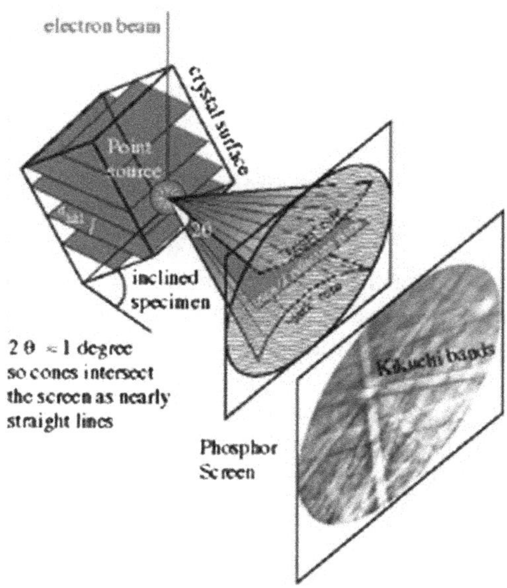

Fig. 6 Diagram of an EBSD.

associated with Raman spectroscopy, for example by Ishihara *et al.*[27] who use Raman spectroscopy for the characterization of graphene layer growth and EBSD for characterizing the copper films used as the substrate for graphene growth by Chemical Vapour Deposition (CVD).

However, as demonstrated by Wilkinson *et al.*[28] in his study on the performance of EBSD for strain mapping at a nanometric scale, these two techniques have several differences that make each specific for several applications.

Electron microscopy is a powerful technique, but it involves several sample specificities, due to secondary vacuum working conditions and the conducting sample surface (obtained for isolated samples by surface coating with a thin layer of carbon or metal). New SEMs, known as "environmental" (ESEM) or "low vacuum" (LV-SEM) SEMs, have emerged over the last fifteen years. These microscopes allow the user to work without a conductive layer coating (ESEM and LV-SEM) under higher vacuum (several tens or hundreds of Pa) or under water vapour pressure (ESEM). As will be seen later, this type of SEM is strongly recommended for coupling with micro-Raman spectroscopy.

Micro-Raman spectroscopy

Micro-Raman spectroscopy is also a very important analytical method in materials sciences. Indeed, the study of the inelastic light scattering (also known as Raman scattering) spectrum provides valuable information on chemical bonds, molecular structure and the crystallinity of a material or mineral.

Raman spectroscopy consists in illuminating a sample with a monochromatic light beam. Molecular bonds are excited through photonic excitation. Bond vibrations result in a secondary photonic emission, whose wavelength difference (compared to the excitation wavelength and independent of it) characterises the nature of the bond. Micro-Raman spectroscopy offers outstanding performance in the structural characterization of materials at a micrometric scale and can also be used on liquids, gases and fluid inclusions.

Raman spectroscopy is however limited by several technical problems. Raman spectroscopy cannot be used for some metals and crystals such as gold or NaCl. The Raman Effect is very weak and detection requires sensitive and highly optimized instrumentation. Fluorescence of impurities or of the sample itself can mask the Raman spectrum, although this can partly be solved by changing the wavelength of the laser beam. Heating through intense laser radiation can destroy or modify the sample. The resolution of an optical microscope in micro-Raman spectroscopy is limited to around 1 μm (lateral resolution) and depth resolution can vary from one sample to another, due to the optical transmittance of each material. In optical microscopy, contrasts are not always optimized, meaning it is sometimes difficult to identify regions of interest in the sample.

3 Bibliographic review: from SEM and Raman to Raman-in-SEM

Raman and SEM

Since complementary information can be extracted from SEM-EDS and Raman spectroscopy, it is sometimes difficult to combine data due to the

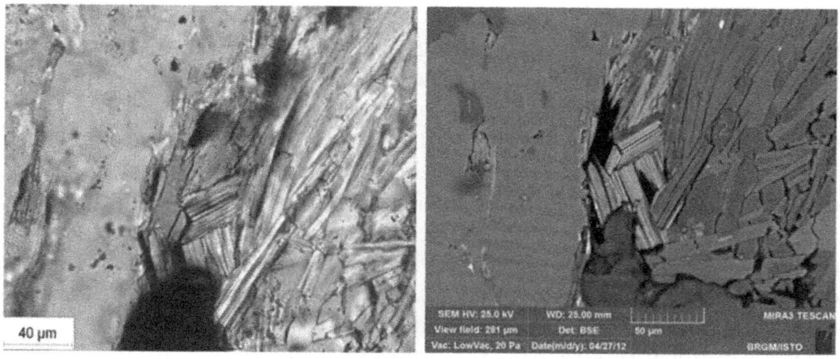

Fig. 7 Comparison of optical microscopy (left) and SEM (right) images of silicates in a polished section of granite from Ploemeur (France).

need for precise location of the analyses in both techniques. For example, it is difficult and time-consuming to locate the same particle by SEM and analysed by EDS under the optical microscope of a micro-Raman spectrometer on a SEM dispersed-particle preparation. Contrasts from optical imaging may be weak or insufficient to differentiate minerals, whereas they can be easy to obtain on SEM using BSE (Fig. 7) or CL detectors. Due to Abbe's law, SEM resolution is up to 1000 times better than an optical image and can help to precisely locate ROI close to 1 μm or less.

Several authors have proposed technical solutions to address the problem of particle location. One of the more efficient solutions is the use of TEM reference grids or centre-marked grids on a SEM stub as particle support.[29–32] This solution is well suited to micrometric particles, the size of which ranges from several microns to several tens of microns and in all cases less than the length of the side of the grid holes (for example, 54 μm for a 300 mesh grid) and with a unique and homogeneous composition. However, the location problem still remains for larger particles (no "large hole" referenced grids), for nanometric particles (too small for optical microscopy), or for heterogeneous particles (which is the most common case, for example, for a crushed rock). Another problem lies in the transport of such a sample from SEM to Raman.[33] Another problematic aspect of this solution is related to the depth of field of SEM images compared to optical microscopy. Thus, particles with strong topography can be easily identified in the SEM but not under the optical microscope. Finally, it is important to note that the SEM image is formed from electron-material interactions localized in a thickness of a few nanometres (SE) to a few micrometres (X-ray) of the surface, while the optical image obtained on a transparent material can provide details lying several microns, or even several millimetres, below the surface. Thus errors can be made in the location of the analysed point.

The great interest of this combination of data and the diversity of analytical problems presented by a combined use of two separate devices is the starting point that led to the integration of these techniques in a unique coupled SEM/EDS/micro-Raman system.Several studies have established the interest of combining SEM-EDS and Raman spectroscopy for characterizing materials. As an example, Nelson *et al.*[33] were among the first to combine these techniques to characterize organic and inorganic particles. These authors and other contributors[29] also noticed the limitations of coupling data from these techniques on separate devices due to optical microscopy limitations in terms of resolution and depth of field, difficulties in locating particles smaller than 100 μm from one device to the other, the difference in the appearance of particles using optical and electron microscopy, *etc.* These analytical problems may lead to inconsistencies between data from the two techniques.[29]

Other applications require the combined use of Raman spectroscopy and SEM-EDS. S. Bruni *et al.*[34] combined SEM-EDS, Raman spectroscopy and IRTF (Infra-Red spectroscopy) to analyse blue pigments in different polychrome works of art. They demonstrated the interest of combining these techniques to establish the presence of several minerals such as lazurite, nosean, hauyne, sodalite or their mixtures, which all are constituent parts of the rock mineral lapis-lazuli. F. Poitrasson *et al.*[35] applied SEM-EDS and Raman spectroscopy together with EPMA and laser ablation—inductively coupled plasma—mass spectrometry (LA-ICP-MS) to the study of the hydrothermal alteration of monazites (rare earth element phosphates). SEM-EDS and Raman spectroscopy were used by M. Hanchen *et al.*[36] to study the influence of temperature and CO_2 pressure in the precipitation of different forms of magnesium carbonate. The combination of SEM-EDS and Raman spectroscopy has also been used in bio-medical studies, such as by the study by J. Chen *et al.*[37] of the mineralization of hydroxyapatite in an electrospun nanofibrous poly(L-lactic acid) matrix.

Raman-in-SEM

Coupled systems have appeared over the last ten years. However, few studies have been published in the open literature regarding the implementation of such systems.

In 2004, Jarvis *et al.*[38,39] reported work on the identification of bacteria using the SERS (Surface-Enhanced Raman Spectroscopy) technique, which consists in analysing bacteria adsorbed on silver colloids, in order to increase Raman signal intensity and reduce the fluorescence effect. Using conventional micro-Raman spectroscopy on *E. Coli* and *Enterococcus* sp.,[38] they noticed a variation in signal intensity on the same type of bacteria. These variations have been associated with the position of the analysed bacteria relative to the silver substrate. SEM images of the samples (Fig. 8) show a heterogeneous distribution of bacteria on the substrate. Thus, depending of the analysed area, the enhancement effect is more or less important. However, the same study conducted by Raman-in-SEM spectroscopy showed good reproducibility of the different measurements.[39] In addition, spectra quality appears to be better, which could be due to the fact that the analyses were performed with a stronger enhancement effect. Thus, the use of SEM imaging to locate regions of interest for Raman analysis is an essential input for this study.

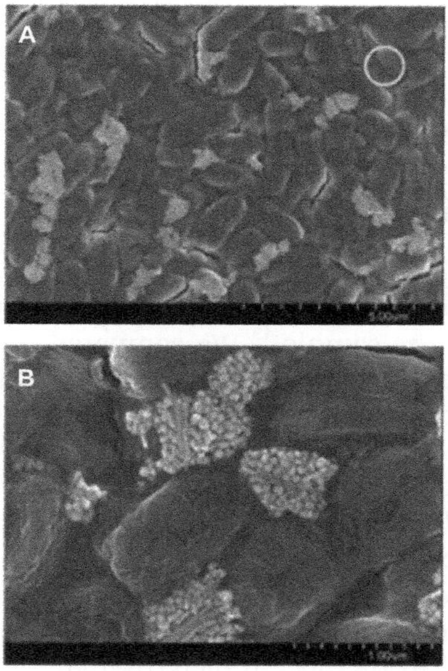

Fig. 8 SEM image of *E. Coli* and aggregated silver colloids (from[39]).

Another application is described by Otieno-Alego *et al.*[40] This work demonstrates the interest of collecting Raman-in-SEM spectra by combination of SEM images, EDS elemental mapping and Raman spectra collected on the same point of the sample for characterizing particles in a heterogeneous mixture. The aim of this study was to characterize and identify the different particles in an explosive compound. The particles could not be differentiated by optical microscopy, but could be identified easily by SEM backscattered electron imaging. SEM was first used to evaluate particle distribution in the mixture using BSE imaging and EDS elemental analysis and mapping. Then, Raman spectra were collected on the same points for phase identification. Thus, BSE image contrasts identified at least eight different particle types randomly distributed in the sample. As an example, medium grey particles in the BSE image were determined by EDS to consist of K, Cl and O. Raman spectroscopy was then used to identify these particles as potassium chlorate ($KClO_3$). Similarly, Raman spectroscopy can be used to differentiate organic compounds in which the BSE contrast and elemental composition are similar. This work clearly demonstrates the interest of coupling SEM, EDS and Raman data for such studies, and the great advantage of using a coupled system instead of two separate devices. However, the case of organic compound identification may also illustrate the interest in the complementary use of Raman-in-SEM and Raman imaging available on conventional Raman systems.

These examples illustrate the potential value of the use of Raman-in-SEM as a complement to conventional Raman spectroscopy (and not in competition with it). However, it is also important to focus on the technical and methodological aspects of the use of such coupled systems.

Recently, Worobiek *et al.*[41,42] compared the Raman spectra collected by conventional spectroscopy and Raman-in-SEM spectroscopy. These authors found a difference in signal intensity (about 10 times less intense in SEM). They mainly attributed this difference to the optic fibres used to transfer the laser beam to the sample and the Raman signal to the spectrometer. This difference requires the use of a higher laser beam power and/or a longer collection time. However, a recent study we conducted on this subject allowed us to establish that the difference is variable, depending on the particular nature of the sample, especially the optical properties of the sample.[43] We compared the signal intensity of Raman spectra collected on the same grains of different minerals by conventional spectroscopy and Raman-in-SEM spectroscopy, using the same laser and spectroscope setup and the same counting time. We observed that the signal ratio is several tens for transparent or partially transparent minerals such as andradite (nesosilicate, garnet group – $Ca_3Fe_2(SiO_4)_3$), but can strongly increase (up to 120) for black minerals such as manganese oxide MnO. These observations therefore imply that the intensity variation also depends on other experimental factors. A difference is noticed in the Spectrometer/Raman-in-SEM intensity ratio from the use of a 514 nm laser to the use of a 785 nm laser beam (the loss of intensity is slightly stronger for the 785 nm for synthetic Cr_2O_3). However, it has been found that the spectra are similar on both systems, both in terms of peak position and peak-to-peak intensity ratios in the spectra.

4 Technical aspects of combined SEM-Raman

Despite articles and patents from Truchet and Delhaye during the 1980s,[6–9] the development of coupled systems is recent. This is mainly due to the technical difficulties associated with the requirements of each technique. Only a few usable systems have been implemented at this time and only a limited number of solutions are commercially available. Two different technical solutions are available at this time from three different Raman spectroscope manufacturers. In this section, we present these systems and their relative advantages.

Retractable specially designed curved mirror and conventional Raman spectrometer (Renishaw SCA system)

The solution presented here is the first that was marketed and which has now been available for 9 years from the manufacturer Renishaw.[44] Several laboratories are currently equipped with this model. It consists of a special interface comprising a parabolic mirror specifically developed for Raman spectroscopy.

The mirror is placed on an automated retractable arm. This system is designed for precisely inserting and retracting the mirror under the pole piece of the SEM, between the SEM electron optics and the sample. A hole in the mirror allows the electron beam to pass, which makes it possible to simultaneously obtain an electronic image (SE). The laser radiation is focused on the sample surface by the mirror. An optical image from a camera is collected together through the mirror, making it possible to precisely locate the laser spot on the sample. This optical image can also be used to check the good agreement between the SEM image and laser beam location and adjust it if necessary, using the SEM image shift function. Scattered light radiation is collected by the mirror and sent to the spectrometer.

This interface is connected to a conventional micro-Raman spectrometer using optic fibres (Figs. 9 and 10). These fibres are used to send laser radiation to the interface installed on the SEM and to send the collected scattered light to the spectrometer. One of the interests of this

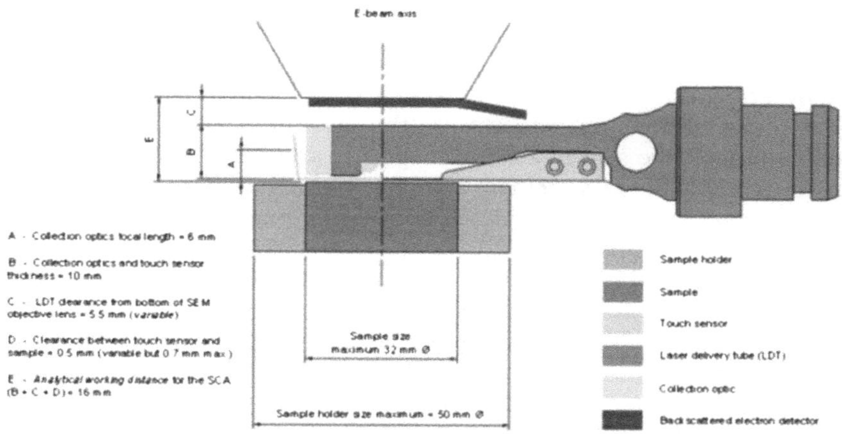

Fig. 9 Schematic diagram showing the laser delivery tube (LDT) of the SEMSCA interface when inserted below the objective lens.

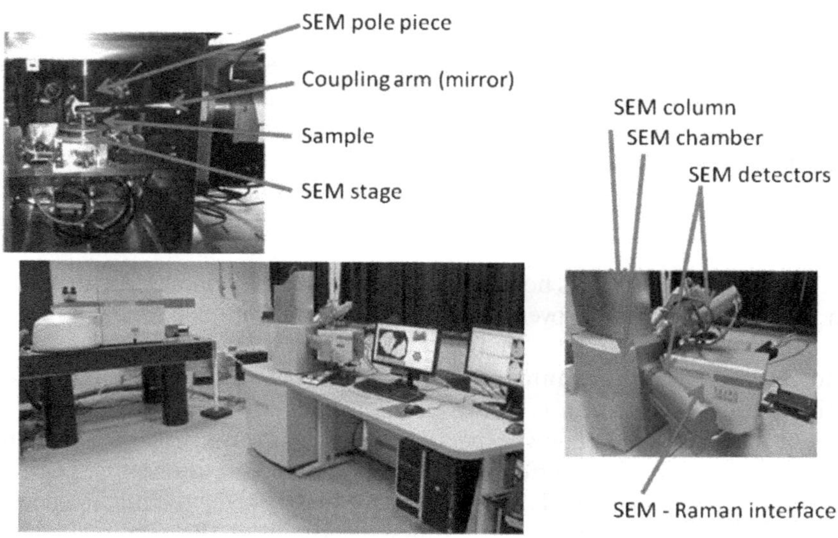

Fig. 10 Raman-in-SEM facility at BRGM Orleans (France): Renishaw InVia with a SEMSCA interface installed on a TESCAN MIRA 3 XMU.

solution is that the coupling system uses the same laser beam radiation and the same spectrometer as the conventional micro-Raman spectrometer. Consequently, it can be assumed to achieve the same performance in terms of spectral range and spectral resolution as a conventional spectrometer. According to the manufacturer, it is equivalent in optical performance to a ×50 long working distance objective.

Due to the same wavelength range, this system can also be used for photoluminescence (visible light emission generated by photonic beam excitation) or cathodoluminescence (visible light emission generated by electron beam excitation) with some limitations as regards the spectral range, depending on the configuration.

Retractable CL mirror and conventional Raman spectrometer (Horiba system)

Another commercial solution is similar to the system described above. The main difference consists in the use of a conventional cathodoluminescence arm instead of a specific arm. This solution was first used in 'home-made' systems[45,46] and is now marketed by Horiba. In this system, the arm is not automated but instead manually inserted.

Off-axis dedicated micro-Raman system (Hybriscan system)

This system consists of a micro-Raman spectrometer positioned laterally on the SEM chamber and in which the Raman optics are arranged outside the electron beam axis. Raman and SEM-EDS data are collected successively, after a calibrated movement of the SEM stage.

Such a system was described in 2005 by Van Apeldoorn *et al.* under the name "CRSEM" (Confocal Raman SEM).[47] Samples are analysed by selecting a location by EM and then shifting the sample into the confocal laser spot at a precise position located "off-axis" (*i.e.* not in the electron beam axis). The laser source is diverted through a side port into the SEM chamber by a beam splitter. The laser is then emitted onto the sample by a set of mirrors through a ×60 objective. The Raman scatter is then collected by the same objective and sent in the opposite direction inside the spectrograph, through a pinhole. A schematic diagram of this system is shown in Fig. 11. One of the main advantages of such a system is its confocality due to a design similar to "conventional" Raman micro-spectrometers equipped with a ×50 long working distance objective.

This kind of solution is now marketed by the manufacturer Hybriscan with a 785 nm laser available over a Raman shift range from 350 to 2500 cm^{-1}.

Advantages and disadvantages of the different systems

The goal of this section is not to promote one or another system. All of the systems have their own specificities and advantages and each one may be best suited to different applications and requirements, both in terms of analytical needs as in terms of cost, size, technical specifications, *etc.* According to the manufacturers, all of the systems can be installed on any SEM (except perhaps tabletop SEMs or SEMs with very small chambers).

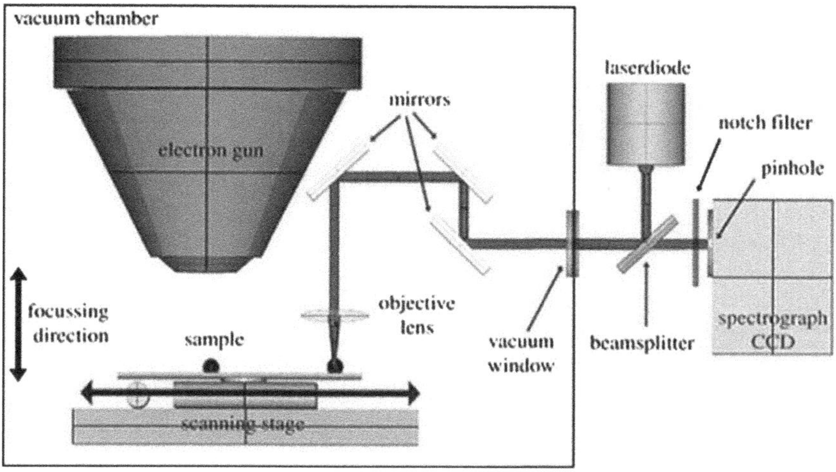

Fig. 11 Schematics of the CRSEM (Confocal Raman SEM) [from[47]].

The main advantage of systems using a retractable mirror is that the analysis can really be obtained simultaneously if specific precautions are taken (this subject will be discussed in the next section, devoted to the technical and metrological aspects of combined analysis). However, the fact that a retractable arm operates under the pole piece of the SEM requires the SEM BSE detector to be retracted, which is positioned in the same place. Similarly, the short distance between the sample and the Raman arm makes EDS analysis difficult or impossible when the Raman arm is inserted. Consequently, the use of a Raman spectrometer and a BSE or EDS detector needs to be done alternately (after retraction of the Raman arm, for EDS and BSE). The use of a secondary Everhart-Thornley detector is always possible, because of its lateral position (subject to a well-adapted collection solid angle).

This limitation is not a problem with the third system, because the micro-Raman spectrometer is "off-axis" and so BSE retraction is not required and the sample is not masked by the Raman arm to the EDS detector. However, with this solution, SEM-EDS and Raman analysis are necessarily performed alternatively.

One point needs to be discussed and checked during a demonstration of the instrument: precise mirror positioning is required to allow perfect reproducibility of the optical path.

Another factor is the spectral range and spectral resolution of the spectrometer. On systems equipped with mirrors, the role of the coupling interface installed on the SEM is to focus the laser radiation onto the sample and to collect the scattered light. The entire analysis is performed using the same spectrometer as the conventional spectrometer. The spectrometer can be used as a conventional Raman spectrometer or as a coupled system, thereby allowing a wider spectral range analysis. This can be a critical aspect for some applications. As an example, the Raman shift of N–H, C–H or O–H bonds in solids appear at around 3000–3700 cm^{-1}. Spectral resolution is another critical point in the choice of a

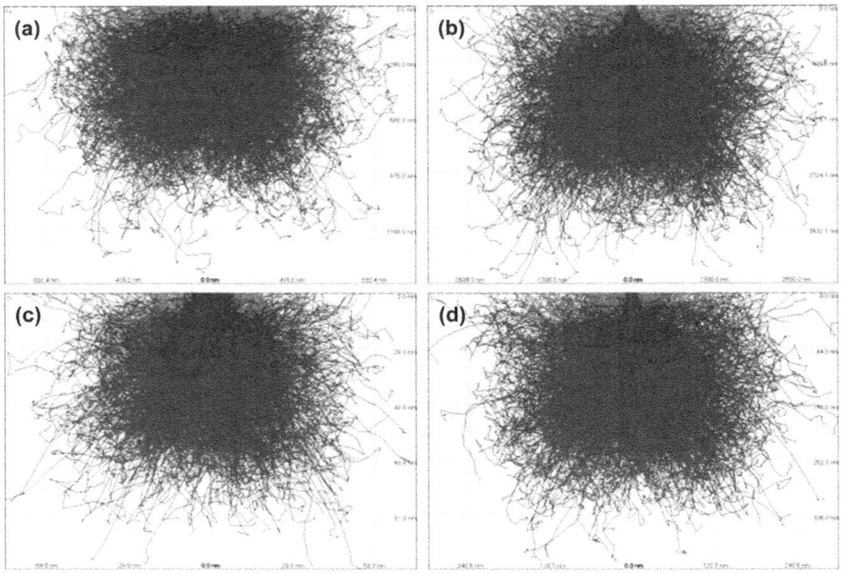

Fig. 12 Penetration and spreading of electrons in iron (a) and silicon (b) at HV = 20 kV, at HV = 2 kV (c) and HV = 5 kV (d) in silicon (beam diameter = 10 nm) (simulation by CASINO 2.48[53]).

Raman system. This is not really a problem for identifying phases using Raman spectroscopy, but it can be for some advanced applications, where precise Raman shift and Raman band shape measurements are required. A good illustration of this point is the use of Raman spectroscopy for geothermometry.[48,49]

Lateral resolution is comparable for Raman-in-SEM and EDS. On the one hand, SEM imaging resolution is nanometric (SEM "ultimate" resolution is given by the manufacturers at around 3 nm for a tungsten emission gun and 1 nm for a Field Emission Gun (FEG- (or FE-) SEM). On the other hand, compositional measurements or determinations carried out by Raman spectroscopy are referred to as an area whose spatial resolution is about 1 micron (due to diffraction). Some specific techniques known as "nano-Raman spectroscopy" have been developed to lower spatial resolution of Raman spectroscopy below the diffraction limit, but it is achieved by significant experimental complexity.[50–52] Except for "nano-Raman spectroscopy", this order of magnitude is fully comparable with microanalysis spatial resolution (*i.e.* EDS and WDS): using 'usual' HV conditions (*i.e.* HV = 15–25 kV), spatial resolution is micrometric (from about 1–5 um^3 for low atomic numbers to 0.2–1 um^3 for high atomic numbers) due to the fact that X-rays are generated from very deep within the interaction volume. 'Nanometric' resolution (in fact hundreds of nm) can only be achieved by EDS/WDS using very low HV conditions (*i.e.* a few kV) that cannot be applied for qualitative/quantitative analysis of many materials by EDS (see Fig. 12).

Sample conductive coating vs. Raman spectrometry. Due to the nature of the beam in a SEM, in other words an electron beam, if a sample is a poor electrical conductor or an insulator (*i.e.* glass, plastic, rock, *etc.*), electrons accumulate on the surface at the electron beam impact point and cannot be removed. This is called a "charging effect", which causes highlights that disturb or prevent observation. Therefore, it is necessary to apply a conductive layer on the sample surface to remove charges from the observation area. Non-conductive samples are usually coated with a conductive layer (of several nanometres thickness). This coating can be a metallic deposit (Au, Pt/Pd, Cr, *etc.*), but when microanalysis (EDS, WDS) is required, a carbon coating is preferred. In fact, a heavy metal coating can lead to the presence of X-ray emission peaks that may overlap emission peaks from elements contained in the sample itself.

However, carbon deposited on the sample surface causes artefacts on the Raman spectrum, which can strongly affect its quality and subsequent data interpretation. An example of this is shown in Fig. 13: Raman spectra were collected on the same andradite ($Ca_3Fe_2(SiO_4)_3$ garnet) grain, before (bottom) and after carbon coating (in top). A broad band can be observed at 1200–1600 cm^{-1} and the baseline level is much higher on the coated sample than on the non-coated sample. The bands are related to the Raman spectrum of the amorphous carbon layer.[54,55] According to these authors, the Raman spectrum of the amorphous carbon layer is characterized by a broad Raman band observed in the range 1000–1800 cm^{-1}. This band can be well fitted by two Gaussian distributions

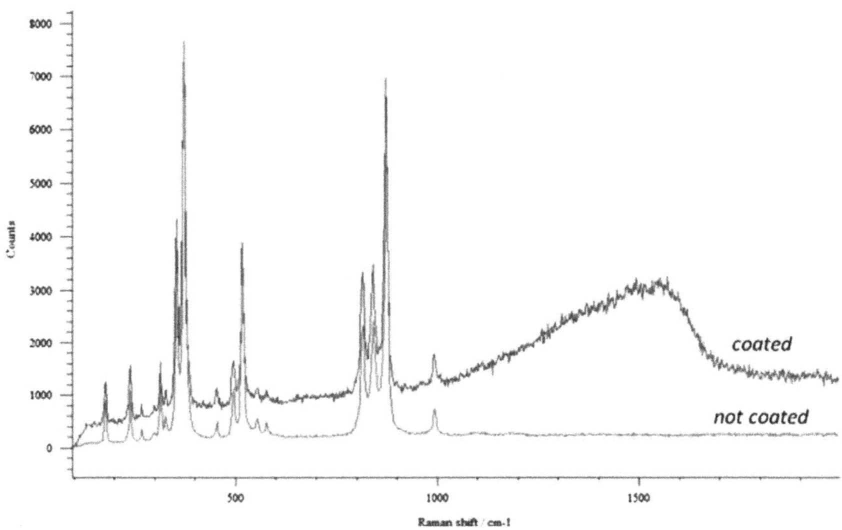

Fig. 13 Raman spectra of the same andradite grain (garnet $Ca_3Fe_2(SiO_4)_3$) carbon coated (top)/not carbon coated (bottom).

known as "D" and "G" bands (respectively 1350 and 1530 cm^{-1} for Scheibe *et al.*[54] and 1380 and 1560 cm^{-1} for Marchon *et al.*[55]).

The baseline level modification is related to the fluorescence phenomenon, probably due to carbon. The carbon coating behaves as a fluorophore, a compound that causes a molecule to absorb energy at a specific wavelength and then re-remit energy at a different wavelength. The amount and wavelength of the emitted energy depend on both the fluorophore itself and on its chemical environment. But, more importantly, the presence of the amorphous carbon layer produces a loss of Raman signal intensity.

Interestingly, since the 1980s, new types of SEMs, known as "environmental" or "low vacuum" SEMs, have appeared on the market. Indeed, conductive layer deposition can be avoided by working with a partial vacuum, in other words a pressure of a few Pa to several thousand Pa (compared to 10^{-3} to 10^{-4} Pa in normal conditions). The electrons accumulated on the sample are neutralized by the positive charges of gas (mainly nitrogen) arising from the incident beam. Starting with the work of the German physicist M. Von Ardenne, early attempts have been reported on the examination of specimens in water or gaseous atmospheres inside specially designed "environmental" cells, in conjunction with conventional and scanning transmission types of electron microscopes.[56] In 1970, Lane reported electron imaging of a wet sample directly in a SEM using a very localized vapour jet to maintain a gaseous gradient just above the specimen.[57] Several developments have led to the introduction of a new type of SEM able to work at room temperature (no sample freezing stage) and at high gas pressure (up to several thousand Pa), and the introduction in 1979 of the term "Environmental SEM", or ESEM, by Danilatos and Robinson.[58] The first commercial ESEM was introduced by Electroscan in 1988. With the expiration of key patents, new commercial instruments have recently been introduced by other SEM manufacturers.

Another type of "partial vacuum" SEM, known as "low vacuum" SEM, has been marketed. These SEMs are able to work at lower pressure than ESEM, usually between a few Pa and a few hundred Pa. This pressure range is not sufficient to work on wet samples (*i.e.* biological samples, wet soils, *etc.*) at room temperature, but it is sufficient to avoid a charging effect on dry samples (ceramics, rocks, glass, *etc.*) or when working on frozen wet samples (using a freezing stage).

Another facility is offered by recent FE-SEMs able to provide good quality images even at low voltage ($<$ a few kV). At low voltage, it is possible to observe non-coated non-conductive samples in the SEM. Unfortunately, under these working conditions, most BSE detectors are not able to work or instead provide poor quality images. In addition, EDS microanalysis is impossible and WDS analysis is complex and limited in these conditions.

Analytical geometry vs. Raman geometry. One of the main selection criteria for a SEM is the 'analytical working distance'. Working Distance (WD) represents the distance from the electron column output (bottom of the pole piece) to the sample surface. The WD value given by SEM software represents the focusing distance and therefore the sample surface position if the image is 'focused'. This parameter is related to the SEM chamber drawing by the SEM manufacturer, for any EDS (or WDS) system from any provider. It represents the optimal distance for X-ray collection in EDS (or WDS) (Fig. 14). Moving away from this position causes a deterioration in X-ray collection and therefore a deterioration or a total loss of the analytical capacities of the SEM.

This fact is illustrated in Fig. 15. In this figure, two SEMs are compared: The analytical working distance is given as 15 mm for SEM 1 and

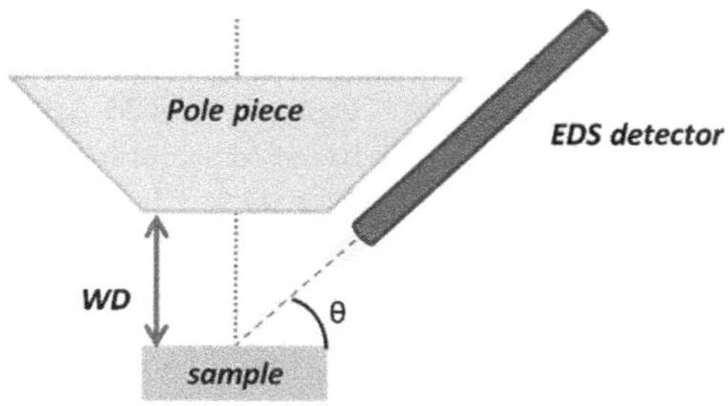

Fig. 14 SEM analytical geometry.

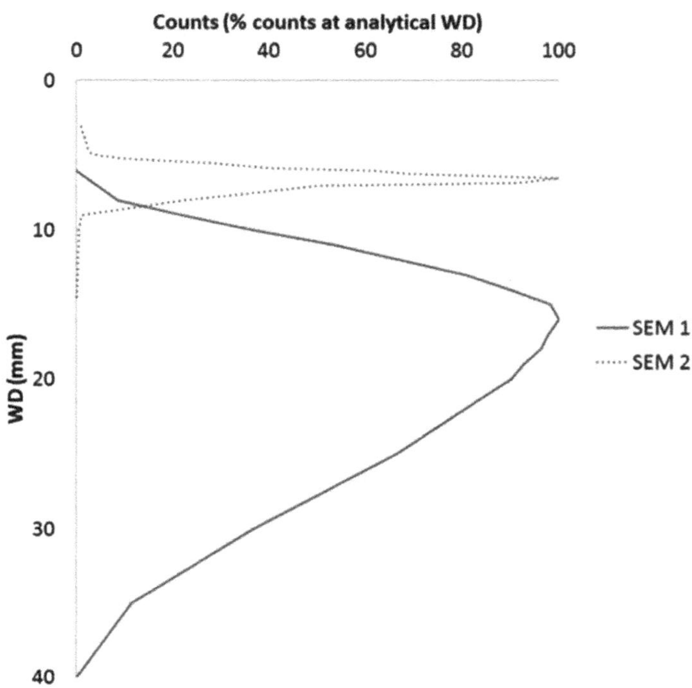

Fig. 15 EDS signal as a function of WD for two SEMs with different analytical WDs (15 mm for SEM 1/5 mm for SEM 2).

5 mm for SEM 2. Both SEMs are able to receive a coupling interface for Raman-in-SEM acquisition. However, if the EDS geometry and Raman-In-SEM geometry are not optimized, use of both for data correlation may be problematic. As an example, with our system, Raman collection is conducted with a sample surface at WD $\approx$ 17.5 mm. With SEM 1, X-ray collection (EDS – 10 mm^2 active surface detector) is 97% of X-ray optimal collection while for SEM 2, X-ray collection is close to 0%. Consequently, SEM 2 would require frequent moves of the stage from EDS WD to Raman WD while X-ray collection is near optimal for 1.

Impact of SEM phenomena on Raman-in-SEM spectroscopy. Coupling techniques related to such different physical phenomena as SEM-EDS and Raman spectroscopy may cause problems directly related to the influence of one on the other. Since Raman-in-SEM spectroscopy can be subject to several influences due to SEM related phenomena, some possible artefacts are documented, which are due to the impact of electron beam induced contamination or cathodoluminescence (CL) contribution to the Raman spectra, especially with geological samples.

Contamination. Electron beam induced specimen contamination has been a well-known problem since the beginning of electron microscopy, both in Transmission Electron Microscopes (TEM) and in Scanning Electron Microscopes (SEM). It results from the development of a dark deposition on the scanning area. Different sources of contamination have been highlighted and discussed as well as different methods and systems to reduce contamination sources and their consequences.[59] The first identified source of contamination is the vacuum system of the SEM.[60] SEMs equipped with diffusion and rotary pumps are often considered as the most sensitive to this form of contamination, due to vapour from the pump oil. SEM manufacturers are increasingly replacing diffusion pumps by turbomolecular pumps and rotary pumps by dry primary pumps in order to reduce this source. Another contamination source is linked to the SEM components. The lubrication of moving parts (stage, retractable detectors, *etc.*) and O-rings are also potentially responsible for hydrocarbon emissions in the SEM chamber. The third main source of contamination is the sample. Several causes can be identified. The first is the sample itself, which can be a source of hydrocarbon molecules due to its composition (polymers, geological samples containing organic matter, biological samples, *etc.*). Unfortunately, this source cannot be reduced and can be a long term problem for the SEM, because carbon molecules evaporated from the sample can be adsorbed on the chamber walls and be desorbed only gradually. The sample may also be linked to other contamination sources that can be more easily reduced by correct sample preparation[61] and handling. Careful cleaning of the sample, use of embedding resins adapted to SEM conditions (secondary vacuum, interaction with electron beam, *etc.*), use of gloves for handling samples and specimen mounts and holders are

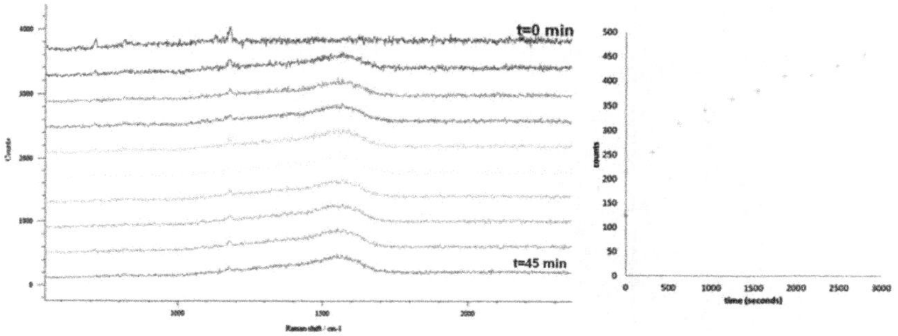

Fig. 16 Evolution of the Raman spectrum of a quartz grain in the range 500–2300 cm^{-1} (left) and signal intensity of the band centered at 1560 cm^{-1} (right) due to increasing contamination.

strongly recommended as "good practices".[62] Another source has appeared more recently with the development of low vacuum SEMs: It involves contamination due to gas purity and the cleanliness of the gas supply network and pipes. Attention must be paid to this point. Several approaches have been developed for reducing hydrocarbon contamination,[59] but the multiple nature and origin of contamination sources mean contamination can be minimized but not completely eliminated.

Since contamination leads to the growth of a thin layer of amorphous carbon on the sample surface scanned by the electron beam, the expected effects are the same as those obtained by conductive carbon layer deposition. An example is shown in Fig. 16: Raman spectra were collected on the same point on a quartz grain (SiO_2) over several minutes of SEM operation at an SEM beam current of 850 pA. Peak positions were determined to be 1347 and 1557 cm^{-1}. Its intensity can be related to the different SEM parameters affecting the development of the contamination on the sample surface: duration of the electron beam scan, beam parameters (HV, beam current), "spot" mode, *etc.* For example, a long time scan of a small area of the sample induces the development of a strong "black square" contamination on the sample surface and a strong broad band in the range 1000–1800 cm^{-1} in the Raman spectrum.

Cathodoluminescence. Cathodoluminescence (CL) is a visible/near-visible light emission caused by the interaction between an electron beam and a luminescent material. It is widely used in geology for providing information on origin, growth fabrics, diagenetic textures and mineral zonation. CL emission can be due to imperfections such as vacancies, loss of ordering or lattice distortion in the crystal lattice (intrinsic luminescence) or to activation or inhibition of CL by impurities in the mineral (extrinsic luminescence).[63,64] Figure 17 shows an example of a CL panchromatic image of a zoned cassiterite (SnO_2) grain (Montbelleux, France). In this example, CL is controlled by crystal growth and orientation and impurities such as Ta, W, Fe, Nb, *etc.*[65–67]

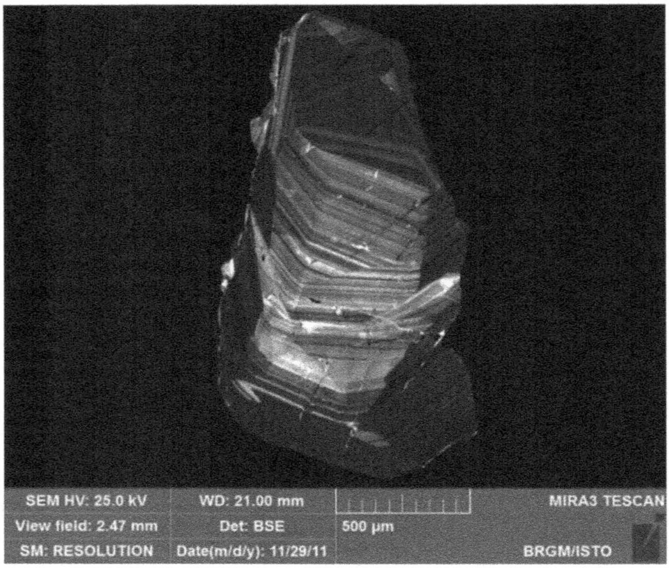

Fig. 17 CL image of a cassiterite grain.

Cathodoluminescence can generate artefacts in Raman spectra if not controlled and taken into account. Indeed, CL issues similar to that of the Raman scattering spectral range occur. Therefore, Raman-in-SEM analysis of a luminescent material can generate electron matter interactions, causing the appearance of CL emission bands that are superimposed on the Raman scattering signal. As for contamination, this phenomenon can be avoided by "blanking" the electron beam during Raman spectra acquisition. To highlight the effect of cathodoluminescence, Raman spectra have been acquired under different conditions (with/without electron beam and/or laser beam) on a cassiterite grain (Fig. 18). Spectrum A shows a flat "baseline" signal of the spectrometer. No band can be observed. Spectrum C shows the Raman spectrum of the cassiterite grain collected with the SEMSCA system in the SEM. Some signal can be observed on spectrum B while the laser is switched off, which means no Raman signal is collected. Spectrum D shows the spectrum collected with the laser and electron beam on. It clearly appears that this spectrum is the result of the combination of spectra B and C.

Nevertheless, this phenomenon can be put to advantage for checking the alignment of the SEM and Raman interface: if the electron beam is focused in the centre of the SEM observation area ("spot mode") on a strongly luminescent material (cassiterite, ZnO, REE oxides, for example), a light spot can be observed on the optical image of the Raman interface (Fig. 19). This method is widely used for checking SEM and optical images on EPMA (Electron Probe MicroAnalysis).

Raman-in-SEM and electron microscopy/microanalysis are really polyvalent techniques for material characterization. However, it is important to note that although these techniques apply equally well to most materials, each has its own specificities. Thus, the analysis of metals and

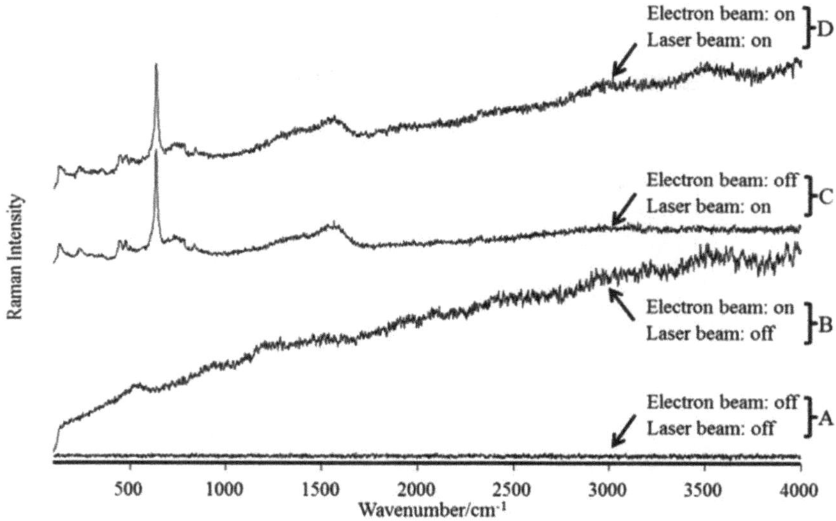

Fig. 18 Effect of CL emission on the Raman spectrum collected by on-axis Raman-in-SEM (A). Raman spectrum recorded without electron and laser beam; (B) Raman spectrum with electron beam only (CL signal); (C) Raman spectrum with laser beam only and (D) Raman spectrum with laser and electron (superimposition of B and C).

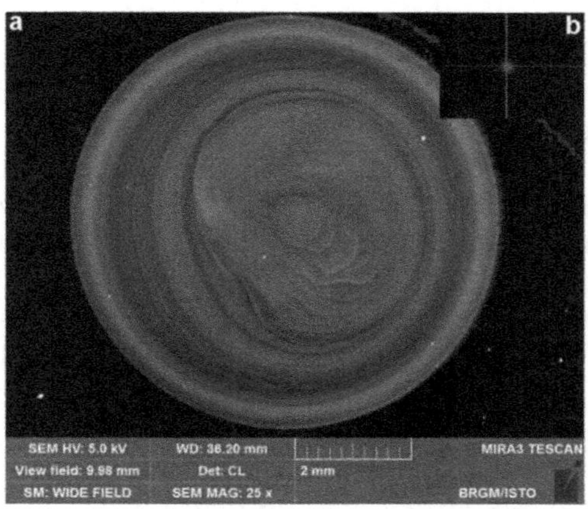

Fig. 19 (a) CL image of a 'milky' ($CaCO_3$ vaterite) pearl (SEM) and (b) signal collected by the SEMSCA camera using SEM 'spot mode'.

alloys by Raman spectroscopy is generally impossible, although SEM, EDS and EBSD are well suited to this type of sample. Instead, wet samples can be analysed in very specific SEMs known as "environmental SEMs" (ESEMs). It may also be noted that some materials pose technical difficulties for analysis by either technique because of their sensitivity to either the electron beam or the laser beam. This point is illustrated in Fig. 20, which shows the electron beam impacts after EPMA analysis of biocarbonate (bivalve).[68]

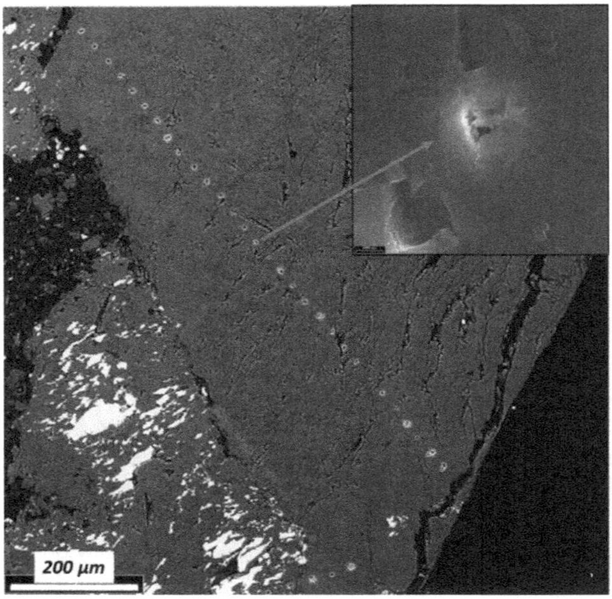

Fig. 20 Electron beam impacts on a polished section of biocarbonate bivalve.

5 Application examples in mineralogy and materials sciences

Characterization of a polymorph at a micrometric scale: milky pearl defect in vaterite

Pearls (*Hyriopsis cumingii*) are produced from a natural biomineralisation process controlled by organic molecules. The pearl biocrystal is a hybrid composite organic/mineral (3%/97%). The regular form of pearl mineralization is aragonite, an orthorhombic polymorph of calcium carbonate. This biocarbonate shape is a stack, of nanometric thickness, of hexagonal plates stacked along the c-axis of the aragonite crystal. It diffracts as a single crystal, or so called "mesocrystal". This mineralization produces the well-known shiny and "pearly" aspect. Sometimes, pearls show a biomineralisation defect characterized by a lack of shine (also called "milky pearl"). It has been established that this defect is related to the change in mineralization form, from orthorhombic aragonite to hexagonal vaterite.[22]

Raman-in-SEM may be applied to the characterization of the "milky" defect of a pearl. SEM imaging, CL, EPMA and Raman spectroscopy in the SEM may be applied to describe the biocrystal structure of both regular and defective nacre of pearls. SEM, SE or BSE imaging can be used to illustrate the change in particle shapes from plates (in the regular aragonite part of the pearl) to needles (in the defective part). CL and EPMA can be used to illustrate the growth ring in the pearl and relate manganese content (few hundreds/thousand ppm of Mn) to growth rings. Then, Raman spectra collected in the SEM are used to characterize the change in crystal structure of the calcium carbonate from aragonite (in the regular part of the pearl) to vaterite (in the defect).

SEM tracking of dispersed micrometric particles: Raman-in-SEM applied to the characterization of pigments of wall paintings
The significant advances made in archaeological sciences through the use of Raman microscopy have been reviewed by Smith & Clark, 2004[69] or summarised in books.[70] As a result, a considerable body of experience has been built up and extensive and reliable spectral libraries have been published in this field.[71–73]

The example reviewed in this paper[74] concerns the 'Grottes de la Garenne', located at Saint-Marcel (France). These caves are a set of Magdalenian living sites[75] discovered in 1848 during the building of the Paris - Toulouse railway. The caves are located on the southern shore of the river Creuse, in a hill known as the Coteau de la Garenne. The walls of some of these caves are covered with a substantial set of red and black paintings, the study of which may allow archaeologists to obtain information or confirm hypotheses about the techniques and tools used by people from the Magdalenian age.[76] We studied samples of pigments taken from the walls of one of these caves, known as the Grotte Blanchard, which allowed us to determine the nature of the main pigments.[74] The red pigment used for the wall paintings is mainly composed of hematite, clays, carbon matter and carbonates, while the black pigments are made of a mixture of cryptomelane, pyrolusite, clays, carbonates and carbon matter.

Raman-in-SEM has been applied to the characterization of samples used for IRTF spectroscopy (pigments dispersed in a KBr pellet). SEM enables tracking of micrometric particles in a complex grain and EDS analysis of these particles. Through the use of Raman-in-SEM spectroscopy, small particles of manganese oxides dispersed in a mixture of minerals and amorphous material can be identified as pyrolusite (MnO_2) and a contribution of carbon matter in addition to black mineral particles (Fig. 21). This can be used to confirm that the painting is not a simple covering of the wall by mineral powder, but by a processed material. The main contribution of Raman-in-SEM is to allow us to precisely locate each particle (by BSE images in SEM) and combine elemental composition (from EDS spectra) and mineral structure (from Raman spectroscopy) for each component of the mixture.

SEM-EDS and micro-Raman data combination applied to the characterization of micrometric aluminium-phosphate-sulphate (APS). The alunite supergroup consists of more than 40 mineral species with the general formula ($AB_3(XO_4)_2(OH)_6$), in which A is occupied by monovalent (*e.g.*, $K+$, $Na+$, NH_4^+, H_3O^+), divalent (*e.g.*, Ca^{2+}, Ba^{2+}, Pb^{2+}) or trivalent (*e.g.*, Bi^{3+}, REE^{3+}) ions, B is typically Al^{3+} or Fe^{3+}, and X is S^{6+}, As^{5+} or P^{5+}.[77] The influence of the ions in the A-, B- and X-sites is highlighted in the Raman spectra by variations in the position of certain vibrations. According to Maubec *et al.*,[78] A-site substitutions are characterized by wavenumber shifts of the vibrations involving hydroxyl groups. The positions of these vibrational bands vary linearly with the ionic radii of the ions in this site. B-site substitutions induce shifts of all bands due to structural modifications that lead to differences in the chemical

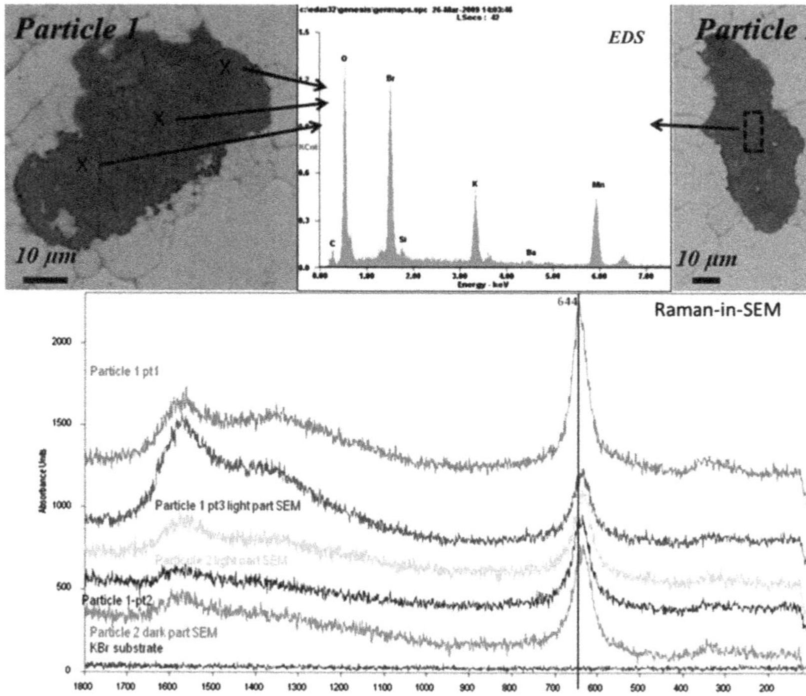

Fig. 21 Raman spectra of pigments of parietal paintings analysed by SEM, EDS and Raman-in-SEM.

environment around the hydroxyl and XO_4 groups and changes in B–O bond lengths. The spectra of compounds containing both sulphate and phosphate groups are described by numerous vibration bands caused by a complex elemental composition and a symmetry change of the XO_4 groups.

A micrometric aluminium-phosphate-sulphate (APS) with an intermediate composition between the minerals woodhouseite ($CaAl_3(PO4)(SO_4)(OH)_6$) and crandallite ($Ca(PO_4)_2(OH)_5.H_2O$) has been identified in a natural sample and characterized using SEM and Raman-in-SEM spectroscopy.[79] As illustrated by a comparison of optical and BSE images, it appears that this kind of grain cannot be located by optical microscopy due to a lack of contrast and resolution (Fig. 22). Using Raman-in-SEM, the backscattered electron imaging mode can be efficiently used to precisely locate the grain and collect Raman spectra of the grain and the alunite matrix. The Raman spectrum obtained can be described by various bands assigned to vibrational modes of sulphate and phosphate units. Some differences can be highlighted by comparison of this mineral Raman spectrum with other APS minerals, with different PO_4/SO_4 ratios, regarding the intensities of some sulphate and phosphate bands. It can be established that the intensity of the sulphate bands (located at around 653 and 1026 cm^{-1}) decreases as a function of the mole fraction of phosphate in the APS minerals. This correlation correctly reflects the effect of the substitution of sulphate anions by phosphate anions.

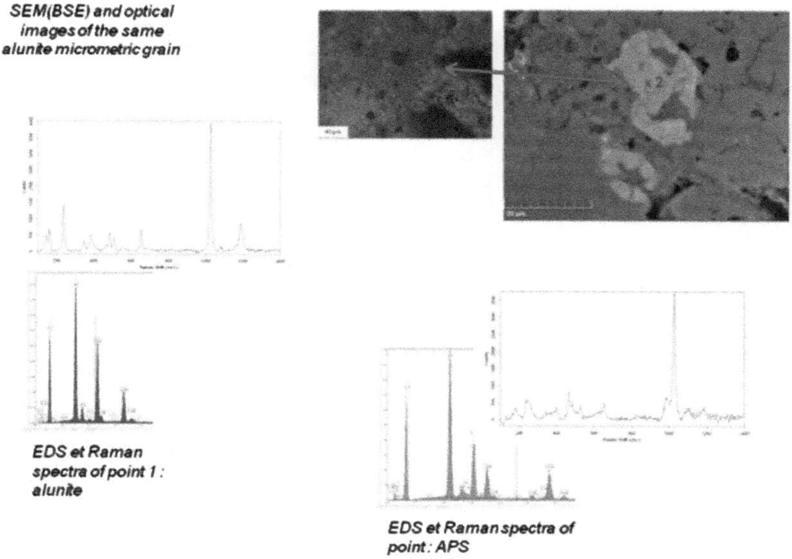

Fig. 22 Raman-in-SEM characterization of alunite and APS minerals (after[79]).

This study also demonstrates that the use of a coupled system is a powerful tool for identifying and characterizing micro-minerals of the alunite supergroup occurring in complex geological samples. The use of BSE images instead of optical imaging can also be very useful when zonation is detected on APS or alunite grains to collect Raman spectra of the different regions of the grain.

Combining SEM, cathodoluminescence and EBSD: application to the study of zonation and twinning in cassiterite. Cassiterite is a mineral species of chemical formula SnO_2. It has been and remains the most important source of tin. Cassiterite is of rutile structure, with a space group of D41h4-P 4/mnm. Natural cassiterite may contain significant amounts of Nb, Ta, W or other metal elements (Fe, Ti, In, Hg, *etc.*).[80,81] Cornwall in England was once a famous cassiterite mining area, but is no longer active today. The most important cassiterite sources now are Malaysia, the Democratic Republic of Congo, Indonesia, Bolivia and Nigeria.

Cassiterite has been extensively studied by Raman spectroscopy. According to Wang *et al.*,[82] the peak intensity ratio and position have been demonstrated to be related to trace element zonation and to crystal orientation. But the main problem of such samples is that crystal orientation and zonation are usually observed at a micrometric scale related to grain growth.

SEM characterization methods have been applied to cassiterite grains from Montbelleux (France). Zonation in a cassiterite grain can be illustrated by elemental mapping of trace elements. In this case, WDS (in EPMA) is preferred to EDS because the better spectral resolution compared to EDS (approx. 5 to 10 times better) allows numerous peak overlaps to be resolved and offers greater sensitivity. Local crystal

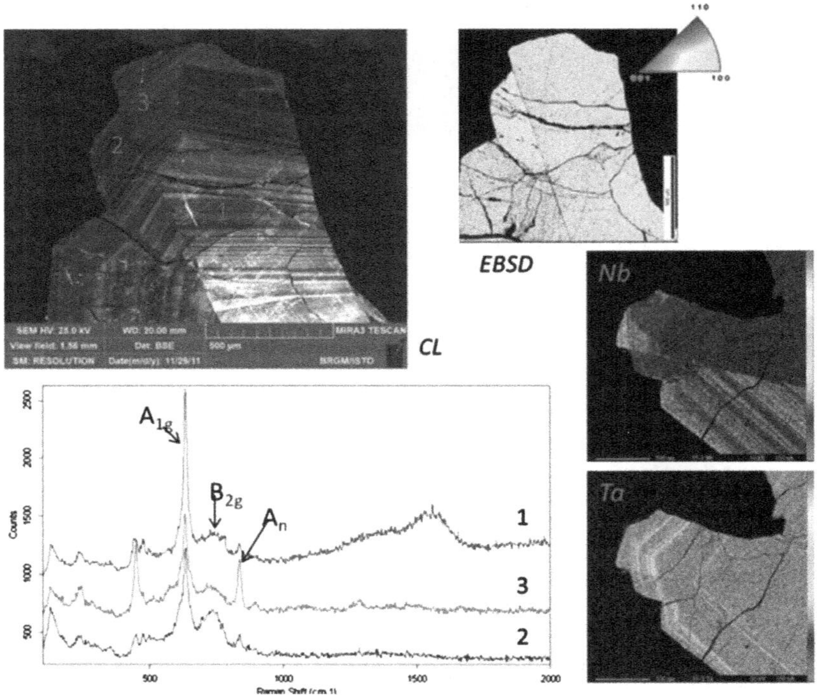

Fig. 23 Coupling SEM-EBSD, EPMA and Raman-in-SEM for characterizing trace element distribution in cassiterite.

orientation can be explored using EBSD in SEM. Cathodoluminescence in the SEM is a very interesting imaging mode for such characterization because the contrasts observed on cassiterite are easily linked to both zonation and crystal orientation at a micrometric scale. The use of these three techniques makes it possible to precisely locate the different regions of interest in the sample, as illustrated in Fig. 23. Two different crystal orientations with a misorientation angle of approximately 68° (twin plane commonly encountered in cassiterite, also known as 'elbow twin') may be observed on this grain. Different zonation regions may be observed on elemental mappings (EPMA), which are characteristic of low and high trace element concentrations.

The use of Raman-in-SEM enables each region of the sample to be located precisely. The SEM-CL detector allows precise positioning and then Raman spectra can be collected on the sample for each area, *i.e.* two different crystal orientations (main orientation and twin) and a strong and a low trace element concentration region. As described by Wang *et al.*,[82] $A1_g$ band shift and An band intensity are related to trace element concentration variations and the intensity of the $B2_g$ band is affected by crystal orientation. Variations observed in this sample for Raman shift and band intensity ratio are observed, correlated with trace element zonation and crystal orientation.

In this example, the combination of SEM detectors and microanalysis with Raman-in-SEM is clearly an advantage for precisely locating the different zones of the sample compared to a conventional micro-Raman spectrometer.

Assessment for a natural environmental risk: natural occurrence of asbestos. Another example is the use of analytical SEM for legal expert assessments on asbestos, which is being standardized in many countries.

Asbestos is a group of six natural mineral fibres belonging to two different mineral groups, serpentines and amphiboles. These fibres have been extensively extracted and used for industrial applications on account of their physical properties. They have mainly been used as insulators, for thermal insulation in buildings, fire retardant coatings, fireproof drywalls, reinforcing fibres in cements, gaskets, *etc.* However, in the early 1900s, a large number of early deaths and lung problems were noticed in the asbestos mining industry. According to the WHO (World Health Organisation), at least 100,000 deaths are caused by asbestos-related diseases each year in the world. For this reason, it is now banned in numerous countries, despite protests from major producing countries.

However, even if asbestos extraction and the production of products containing asbestos are now heavily restricted by this ban, asbestos still remains a public health issue. In fact, asbestos is primarily a group of naturally occurring silicate mineral fibres that can be encountered in nature in different geological settings. Therefore, good knowledge of the distribution of these sites is necessary, both for populations living in these areas and for human activities known to disperse fibres in the atmosphere, such as mining activities in favourable geological contexts.

As represented in Fig. 24, three items of information are needed, namely: fibre morphology, composition and crystalline structure. Now, all three can be obtained at the same time with Raman-in-SEM. Protocols have been developed for asbestos as a manufactured material, which

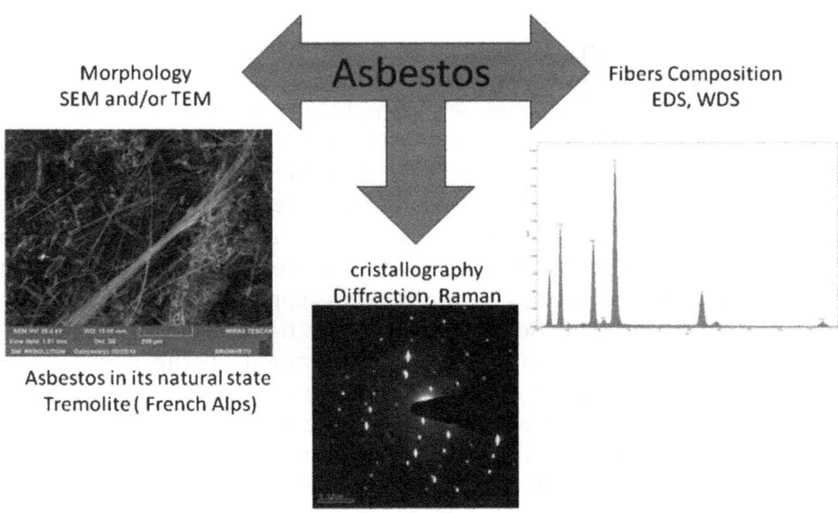

Fig. 24 Overview of the characterization of asbestos in fibre form.

mainly contains extracted and cleaned asbestos fibres (mainly chrysotile, about 90% of the total asbestos used in industry). Standard analytical protocols are routinely applied in specialized laboratories. These are based on the use of Transmission Electron Microscopy coupled with EDS microanalysis. Indeed, the contribution of TEM for this application has been highlighted[83,84] because TEM provides, on a unique sample preparation, morphological information (TEM imaging), chemical information (by EDS microanalysis) and structural information (electron diffraction on unique fibres). But the problem is much more complex for natural fibres in their geological context because of the presence of other minerals that can occur in the same mineral group (*i.e.* serpentine or amphibole) but cannot be classified as asbestos because of one or several criteria (non-fibrous, fibrous but with different crystallisation growth, cleavage fragments, *etc.*). For these samples, the combination of high resolution FE-SEM, EDS microanalysis and Raman-in-SEM can be of great interest, with SEM providing morphological information, EDS providing chemical information and Raman providing structural characterization.

Application of Raman-in-SEM to the characterization of components of cosmetics. Cosmetics such as make-up or sun creams are a complex mixture of different components such as minerals, metallic powders and organic compounds. Many products use submicronic powders for the potential value of their size in terms of optical or photochemical properties.[85–87] Moreover, advanced nanotechnology allows the cosmetics industry to customize multiscale textures with mineral ingredients to extend the limits of the aesthetic and functional properties of cosmetics.[85] Their chemical and physical properties often differ from macrometric materials, which is why they require a specific risk assessment and a specific characterization approach. This should cover both health risks of workers and consumers[88] and potential environmental risks. The characterization of such mixtures requires the ability to determine both the elemental and structural chemistry of micrometric and submicrometric materials.

The characterization of commercial make-up from a well-known brand was conducted using SEM and EDS. The make-up was deposited on a silicon plate, using the brush supplied with the product, as it would be done by the consumer. This analysis of the different particles that make up the powder provides information on the morphology and chemical composition of the individual components of the mixture (Fig. 25). Fillers (talc, mica, alumina sheets, *etc.*), pigments (copper plates, mica sheets coated with bismuth nanoparticles, *etc.*) or nanometric spheres were identified. But no information could be obtained on the phase of the particles. Then, structural chemistry information could be obtained by Raman spectroscopy. However, the size of these powder particles means conventional micro-Raman spectroscopy cannot be used. As an example, the large plates (n° 1, 2, 3) in Fig. 25 are visible using optical microscopy, but white submicronic particles (n° 4) cannot be observed and cannot be analysed by micro-Raman spectroscopy.

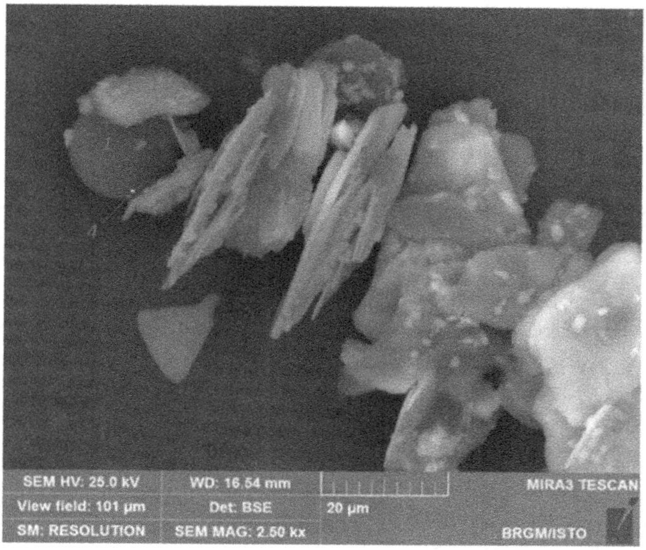

Fig. 25 SEM (BSE) image of a particle agglomerate in a make-up.

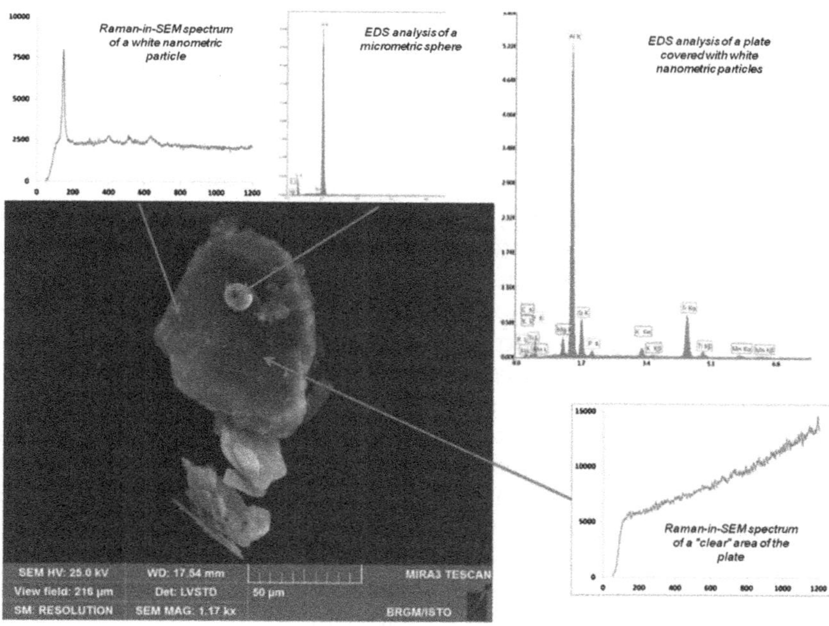

Fig. 26 SEM/Raman-in-SEM analysis of micrometric and sub-micrometric particles in a make-up.

Raman-in-SEM was then applied to the characterization, enabling the coupling of information from SEM imaging, chemical composition and mineral phase identification.

Several items of information can be obtained in this way (Fig. 26). Raman-in-SEM on the plate, the EDS analysis of which indicates that it appears to contain Al in its metallic form (very low oxygen peak), confirms that it is not made of alumina. The Raman spectra of the very small white particles make it possible to identify them as anatase particles. This information also helps to identify the origin of the titanium peak observed in the EDS spectra of the aluminium plates.

6 Evolution of Raman-in-SEM and combination of Raman spectroscopy and other techniques in the SEM

To date, commercial micro-Raman spectroscopy systems have been coupled with SEM as 'Point analysis Raman-in-SEM'. Many problems can be solved in terms of chemical and structural material characterization. However, if we look at the way elemental microanalysis has evolved, EDS/WDS microanalysis has expanded from a point analysis on the first electron microprobe to several other applications today (mapping, line profiles, automatic particles/phases analysis, *etc.*), thanks to a stronger combination with the SEM (development of SEM column and stage control by the microanalysis software, fully integrated EDS/SEM systems in 'integrated mineral analyser' systems). Then, the next step in the development of Raman-in-SEM may be stronger integration, with Raman mapping and line profiles in the SEM by controlling the SEM stage in the Raman-in-SEM software (or Raman spectrum collection in the SEM software). This function was first tested by A. Jambon and O. Boudouma[46] on their 'home-made' on-axis system, but many problems arose, such as focusing control, Raman spectra collection times, *etc.*

Another (and very interesting) perspective in the evolution of Raman-in-SEM spectroscopy is also a stronger combination with elemental microanalysis (*i.e.* EDS/WDS) for phase identification. For this aspect of the technique, one can refer to the evolution of EBSD. At the beginning of its development as a characterization method,[23] EBSD was developed as an electron diffraction tool for mapping well-known phases. This meant that the phases were known before mapping and no confusion was possible. Identification of phases by EBSD without prior EDS analysis was difficult or impossible and in some cases phase mapping was impossible, such as the mapping of materials made of a mixture of cubic phases, which are difficult to differentiate by EBSD, especially when lattice parameters are similar. Then fully integrated systems coupling EDS and EBSD were developed and have proven their efficiency in phase identification and mapping.[89] In such systems, EDS data collection is controlled by EBSD software and elemental identification or elemental mapping by EDS / WDS is used for EBSD pattern indexation. The transfer of this idea to Raman-in-SEM spectroscopy is also a major step to a more integrated and user-friendly analytical system, a sort of 'analytical expert-system'.

A final evolution may be considered regarding both SEM and Raman spectroscopy development separately: Both SEM and Raman spectroscopy can now be associated with other techniques such as Atomic Force Microscopy (AFM), micro-mechanical testing machines, *etc.* Such systems are commercially available and their interest has been demonstrated in recent publications (see for example[90–94]). In addition to specific SEM technical developments (resolution, detection, *etc.*), an important part of development has been focused on the development of the SEM as a micro-laboratory. Integration of coupled AFM-micro-Raman spectroscopy in this analytical 'micro-lab' would be a coherent development.

7 Conclusions and perspectives

As established by different authors, the use of Raman spectroscopy associated with electron microscopy and microanalysis offers many opportunities for characterizing both natural and synthetic materials. However, the use of separate devices can be complicated in some cases, especially for micrometric objects, dispersed particles or when images from the optical microscope available on the conventional micro-Raman spectrometer cannot be connected to the different imaging modes from the SEM. The possibility of focusing a laser beam inside the SEM chamber at the same point as the electron beam and with the same resolution as EDS or WDS X-ray spectrometry (elemental microanalysis) offers new opportunities for the characterization of such samples. It strongly increases the precision of analyses using SEM imaging capacities (resolution, variety of imaging modes) and combined elemental and structural chemical data. For numerous applications, the use of coupled Raman-in-SEM systems makes it possible to collect combined data where separate devices would not be suitable for conducting such analyses.

Studies carried out with coupled systems of this type demonstrate the diversity of applications and samples that can benefit from such characterization approaches compared to the conventional approach consisting of two separate systems for SEM-EDS and micro-Raman spectroscopy. Only a few studies showing the contribution of coupled systems have been published to date, which mainly relate to the study of biological and synthetic materials. However, the potential for applications in other areas, including geosciences, is important: micro-grains or fibres, micro-inclusions in rocks, mineral alterations, zonation, polluted soils, *etc.* These issues require fine chemical and structural characterization at a micrometric scale and a multi-technique approach based on separate systems is frequently not suitable. The use of combined techniques based on Raman-in-SEM spectroscopy offers new opportunities.

In addition, specific capacities of modern SEM could offer new perspectives by using or adapting other detectors (cathodoluminescence, Transmitted Electrons detectors (STEM), micro-characterisation techniques (EBSD) or *in-situ* applications (freezing and heating stages, micro-mechanical testing systems, *etc.*).

Micro-Raman spectroscopy available in the SEM makes it possible to precisely locate the point of interest using the different display modes (SE, SBE, CL, *etc.*) available in scanning electron microscopy and SEM nanometric resolution together with elemental microanalysis from EDS/

WDS. This study has also demonstrated the perfect match between EDS and Raman-in-SEM lateral resolution (micrometric). A more difficult question is depth resolution for Raman spectroscopy, which is strongly dependent on the optical system. If an off-axis system is equipped with well-known objectives (×50 long working distance), then it is possible to connect the confocality of such systems to conventional micro-Raman spectroscopy with such objectives. For on-axis systems, it is more difficult to evaluate, and studies need to be continued to evaluate this aspect.

However, this study demonstrates the considerable analytical potential of such combined systems for material characterization: in materials with microscopic/submicroscopic heterogeneity, SEM imaging provides morpho-chemical information through SE/BSE or CL imaging modes, EDS/WDS provide elemental chemical information (qualitative/quantitative analyses), and thanks to Raman-in-SEM, one has access to information concerning structural chemistry, atomic structure, molecular and crystal structure at the same scale as elemental microanalysis. We can also conclude on the considerable potential in the combination of Raman-in-SEM with crystallographic information from cathodoluminescence or EBSD available in the SEM. With a specific sample holder drawing, Raman-in-SEM could be applied to STEM-in-SEM microscopy. Finally, connecting Raman spectroscopy to FIB-SEM[95] could lead to a new type of 3D analysis.

Acronyms used in this chapter

BSE	BackScattered Electrons (SEM imaging mode)
BF	Bright Field (STEM imaging mode)
CRSEM	Confocal Raman SEM (Hybriscan™ commercial system)
DF	Dark Field (STEM imaging mode)
EBSD	Electron BackScattered Diffraction
EDS	Energy Dispersive Spectroscopy
EPMA	Electron Probe Microanalyzer (equipped with WDS)
ESEM	Environmental Scanning Electron Microscope (allowing for a gaseous environment in the specimen chamber – working under dry gas or H_2O vapour pressure from a few Pa up to more than 1 kPa allows for the option of collecting electron micrographs of specimens that are "wet", uncoated or both)
LV-SEM	Low Vacuum Scanning Electron Microscope (allowing for a gaseous environment in the specimen chamber – working under dry gas pressure from a few Pa to several hundred Pa)
Raman	often used as Raman Spectroscopy
SCA	Structural and Chemical Analyser, interfaced with SEM from Renishaw
SEM	Scanning Electron Microscope
SE	secondary electron (signal used to image the surface by scanning the probe)

STEM	Scanning Transmission Electron Microscope (transmission configuration in the SEM)
TEM	Transmission Electron Microscope
WDS	Wavelength Dispersive X-ray Spectroscopy

References

1 L. Brillouin, *Annales de Physique*, 1922, **17**, 88–122.
2 A. Smekal, *Naturwissenschaften*, 1923, **11**, 873–875.
3 C. V. Raman and K. S. Krishnan, *Nature*, 1928, **121**, 501–502.
4 G. Plaezek, *Zeitschrift für Physik*, 1929, **58**, 585–594.
5 R. Castaing and J. Descamps, *J. Phys. Radium*, 1955, **16**, 304–317.
6 M. Truchet and M. Delhaye, *Journal de microscopie et de spectroscopie electronique*, 1988, **13**, 167–175.
7 M. Truchet and M. Delhaye, *France Pat. FR2596863*, 1986.
8 M. Truchet, M. Delhaye, C. Ballan-Dufrancais and A.-Y. Jeantet, *Biology of the Cell*, 1988, **63**, 26.
9 P. Dhamelincourt, M. Delhaye, M. Truchet and E. Dasilva, *Journal of Raman Spectroscopy*, 1991, **22**, 61–64.
10 T. P. Mernagh and L. G. Liu, *Physics and Chemistry of Minerals*, 1991, **18**, 126–130.
11 F. Maurice, L. Meny and R. Tixier, eds., *Microanalysis and scanning electron microscopy*, Les editions de physique, 1978.
12 T. E. Everhart and R. F. M. Thornley, *Journal of Scientific Instruments*, 1960, **37**, 246.
13 M. Jacka, M. Zadrazil and F. Lopour, *Scanning*, 2003, **25**, 243–246.
14 J. Jirak, V. Nedela, P. Cernoch, P. Cudek and J. Runstuk, *J Microsc*, 2010, **239**, 233–238.
15 G. D. Danilatos, *Micron and Microscopica Acta*, 1983, **14**, 307–318.
16 P. Gnauck, V. Drexel, D. Bate, E. Essers, patent US 7462839, 2008
17 G. Koschek, *Journal of Microscopy*, 1993, **171**, 223–232.
18 M. R. Landtwing and T. Pettke, *American Mineralogist*, 2005, **90**, 122–131.
19 T. Sekiguchi, X. L. Yuan and J. Niitsuma, *Scanning*, 2005, **27**, 103–104.
20 F. Millot, J.-C. Rifflet, G. Wille and V. Sarou-Kanian, *High Temperatures-High Pressures*, 2009, **38**, 245–257.
21 G. Wille, F. Millot and J. C. Rifflet, *International Journal of Thermophysics*, 2002, **23**, 1197–1206.
22 X. Bourrat, L. Qiao, Q. L. Feng, M. Angellier, A. Dissaux, J. M. Beny, V. Barbin, P. Stempfle, M. Rousseau and E. Lopez, *Materials Characterization*, 2012, **72**, 94–103.
23 D. J. Dingley, *Scanning Electron Microscopy*, 1984, 569–575.
24 D. Dingley, *Journal of Microscopy-Oxford*, 2004, **213**, 214–224.
25 D. Mainprice, J. Bascou, P. Cordier and A. Tommasi, *Journal of Structural Geology*, 2004, **26**, 2089–2102.
26 D. J. Prior, A. P. Boyle, F. Brenker, M. C. Cheadle, A. Day, G. Lopez, L. Peruzzo, G. J. Potts, S. Reddy, R. Spiess, N. E. Timms, P. Trimby, J. Wheeler and L. Zetterstrom, *American Mineralogist*, 1999, **84**, 1741–1759.
27 M. Ishihara, Y. Koga, J. Kim, K. Tsugawa and M. Hasegawa, *Materials Letters*, 2011, **65**, 2864–2867.
28 A. J. Wilkinson, G. Meaden and D. J. Dingley, *Superlattices and Microstructures*, 2009, **45**, 285–294.
29 E. A. Stefaniak, A. Worobiec, S. Potgieter-Vermaak, A. Alsecz, S. Torok and R. Van Grieken, *Spectrochimica Acta Part B-Atomic Spectroscopy*, 2006, **61**, 824–830.

30 E. A. Stefaniak, A. Alsecz, R. Frost, Z. Mathe, I. E. Sajo, S. Torok, A. Worobiec and R. Van Grieken, *Journal of Hazardous Materials*, 2009, **168**, 416–423.
31 F. Pointurier and O. Marie, *Spectrochimica Acta Part B-Atomic Spectroscopy*, 2010, **65**, 797–804.
32 R. H. M. Godoi, S. Potgieter-Vermaak, J. De Hoog, R. Kaegi and R. Van Grieken, *Spectrochimica Acta Part B-Atomic Spectroscopy*, 2006, **61**, 375–388.
33 M. P. Nelson, C. T. Zugates, P. J. Treado, G. S. Casuccio, D. L. Exline and S. F. Schlaegle, *Aerosol Science and Technology*, 2001, **34**, 108–117.
34 S. Bruni, F. Cariati, F. Casadio and L. Toniolo, *Vibrational Spectroscopy*, 1999, **20**, 15–25.
35 F. Poitrasson, S. Chenery and T. J. Shepherd, *Geochimica Et Cosmochimica Acta*, 2000, **64**, 3283–3297.
36 M. Hänchen, V. Prigiobbe, R. Baciocchi and M. Mazzotti, *Chemical Engineering Science*, 2008, **63**, 1012–1028.
37 J. L. Chen, B. Chu and B. S. Hsiao, *Journal of Biomedical Materials Research Part A*, 2006, **79A**, 307–317.
38 R. M. Jarvis and R. Goodacre, *Analytical Chemistry*, 2004, **76**, 40–47.
39 R. M. Jarvis, A. Brooker and R. Goodacre, *Analytical Chemistry*, 2004, **76**, 5198–5202.
40 V. Otieno-Alego, *Journal of Raman Spectroscopy*, 2009, **40**, 948–953.
41 A. Worobiec, S. Potgieter-Vermaak, A. Brooker, L. Darchuk, E. Stefaniak and R. Van Grieken, *Microchemical Journal*, 2010, **94**, 65–72.
42 A. Worobiec, L. Darchuk, A. Brooker, H. Potgieter and R. Van Grieken, *Journal of Raman Spectroscopy*, 2011, **42**, 808–814.
43 G. Wille, X. Bourrat, N. Maubec and A. Lahfid, ed. BRGM, under review for publication in microscopy and microanalysis.
44 R. Bennett, A. M. Woolfrey, J. C. C. Day and A. Bewick, Patent 6885445, 2005.
45 A. Laroussi, A. Jambon, O. Boudouma and H. Chennaoui-Aoudjehane, 74th annual meteoritical society meeting, 2011.
46 A. Jambon and O. Boudouma, in *journées thématiques GN-MEBA*, Strasbourg - France, 2011
47 A. A. van Apeldoorn, Y. Aksenov, M. Stigter, I. Hofland, J. D. de Bruijn, H. K. Koerten, C. Otto, J. Greve and C. A. van Blitterswijk, *Journal of the Royal Society Interface*, 2005, **2**, 39–45.
48 A. Lahfid, O. Beyssac, E. Deville, B. Goffe and C. Chopin, *Geochimica Et Cosmochimica Acta*, 2008, **72**, A511–A511.
49 A. Lahfid, O. Beyssac, E. Deville, F. Negro, C. Chopin and B. Goffe, *Terra Nova*, 2010, **22**, 354–360.
50 L. Zhu, C. Georgi, M. Hecker, J. Rinderknecht, A. Mai, Y. Ritz and E. Zschech, *Journal of Applied Physics*, 2007, **101**.
51 C. L. Jahncke, H. D. Hallen and M. A. Paesler, *Journal of Raman Spectroscopy*, 1996, **27**, 579–586.
52 Y. D. Suh, G. K. Schenter, L. Y. Zhu and H. P. Lu, *Ultramicroscopy*, 2003, **97**, 89–102.
53 D. Drouin, A. R. Couture, D. Joly, X. Tastet, V. Aimez and R. Gauvin, *Scanning*, 2007, **29**, 92–101.
54 H. J. Scheibe, D. Drescher and P. Alers, *Fresenius Journal of Analytical Chemistry*, 1995, **353**, 695–697.
55 B. Marchon, J. Gui, K. Grannen, G. C. Rauch, J. W. Ager, S. R. P. Silva and J. Robertson, *Ieee Transactions on Magnetics*, 1997, **33**, 3148–3150.
56 M. Von Ardenne and D. Beischer, *Zeitschrift für Elektrochemie und angewandte physikalische Chemie*, 1940, **46**, 270–277.
57 W. C. Lane, *Scanning Electron Microsc.*, 1970, 43–48.

58 G. D. Danilatos and V. N. E. Robinson, *Scanning*, 1979, **2**, 72–82.
59 M. T. Postek, *Scanning*, 1996, **18**, 269–274.
60 U. R. Bance, I. W. Drummond, D. Finbow, E. H. Harden and P. Kenway, *Vacuum*, 1978, **28**, 489–496.
61 R. V. Krstic and P. A. Milliquet, *Scanning*, 1980, **3**, 47–48.
62 F. Brisset, *Microscopie électronique à balayage et Microanalyses*, EDP Sciences, 2008.
63 A. N. Mariano, *Reviews in Mineralogy*, 1989, **21**, 339–348.
64 A. Baumer, P. Blanc, F. Cesbron and D. Ohnenstetter, *Chemical Geology*, 1997, **138**, 73–80.
65 C. B. Farmer, A. Searl and C. Halls, *Mineralogical Magazine*, 1991, **55**, 447–458.
66 G. Remond, F. Cesbron, R. Chapoulie, D. Ohnenstetter, C. Roquescarmes and M. Schvoerer, *Scanning Microscopy*, 1992, **6**, 23–68.
67 G. Remond, *Bull. Soc. franc. Mineral. Crist.*, 1973, **96**, 183–198.
68 C. Lerouge, F. Claret, M. A. Denecke, G. Wille, G. Falkenberg, C. Ramboz, C. Beny, E. Giffaut, T. Schaefer, E. C. Gaucher and C. Tournassat, *Physics and Chemistry of the Earth*, 2010, **35**, 271–277.
69 G. D. Smith and R. J. H. Clark, *Journal of Archaeological Science*, 2004, **31**, 1137–1160.
70 F. R. Pérez, *Selected Topics in Raman Spectroscopic Applications Geology, Biomaterials, Art*, Departamento de Física de la Materia Condensada Cristalografía y Mineralogía, Universidad de Valladolid, 2007.
71 I. M. Bell, R. J. H. Clark and P. J. Gibbs, *Spectrochimica Acta Part a-Molecular and Biomolecular Spectroscopy*, 1997, **53**, 2159–2179.
72 L. Burgio and R. J. H. Clark, *Spectrochimica Acta Part a-Molecular and Biomolecular Spectroscopy*, 2001, **57**, 1491–1521.
73 M. Bouchard and D. C. Smith, *Spectrochimica Acta Part a-Molecular and Biomolecular Spectroscopy*, 2003, **59**, 2247–2266.
74 P. Jezequel, G. Wille, C. Beny, F. Delorme, V. Jean-Prost, R. Cottier, J. Breton, F. Dure and J. Despriee, *Journal of Archaeological Science*, 2011, **38**, 1165–1172.
75 J. Despriee and S. Tymula, eds., *Bulletin de l'association pour la sauvegarde du site archeologique d'Argentomagnus et Amis du Musee - Le Coteau de la garenne, projet collectif de recherche, étude 1999–2001 (Bulletin of the Association for the Protection of archeological site Argentomagnus and Friends of the Museum, slopes of La Garenne - collaborative research project 1999–2001) - (in french)*, ASSAAM, 2001.
76 C. Vignaud, H. Salomon, E. Chalmin, J.-M. Geneste and M. Menu, *L'Anthropologie*, 2006, **110**, 482–499.
77 J. L. Jambor, *Canadian Mineralogist*, 1999, **37**, 1323–1341.
78 N. Maubec, A. Lahfid, C. Lerouge, G. Wille and K. Michel, *Spectrochimica acta. Part A, Molecular and biomolecular spectroscopy*, 2012, **96**, 925–939.
79 N. Maubec, C. Lerouge, A. Lahfid, G. Wille, K. Michel and X. Bourrat, *Journal of Molecular Structure*, 2013, **1048**, 33–40.
80 P. Cerny and T. S. Ercit, *Bulletin De Mineralogie*, 1985, **108**, 499–532.
81 L. Izoret, G. Marnier and Y. Dusausoy, *The Canadian Mineralogist*, 1985, **23**, 221–231.
82 R. Wang, J. Wu, J. Dubessy and P. Monchoux, *Chinese Journal of Geochemistry*, 1993, **12**, 353–360.
83 D. Steen, M. P. Guillemin, P. Buffat and G. Litzistorf, *Atmospheric Environment*, 1983, **17**, 2285–2297.
84 C. Malami and M. Grasserbauer, *Fresenius Zeitschrift Fur Analytische Chemie*, 1982, **311**, 1–6.

85 X. Bourrat, G. Wille and M. Bizi, in *Cosm'innov 2013 - Cosmetic innovation days*, Orleans - France, 2013.
86 M. Auffan, M. Pedeutour, J. Rose, A. Masion, F. Ziarelli, D. Borschneck, C. Chaneac, C. Botta, P. Chaurand, J. Labille and J. Y. Bottero, *Environmental Science & Technology*, 2010, **44**, 2689–2694.
87 N. A. Monteiro-Riviere and J. E. Riviere, *Nanotoxicology*, 2009, **3**, 188–193.
88 A. Nel, T. Xia, L. Madler and N. Li, *Science*, 2006, **311**, 622–627.
89 S. I. Wright and M. M. Nowell, *Microscopy and Microanalysis*, 2005, **11**, 672–673.
90 G. D. West and R. C. Thomson, *Journal of Microscopy*, 2009, **233**, 442–450.
91 F. Foucher and F. Westall, *AIP Conference Proceedings*, 2009.
92 F. Hang, D. Lu and A. H. Barber, *MRS Proceedings*, 2009.
93 E. M. van Schrojenstein Lantman, T. Deckert-Gaudig, A. J. Mank, V. Deckert and B. M. Weckhuysen, *Nature nanotechnology*, 2012, **7**, 583–586.
94 F. Adar, M. Delhaye and E. DaSilva, *Journal of Chemical Education*, 2007, **84**, 50–60.
95 R. Wirth, *Chemical Geology*, 2009, **261**, 217–229.

Structure-property correlations of inorganic nanomaterials by different X-ray related techniques

Kanishka Biswas

DOI: 10.1039/9781782621485-00117

Substantial development in the nanoscience and nanotechnology has been evinced in the last few years due to the availability of sophisticated physical methods to characterize nanomaterials. Among the various physical techniques, X-ray related characterizations are superior for the understanding of the crystal structure, size, shape, composition and electronic structure of inorganic nanomaterials. These techniques include X-ray diffraction, small angle X-ray scattering, X-ray reflectivity, pair distribution function analysis, X-ray photoelectron spectroscopy, energy dispersive X-ray analysis and others. Characterization of the nanomaterials includes not only the determination of size and shape but also the atomic and electronic structures as well as other important properties. In this article we describe some of the important X-ray related methods employed for characterization of nanostructures. In order to provide a feeling for the use of these methods, a few case studies are given.

Introduction

Characterization of the inorganic nanomaterials is carried out at various levels. The various aspects of characterization are: (a) structure, revealing the crystallinity or otherwise of the specimen, crystal system, unit cell and where possible (or necessary) atomic coordinates, bonding and electronic structure, (b) chemical composition and compositional homogeneity of the specimen, (c) impurities or doping that may affect the properties, and (d) the nature and concentration of imperfections (defects) influencing the properties. Generally, X-ray related techniques are superior than other characterization methods to understand the structure-property correlation in the inorganic nanomaterials. X-ray related characterizations of nanomaterials has a wide diversity, which includes: powder X-ray diffraction, single crystal X-ray diffraction, small angle X-ray scattering, pair distribution function analysis of high angle powder X-ray diffraction data, X-ray reflectivity, X-ray photoelectron spectroscopy, energy dispersive X-ray analysis and extended X-ray absorption fine structure. The structure of materials can be studied at different levels of sophistication, which include crystal structure, microstructure, atomic-level structure and electronic structure. X-ray diffraction is routinely used for determination of crystal structure, but the use of synchrotron X-ray radiation, along with Rietveld analysis have made these techniques more powerful. X-ray diffraction is also used to obtain the average size of the nanocrystals by the use of the Scherrer formula. Temperature and pressure dependent X-ray diffraction provides important information of the phase transitions in inorganic nanomaterials. X-ray scattering is an important technique to study the

New Chemistry Unit, Jawaharlal Nehru Centre for Advanced Scientific Research (JNCASR), Jakkur P.O., Bangalore 560064, India. E-mail: kanishka@jncasr.ac.in

distribution of shapes and sizes of the nanocrystals. Pair distribution function analysis of powder diffraction provides the bonding information in amorphous solid and nanomaterials. Several techniques including X-ray photoelectron spectroscopy (XPS) provide information on the electronic structure of nanomaterials. This chapter discusses the different X-ray related characterization techniques, and how these techniques are used to study the structure and property of various inorganic nanostructures.

Powder X-ray diffraction

Powder X-ray diffraction (PXRD) is the most essential tool employed for characterizing the crystal structures. Evaluation of the crystal structure is important even for the nanoscale materials.[1,2] While the properties might be affected by the structures on the nanometer scale, the crystal structures are determined by the arrangement of atoms separated by about 0.1 nm. The principle of powder XRD is based on the well known Bragg's law ($\lambda = 2d\sin\theta$). The most commonly used database for the identification of crystal structures is the JCPDS-ICDD system. Rietveld profile analysis[3] of the experimental PXRD patterns gives information about the space group and the structural parameters of materials.

ReO_3 is an unusual metal oxide with a cubic structure, exhibiting metallic conductivity.[4] It is comparable to copper both in its appearance and electronic properties. Metallic ReO_3 nanocrystals exhibit a size dependent surface plasmon band , similar to gold nanoparticles.[5] The structure of the ReO_3 nanocrystals have been determined by powder XRD followed by Rietveld profile analysis.[5] It crystallizes in the cubic perovskite (ABO_3 with A cation vacant) structure with Pm-3m space group. Hexagonal rod-shaped CoO nanocrystals are unstable in the bulk phase, and have been identified by powder XRD (Fig. 1).[6] Kinetic hexagonal CoO nanocrystals synthesized by a low temperature solution phase synthesis,

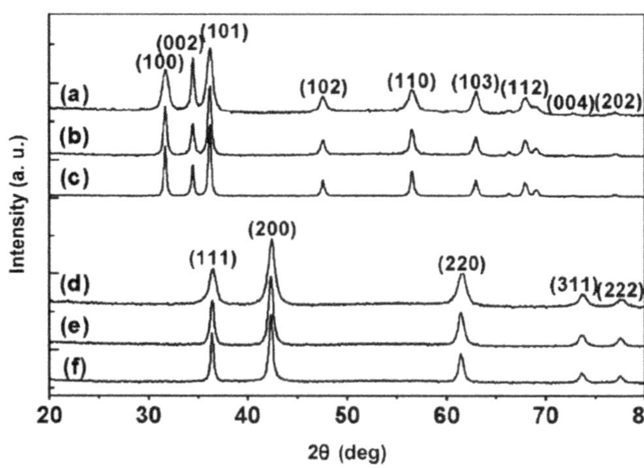

Fig. 1 XRD patterns of (a) 11×40 nm rod-shaped and (b) 47 nm and (c) 83 nm pyramid-shaped hexagonal CoO nanocrystals. XRD patterns of (d) 13 nm, (e) 24 nm, and (f) 33 nm cubic CoO nanocrystals. Reprinted with permission from Seo *et al. J. Am. Chem. Soc.*, 2005, **127**, 6188 © 2005 American Chemical Society.

while thermodynamically controlled reactions resulted cubic CoO nanocrystals. Park *et al.*[7] have used metal-oleates as precursors for the preparation of monodisperse Fe_3O_4, MnO and CoO nanocrystals. 1-octadecene, octyl ether and trioctylamine have been used as solvents. These metal oxide nanocrystals are initially characterized by PXRD. Cubic form of $Pb_{2-x}Sn_xS_2$ have been stabilized through size reduction to the nanoscale recently by Kanatzidis and coworkers.[8] The cubic $Pb_{2-x}Sn_xS_2$ nanocrystals have been synthesized using a modified hot injection colloidal synthetic route. The x value is in the range $0.40<x<1$. Even though these compositions lie in a region of the PbS–SnS phase diagram where no single phase exists, and despite the fact that $PbSnS_2$ is a distorted orthorhombic phase, the $Pb_{2-x}Sn_xS_2$ nanocrystals are single phase solid solutions with cubic NaCl-type structure. The phase was characterized by PXRD studies as well. PXRD patterns of the $Pb_{2-x}Sn_xS_2$ nanocrystals revealed the long-range crystalline nature of the particles with cubic rock salt symmetry ($Fm\bar{3}m$ space group), Fig. 2. This result is in contrast with the bulk $PbSnS_2$ that normally crystallizes in an orthorhombic space group (*Pnma*). The diffraction patterns are shifted to larger 2θ values with respect to the powder pattern of pure cubic PbS suggesting a cell contraction consistent with the incorporation of Sn into the PbS lattice. Furthermore, a series of novel rock-salt-type

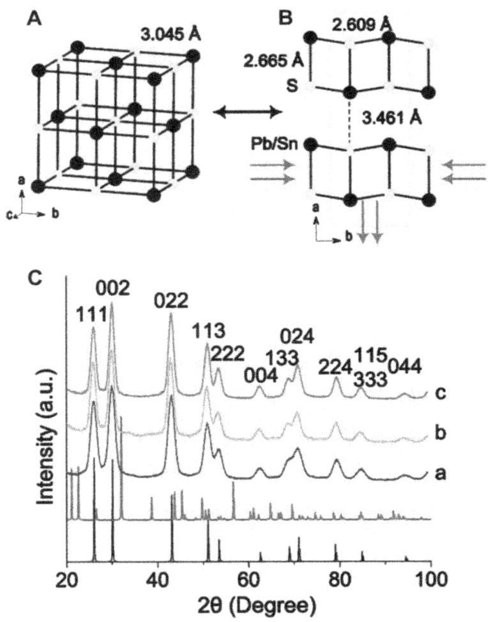

Fig. 2 Cubic (A) and orthorhombic (B) crystal polymorphs of $PbSnS_2$ showing the structural relationship through a compression along the *b*–*c* plane and an elongation along the *a*-axis (arrows). The 3.045 Å distance is the average Pb/Sn–S bond length found in the cubic structure. (C) Powder X-ray diffraction pattern of the $Pb_{2-x}Sn_xS_2$ nanocrystals prepared at different lengths of reaction time: (a) 30 s, (b) 60 s, and (c) 120 s. Black lines represent the simulated PXRD pattern of bulk PbS. Line below (a) represents the simulated PXRD pattern of bulk $PbSnS_2$. Reprinted with permission from Soriano *et al. J. Am. Chem. Soc.*, 2012, **134**, 3228 © 2012 American Chemical Society.

$Pb_mSb_{2n}Te_{m+3n}$ nanocrystals ($m = 2$, 3, 4, 6, 8, and 10; $n = 1$ and 2) have been synthesized using a colloidal synthesis route.[9] These materials are stable only on the nanoscale and have no bulk analogues. The powder X-ray diffraction patterns of the as-synthesized $Pb_mSb_{2n}Te_{m+3n}$ nanocrystals show high crystallinity with the characteristic cubic rock-salt structure belonging to the $Fm\bar{3}m$ space group. Ternary silver bismuth sulfide, $AgBiS_2$, is a typical member of the I–V–VI_2 family. At room temperature, bulk $AgBiS_2$ crystallizes in hexagonal phase (space group, *P*-3*m*1) and transforms to cubic rocksalt structure (space group, *Fm*-3*m*) at around ~473 K. High temperature cubic bulk $AgBiS_2$ possesses disordered Ag and Bi atoms (Fig. 3(a)). Cubic, bulk $AgBiS_2$ possesses very low thermal conductivity, which has immense importance in the field of thermoelectric energy conversion. High temperature rocksalt phase of $AgBiS_2$ and $AgBiS_{2-x}Se_x$ (x = 0.05–0.1) have been kinetically stabilized at room

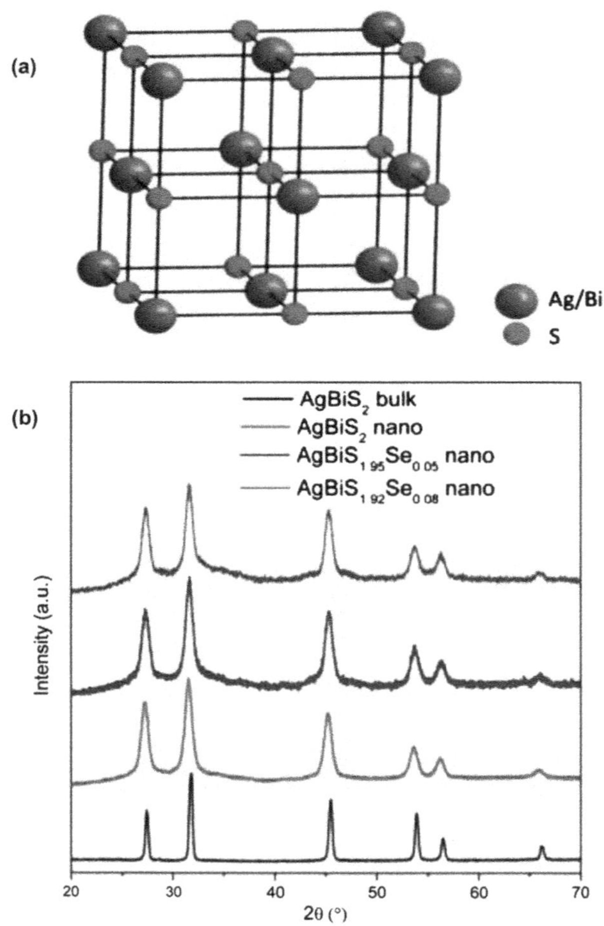

Fig. 3 (a) Crystal structure of the high temperature cubic phase of $AgBiS_2$ with disordered Ag/Bi positions. (b) XRD patterns of as prepared $AgBiS_2$ and $AgBiS_{2-x}Se_x$ nanocrystals along with cubic bulk $AgBiS_2$. Reprinted with permission from Guin *et al. Chem. Mater.*, 2013, **25**, 3225 © 2013 American Chemical Society.

temperature in the nanocrystals (~11 nm) by simple solution based synthesis.[10] Experimental evidence for this derives from PXRD (Fig. 3 (b)) and many other techniques. Similar kinetic stabilization of polytype phases are rare and have been observed only in few semiconductors where a metastable phase is accessed on the nanoscale as, for example, in wurzite MnSe,[11] Cu_2SnSe_3,[12] $CuInS_2$,[13] and orthorhombic MnAs.[14]

In general, the XRD peaks of nanocrystals are much broader compared to those of the bulk. Evaluation of the average crystallite size by using the Scherrer formula is one of the important applications of powder XRD in the study of nanomaterials:

$$t = \frac{0.9\lambda}{B\cos\theta_B} \tag{1}$$

where, t is the size of nanocrystals (in Å), λ the X-ray wavelength and θ_B the Bragg angle. B is the measured peak width in radians at the full width of half maxima. For example, in the case of nanocrystals of MnO,[15] CoO,[16] NiO, CdS, CdSe,[17] PbS and GaN,[18] Rietveld profile analysis and the Scherrer formula have been employed to determine the structure and average particle size. In the case of doped nanomaterials, it is often possible to obtain the extent of doping by the shifts in the XRD reflections. While XRD is a routine technique, carried out with standard laboratory diffractometers, it is necessary to use synchrotron X-rays to obtain better resolution and sensitivity for many of the materials such as metal oxides.

An indication of the anisotropic growth of the nanostructure can be obtained through the analysis of the intensity difference in the XRD peaks. If the nanostructures prefers to grow in the one dimension (nanorod or nanowire), crystal planes involved in such growth show high intensity peak in PXRD. This was observed in the case of ZnO and GaN nanostructures. Interestingly, 2–3 layered MoS_2 does not show any (002) reflection in PXRD.[19] A similar kind of lowering of the intensity of (001) plane in XRD has been observed for a few layer Bi_2Se_3 two-dimensional sheet.[20] The low intensity of (001) reflections relative to those of bulk counterpart indicate reduced periodicity along the crystallographic c-axis.

Nanoscale inclusions in bulk materials can dramatically suppress the lattice thermal conductivity by scattering the longer wavelength heat-carrying phonons, as in PbTe-SrTe[21,22] or PbTe-PbS,[23,24] which resulted in a high thermoelectric figure of merit. The system $(PbTe)_{1-x}$-$(PbS)_x$ is ,in fact, nanostructured (and not a solid solution) over a wide range of concentrations. It exhibits spinodal decomposition and or nucleation and growth, depending on x. Analysis of this phase separation behavior have examined by high resolution temperature dependent PXRD using synchrotron X-ray sources (Fig. 4).[25] For the PbTe–PbS 8% sample, PbS reflections become visible above 250 °C (Fig. 4(a)). The region of immiscibility is roughly in the range of 200 to 500 °C. The temperature of dissolution of PbS, at approximately 500 °C, is in close agreement with the phase diagram. For the PbTe–PbS 30% sample, a significant precipitation of PbS is observed between 250 and 700 °C (Fig. 4 (b)). Additionally, above ~550 °C the movement of the sample from the spinodal to the nucleation and growth region results in increased dissolution of S into PbTe

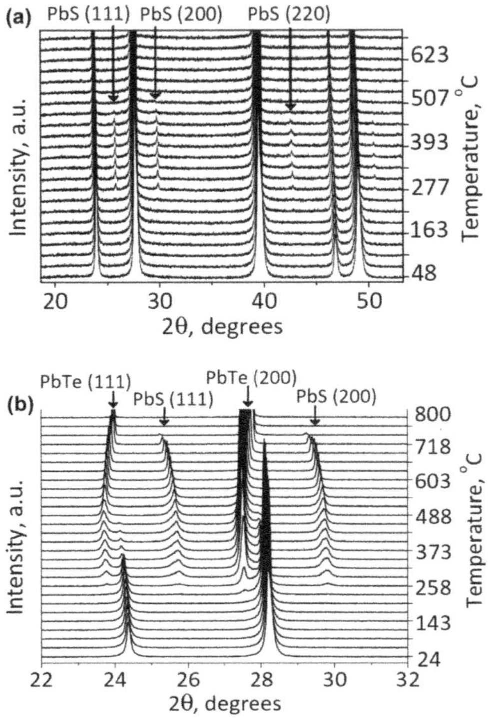

Fig. 4 (a) *In situ* high-temperature PXRD of PbTe–PbS 8% solid-solution alloys. The precipitation of PbS is evidenced by the appearance of PbS (111), (200), and (220) reflections between 200 and 500 °C. At temperatures >500 °C, the PbS redissolves into the PbTe matrix, re-forming a solid solution. (b) *In situ* high-temperature PXRD of PbTe–PbS 30% solid-solution alloys. The higher PbS concentration produces a region of immiscibility between 250 and 750 °C. At temperatures >750 °C, the PbS redissolves into the PbTe matrix, re-forming a solid solution. Reprinted with permission from Girard *et al. Adv. Func. Mater*, 2013, **23**, 747 © 2013 Wiley-VCH Verlag GmbH & Co. K GaA, Weinheim.

and Te into PbS, contracting the PbTe and expanding the PbS lattices, in agreement with the lever rule.[23,25]

In-situ temperature dependent PXRD is a valuable tool to monitor the phase transition in nanocrystalline solids. $AgBiSe_2$, as a typical member of I–V–VI_2 compounds with phase transition behavior,and it's temperature-dependent structural evolution is well-known as shown in Fig. 5(a).[26] At room temperature, $AgBiSe_2$ is a *p*-type semiconductor and crystallizes in the hexagonal phase with lattice parameters $a = 4.18$ Å and $c = 19.67$ Å (space group $P\bar{3}m1$). As the temperature increases, $AgBiSe_2$ is observed to undergo continuous phase transition to rhombohedral phase around 410 K and then to cubic phase around 580 K. This reversible phase transition is undergone, from a high temperature cubic phase, to an intermediate temperature rhombohedral structure ~560 K and then to a low-temperature hexagonal phase ~393 K as temperature decreases. In the intermediate temperature rhombohedral phase (space group $R\bar{3}m$ with lattice constants $a = 7.022$ Å and $\alpha = 34.5°$), ordering of Ag and Bi atoms takes place in quite distinguishable positions, while in the high temperature cubic phase (space group $Fm\bar{3}m$ with lattice constants

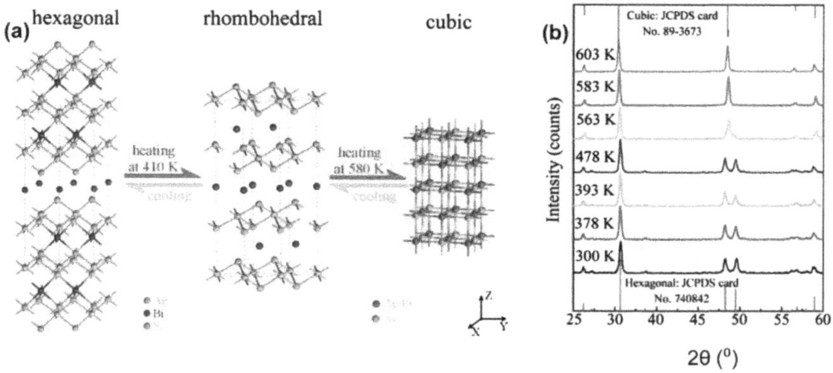

Fig. 5 Crystal structural evolution among hexagonal, rhombohedral, and cubic phase of $AgBiSe_2$. (b) Temperature-dependent XRD patterns for $AgBiSe_2$ nanocrystals. Reprinted with permission from Xiao *et al. J. Am. Chem. Soc.*, 2012, *134*, 18460 © 2012 American Chemical Society.

$a = 5.832$ Å), the Ag and Bi atoms are fully disordered. X-ray diffractograms from synthetic $AgBiSe_2$ nanocrystals (Fig. 5(b)) exhibit peaks corresponding to the hexagonal structure with space group of $P\bar{3}m1$.[26] Temperature-dependent XRD patterns show the structural transition from the rhombohedral to the cubic phase. While the hexagonal-rhombohedral phase transition was not differentiated in the temperature-dependent XRD patterns. This proves that the atoms slightly shift, rather than rearrange during this phase transition. Ag_2Se crystallizes in the orthorhombic phase at room temperature,and is a narrow-band gap semiconductor with two crystallographically distinct silver atoms,In the high-temperature cubic phase, α-Ag_2Se, the selenium sublattice is ordered in a bcc lattice while silver atoms are statistically distributed over several interstitial sites, through which Ag^+ cations can move easily and show superionic conductivity.[27] Detailed insight into the phase transition behavior of Ag_2Se nanocrystals has also been obtained from the temperature-dependent PXRD.[28]

Pressure induced phase transition in nanocrystalline materials is an exciting topic in solid state chemistry and can be monitored by studying the pressure dependent powder X-ray diffraction. CdSe nanocrystals show a wurtzite to zinc blende phase transition at an increasing pressure of 6.3 GPa, which is twice the pressure increase of about 3 GPa for bulk CdSe.[29] Pressure-induced phase transitions in the nanocrystals of ReO_3 with an average diameter of ~12 nm have been investigated in detail by using synchrotron x-ray diffraction . The results compared with the bulk samples of ReO_3.[30] The study shows that at ambient-pressure, cubic I phase (space group *Pm*-3*m*) transforms to a monoclinic phase (space group *C* 2/*c*), then to a rhombohedral I phase (space group *R*-3*c*), and finally to another rhombohedral phase (rhombohedral II, space group *R*-3*c*) with increasing pressure over the 0.0–20.3 GPa range (Fig. 6(A)). The cubic I to monoclinic transition is associated with the largest volume change (~5%), indicative of a reconstructive transition (Fig. 6(B)). The transition pressures are generally lower than those known for bulk ReO_3. The cubic II (*Im*-3) or tetragonal (*P*4/*mbm*) phases do not occur at lower

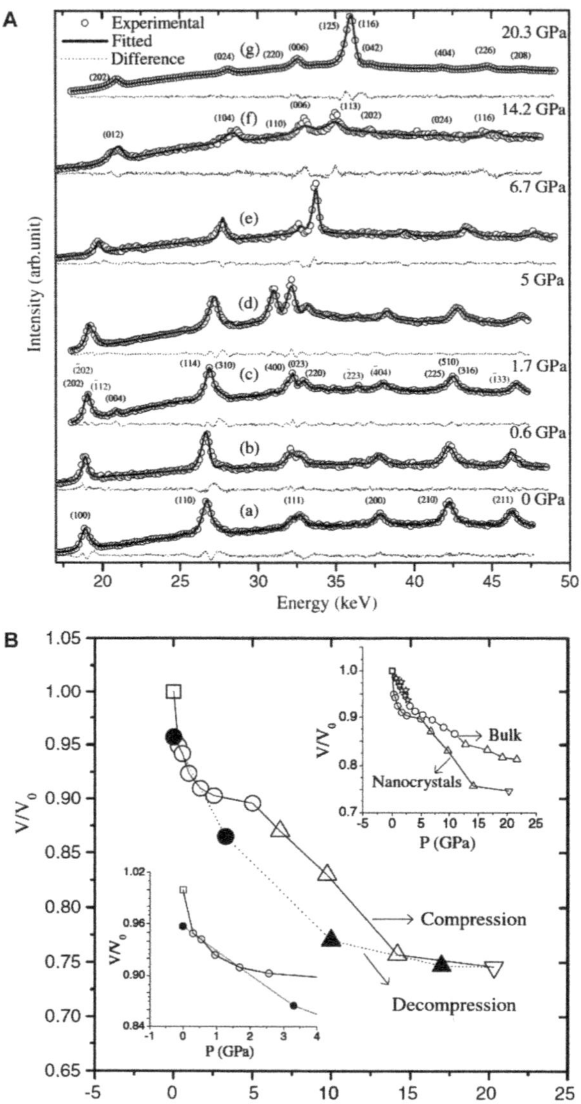

Fig. 6 (A) High-pressure energy dispersive experimental XRD patterns of the ~12 nm ReO_3 nanocrystals along with Rietveld fits and difference patterns at (a) 0.0, (b) 0.6, (c) 1.7, (d) 5, (e) 6.7, (f) 14.2 and (g) 20.3 GPa respectively in the compression process. (B) Relative volume *vs.* pressure curve for different phases of ReO_3 nanocrystals in the compression (filled symbols) and decompression (open symbols) experiments in the 0.0–20.3 GPa pressure range. Cubic I (squares), monoclinic (circles), rhombohedral I (triangles) and rhombohedral II (inverted triangles). The top inset compares the phase transition data of the nanocrystals under compression with the data on bulk ReO_3. Stars are used to represent the cubic II phase identified in the case of bulk sample. The bottom inset shows the magnified 0.0–4.0 GPa region. Reprinted with permission from Biswas *et al. J. Phys.: Condens. Matter*, 2007, **19**, 436214 © 2007 IOP Publishing.

pressures. The nanocrystals are found to be more compressible than bulk ReO_3. On decompression to ambient pressure, the structure does not revert back to the cubic I structure.

Small angle X-ray scattering

Small angle X-ray scattering (SAXS) enables the evaluation of the size of the small particles or the modulation of the electron density in the scale range from several nm to 100 nm, by analyzing the scattered intensity of X-rays in a small angular range from about 0.1 to 3°.[31,32] In contrast to XRD, SAXS can be applied not only to the crystalline materials but also to the amorphous materials, polymers and biomolecules. The SAXS intensity depends on the electron density (ρ), the structure factor [$S(q)$], and the form factor ($F(q)$) of the particle, where $q = 4\pi \sin\theta/\lambda$ is the scattering vector, θ is the scattering angle and λ is the wavelength of the X-ray.[31,32] The most basic application of SAXS is in the evaluation of the particle size, shape, orientation and size distribution in dilute dispersions, where the particle-particle correlation factor [$S(q)$] is negligible. In order to evaluate the particle size distribution, experimental SAXS data are generally fitted to a theoretical model.

SAXS can be used as an effective means to track the *in-situ* growth of nanocrystals in solution. The growth of gold nanocrystals prepared by the reduction of tetrachloroauric acid ($HAuCl_4$) by tetrakis(hydroxymethyl)phosphonium chloride which allows slow reduction, has been investigated by SAXS.[33,34] *In-situ* SAXS measurements on the gold hydrosols prepared with different concentrations of $HAuCl_4$, have been carried out over the 10–700 min range with an exposure time of 600 sec in order to obtain good signal-to-noise ratios. Figure 7(a) and (b) show intensity *vs.* scattering vector plots in the logarithmic scale for different times of growth in the case of gold hydrosols prepared respectively with 125 μL (0.006 mmol) and 375 μL (0.018 mmol) of 50 mM $HAuCl_4$ solutions. In order to estimate the average diameter and diameter distribution of gold nanocrystals, the experimental data has been fitted to the spherical form factor model. The scattering contrast for X-rays is given by the electron density difference between the particle and the solvent. Since gold has higher contrast than the solvent, only the gold particles were considered for the analysis. In the case of a diluted assembly of spherical particles, neglecting particle-particle interaction, the scattering intensity is given by equation 2,

$$I(q) \propto \int f(R)V(R)^2 P(q,R)dR \qquad (2)$$

where $V(R)$ and $P(q, R)$ are the volume and form factor respectively of a sphere of radius R. The form factor of the sphere is given by equation 3,[35]

$$P(q,R) = \left[\frac{3\{\sin(qR) - qR\cos(qR)\}}{(qR)^3}\right]^2 \qquad (3)$$

where, q is the scattering vector, and R the radius of the sphere. To obtain the particle size distribution, Gaussian distribution, $f(R)$ has been used (see equation 4),

$$f(R) = \frac{1}{\sigma\sqrt{2\pi}} e^{-\left[(R-R_0)^2/2\sigma^2\right]} \qquad (4)$$

Least square refinement yielded two parameters, R and σ where the latter is the standard deviation. The solid curves in the Fig. 7 are the

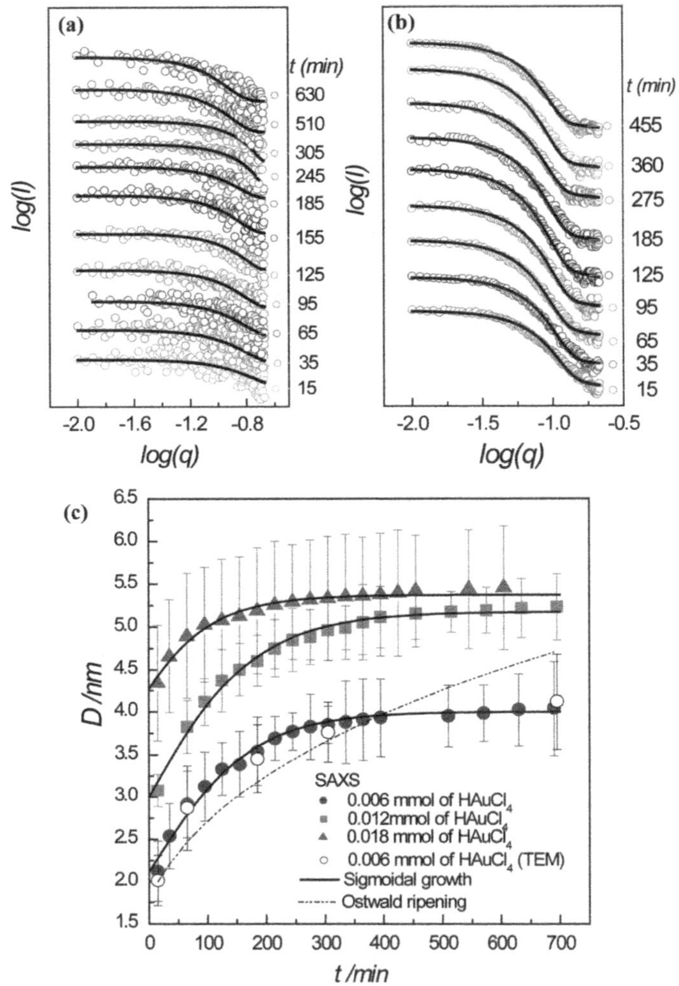

Fig. 7 SAXS data for the growth of gold nanocrystals prepared from the solution of (a) 125 μL (0.006 mmol), and (b) 375 μL (0.018 mmol) of 50 mM $HAuCl_4$ after different times of the reaction. Solid lines are the spherical model fits to the experimental data. (c) Time evolution of the average diameter (*D*) of the gold nanocrystals obtained from SAXS for different concentrations of $HAuCl_4$ (solid symbols). Solid curves represent the sigmoidal growth model fits to the experimental data. Dotted curve is the Ostwald ripening model fit to the experimental data. Reprinted with permission from Biswas *et al. Small*, 2008, **4**, 649 © 2008 Wiley-VCH Verlag GmbH & Co. K GaA, Weinheim.

sphere model fits of the experimental SAXS data. The fits with the experimental patterns yielded the average diameter and the diameter distributions for different growth times. Average diameter of gold nanocrystals prepared with three different concentrations of $HAuCl_4$ against time has been given in Fig. 7(c). The growth of the nanocrystals does not follow the diffusion-limited Ostwald ripening,[33,34] and instead follows a sigmoidal rate curve. The growth of capped CdSe and CdS nanocrystals formed by the reaction of selenium or sulphur with cadmium stearate in toluene solution in the presence of dodecanthiol or trioctylphosphine oxide and tetralin has also been investigated by using time a dependent SAXS study.[36] Growth of CdSe and CdS nanocrystals having contributions from both diffusion and surface reaction, with a D^3+D^2 type behavior, independent of the capping agent. Sarma and coworkers[37] report the growth kinetics of CdS and ZnS nanocrystals by small angle X-ray scattering (SAXS) which do not follow the Ostwald ripening model. These workers also report the growth kinetics of the ZnO nanocrystals in the absence of any capping agent to deviate from diffusion-controlled Ostwald ripening.[38,39]

Capping agent dependent growth kinetics of ZnO nanorods in solution by time dependent SAXS measurement has been investigated by Rao and co-workers.[40] Under the reaction conditions employed, the length of the nanorods varies significantly with time, while the diameter varies only slightly. Based on this observation,a cylindrical shape for the nanorods was assumed to analyze the SAXS data. A cylinder is a close approximation of a hexagonal rod. In order to estimate the average length and diameter of ZnO nanorods, experimental data were fitted to cylinder model. The form factor of the cylinder is that due to Fournet:[41,42]

$$P(q,R,L)=\int_0^{\pi/2}\left[\frac{2J_1(qR\sin\alpha)}{qR\sin\alpha}\frac{\sin(qL\cos\alpha/2)}{qL\cos\alpha/2}\right]^2\sin\alpha d\alpha \tag{5}$$

Here, q is the scattering vector, R the radius of the cylinder, L the length of the cylinder, α the orientation dependent parameter (angle between scattering vector and cylinder long axis) and $J_1(x)$ is the first order Bessel function of the first kind. Length and diameter distribution in the scattering cross section were not introduced in the equation because the form factor amplitude of a cylinder is a more complicated function of length, diameter and the orientation of the cylinder than the form factor amplitude of a spherical model where the diameter is the only variable. The growth of the nanorods mainly occurs by the diffusion of monomers from solution to the nanorod surface or by the reaction at the surface where units of the diffusing particles get assimilated into the growing nanorods. Diffusion and surface reaction are the two limiting cases in the growth of nanorods. In the absence of a capping agent, the nanorod growth is essentially controlled by diffusion, supported by the goodness of fit of $L(t)$ data to diffusion-limited growth model. Presence of capping agent gives rise to a diffusion barrier. As a result, the contribution of the surface reaction becomes more prominent. The growth of capped nanorods occurs through a combination of diffusion and surface reaction processes. There is no barrier present to limit the diffusion process in the case of uncapped nanorods.

In order to understand the particle-particle correlation (where $S(q)$ has major role) in nanoregime SAXS has recently been used extensively. In the case of formation of nanocrystalline gold films at the toluene–water interface, SAXS has been utilized to study the aggregation process.[43] The measurements provided direct evidence for the formation of 1.2 gold nm particles with a particle-particle distance of 2.3 nm. The fractal structure of these organic-capped nanoparticles was found to change as the reaction progresses. Nanoparticles can be combined with nucleic acids to programme the formation of three-dimensional colloidal crystals where the particles' size, shape, composition and position can be independently controlled.[44,45] SAXS has been used extensively to determine the crystal structure of these three-dimensional colloidal crystals (Fig. 8).[44]

Pair-distribution function analysis

In many of the materials, especially those with intermediate range order, structural coherence dies out on a nanometer scale, making it difficult to obtain structural solutions by standard crystallographic techniques. Significant advances have been made recently on the use of atomic pair distribution function (PDF) analysis of powder diffraction data coupled with the use of advanced high energy X-rays and fast computers.[46] The PDF technique makes use of the Fourier relationship between the diffraction intensities and the real-space arrangement of pairs of atoms.

The pair distribution function can be defined in real-space in terms of the atomic coordinates. Since the peaks in the PDF arise from pairs of atoms in the material, a peak shifted to a lower-r indicates that the pair of atoms has a shorter bond. PDF analysis of X-ray diffraction data which account for both the Bragg and the diffuse components of scattering are used to solve structures of nanomaterials, amorphous materials and gels which cannot otherwise be examined by traditional crystallographic techniques. PDF analysis gives local structural information on a material which cannot be obtained by crystallography.

In Fig. 9 shows the PDF data of fluorothiol-capped gold nanparticles of 4 nm diameter.[47] The PDF pattern is attenuated compared with bulk gold, due to the finite size of the nanocrystals. In the nearest region of the PDF (lower part of Fig. 9), we see two distinct peaks. The sharp intense peak at 2.8 Å is due to the shortest Au–Au distance and the peak at 2.4 Å is due to the Au–S (gold to capping thiol) distance. It is possible to obtain other structural information such as the lattice parameters based on model fits to the experimental data. PDF analysis can be used to determine the internal atomic ordering, and also to assess the geometry of nanocrystals. In particular, they provided very detailed structural information for γ-Fe_2O_3 spherical and tetrapod-shaped nanocrystals, which can be used to explain their growth mechanism as well as their magnetic properties.[48] PDF analysis on the $PbSnS_2$ nanocrystals have been performed in order to verify that their short and medium range structure was indeed cubic and not locally distorted as in orthorhombic $PbSnS_2$.[8] It has been pointed out previously that it is possible for the long-range structure suggested by PXRD to be different from the short-range local structure and PDF analysis is key to resolving such issues.[46,49] The experimental PDF plots and the corresponding fits of the cubic model of the $Pb_{2-x}Sn_xS_2$ nanocrystals prepared in 30, 60, and 120 s have been shown in Fig. 10. All three reaction conditions gave nanocrystals with the same cubic structure as indicated from the excellent agreement of the fits

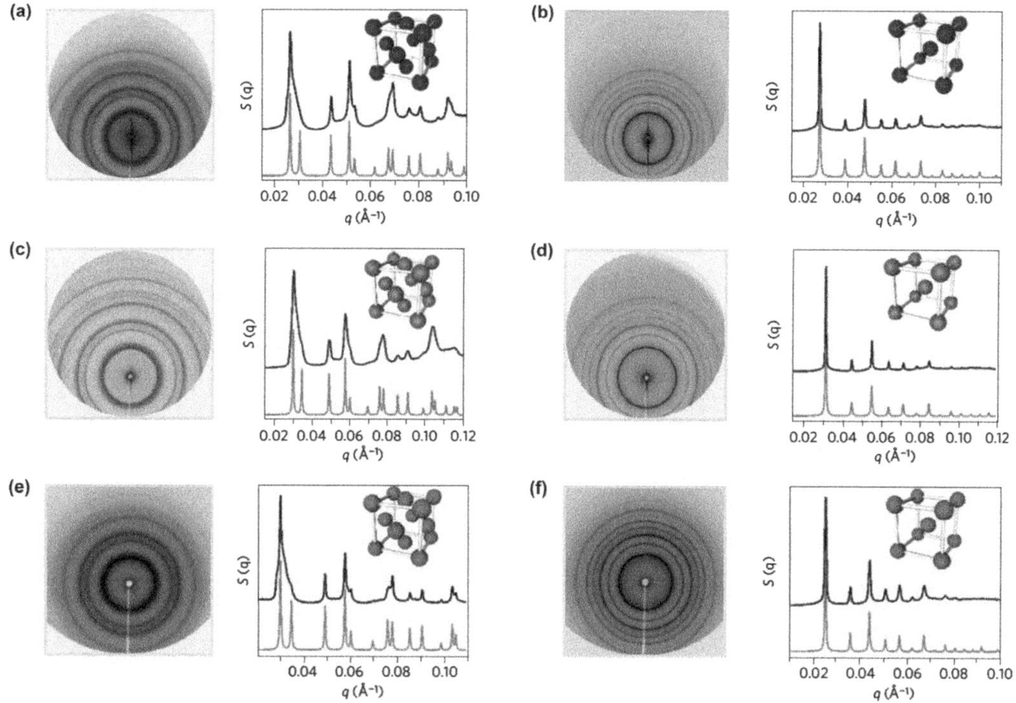

Fig. 8 a–f, 2D and radially averaged 1D SAXS patterns of: Quantum dot-programmable atom equivalent (PAEs) in (a) fcc and (b) bcc structure; dodecanthiol capped Au-PAEs in (c) fcc and (d) bcc arrangements; Fe_3O_4-PAEs in (e) fcc and (f) bcc arrangements. Experimental data are shown in upper panel, and predicted scattering patterns are shown in lower panel. For each pattern, a unit cell is shown of the corresponding crystal symmetry and particle composition. Reprinted with permission from Zhang *et al. Nature Mater.*, 2013 , **12** , 741. © 2013 Nature Publishing Group.

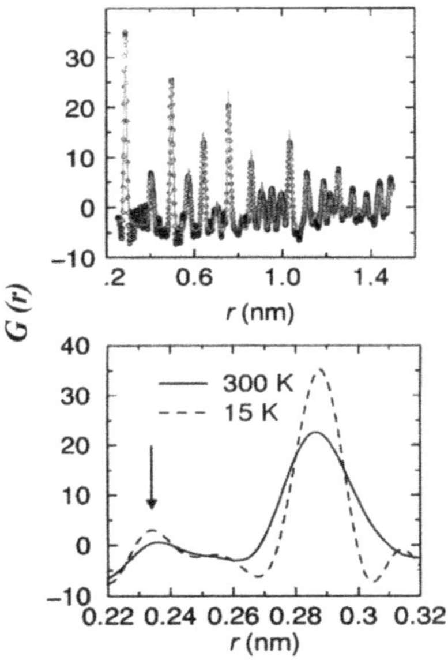

Fig. 9 (Upper) PDF of gold nanocrystals at 15 K. The filled circles represent the experimental data and the solid gray line the model PDF. (Lower) shows the nearest neighbor region of the PDF. A structural feature can be seen at $r = 0.235$ nm marked by arrow. Reprinted with permission from Page *et al. Chem. Phys. Lett.*, 2004, **393**, 385 © 2004 Elsevier.

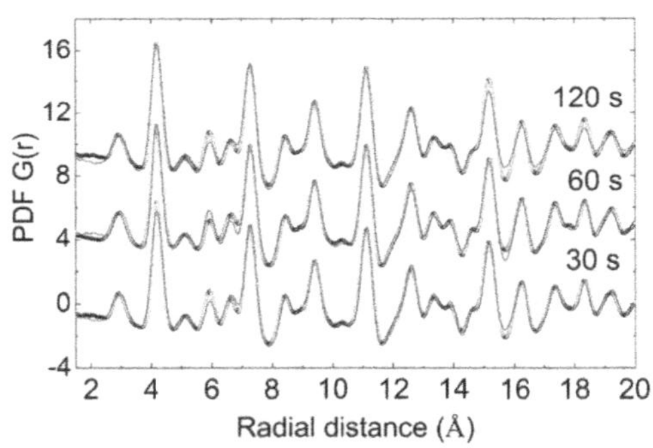

Fig. 10 PDF plots of the $Pb_{2-x}Sn_xS_2$ nanocrystals prepared in 30, 60, and 120 s. Experimental data are shown in open circles, and the corresponding fits of the cubic NaCl-type model are shown in solid lines. The fit of the cubic structure is in excellent agreement with the experimental data. Reprinted with permission from Soriano *et al. J. Am. Chem. Soc.*, 2012, **134**, 3228 © 2012 American Chemical Society.

(solid lines) with the experimental PDF data (open circles). The first observed interatomic vector in the experimental PDF at around 3 Å is characteristic of the M–S bond (M = Pb/Sn) in the rock salt structure while the corresponding M–S bonding distances in the distorted orthorhombic model are much shorter at ~2.61 and 2.66 Å (Fig. 2(a)). Local structure of $Pb_mSb_{2n}Te_{m+3n}$ nanocrystals have also been found by PDF analysis.[9]

MoS_2 is the key catalyst for the removal of sulfur from crude oil (hydrodesulfurization). Pristine MoS_2 is perfectly crystalline and consists of layers of Mo-S_6 trigonal prisms held together by van der Waals forces. $LiMoS_2$ has Li intercalated between the MoS_2 layers. It is important as a precursor of stable MoS_2 colloids used to prepare a variety of lamellar nanocomposites. Despite being extensively studied the structure of $LiMoS_2$ has not been determined. The reason is that, on Li intercalation, pristine MoS_2 is dramatically modified resulting in a product that is too poorly diffracting to allow a traditional structural determination. The approach of atomic PDF analysis have been used because of the lack of well-defined Bragg peaks due to the short structural coherence ~50 Å in this intercalation compound.[50] The reduction of Mo by Li results in Mo–Mo bonding with the formation of chains of distorted Mo-S_6 octahedra.

Aerogels are a unique class of porous materials that are largely composed of randomly interconnected nanoparticles. Aerogels possess low density and high internal surface area, which leads to potential applications such as catalysis, separations, sensing devices, and charge storage. Chalcogenide cluster-based aerogels, nicknamed chalcogels, are a new class of porous materials and have been made using a simple metathesis reaction between cationic transition-metal linkers and anionic chalcogenide clusters.[51] Understanding the structure by normal PXRD of is really difficult task due to absence of long rage ordering in this class of materials. PDF analysis of the total X-ray scattering pattern have been provided useful information of the local structure and building blocks (molecular fragments) such as the ZTS-cg materials.[52]

X-ray reflectivity

A toluene–water interface has been exploited to form two-dimensional aggregates of nanocrystals of metals, metal sulfides, and other materials at the interface by clever choice of interfacial chemistry.[53–55] The film of nanoparticles produced at the liquid-liquid interfaces is expected to be very thin because it is known that water "fingers" and organic "fingers" protrude into one another only by around 10 Å, lasting for tens of picoseconds, to facilitate interfacial chemical reactions. Sanyal and coworkers have studied the *in situ* formation and ordering of gold nanoparticles at the water–toluene interface using high-energy synchrotron X-ray reflectivity techniques.[56] Gold nanocrystals have been formed at the toluene–water interface by the reaction between a metal-organic compound triphenylphosphine gold chloride, $Au(PPh_3)Cl$ (P≡phosphorus, Ph≡phenyl), in the toluene layer and the reducing agent, tetrakishydroxymethylphosphonium chloride (THPC), in the aqueous layer. Six reflectivity curves collected at different time after the initiation of the reaction are shown in Fig. 11(a). Insets of Fig. 11 (a) and (b) show the reflectivity data and electron density profile of the controlled toluene–water interface, respectively. The oscillations in the measured reflectivity curves indicate the presence of a thin film at the toluene–water interface. These curves have been fitted using an iterative inversion technique based on the Born approximation.[56] The inversion scheme is based on recursive Fourier transform (FT) and inverse Fourier transform (FT^{-1}) of

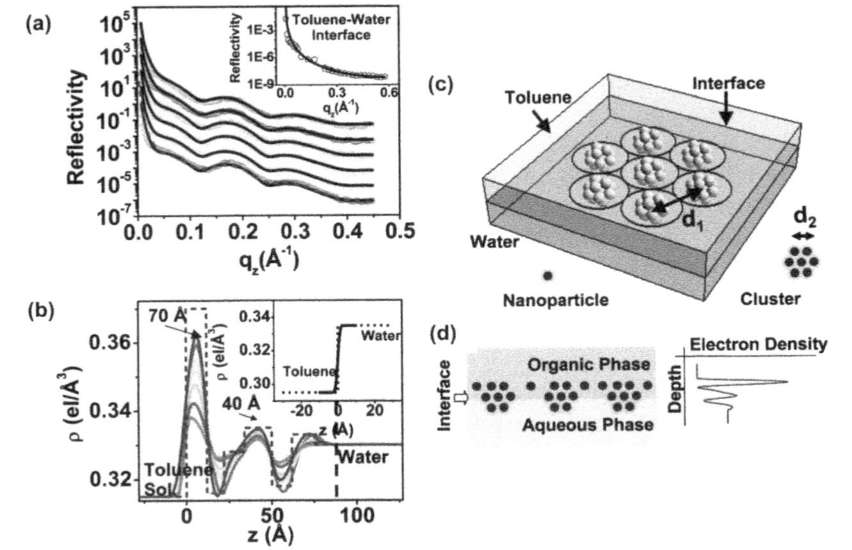

Fig. 11 (a) Variation of reflectivity and fits as function of q_z after initiation of the toluene–water-interface reaction (194 min; 224 min; 253 min; 283 min; 312 min and 364 min) and the fits (solid lines). The reflectivity (dashed line) generated from the simple box model is also shown. Inset: The reflectivity data from the toluene–water interface and its fit (solid line). (b) The electron density profiles (EDPs) as function of depth obtained from fitting. Positions of the two upper peaks as measured from the water interface (dashed line) 70 Å and 40 Å are marked. A simple model without (dashed line) and with roughness convolution (solid line) are also shown. (c) Three-dimensional schematics of the proposed model that involve a 13-member cluster of organic capped gold nanoparticles at the toluene–water interface are shown along with the two-dimensional schematics of an individual nanoparticle and a cluster. (e) Two-dimensional schematics for the same model is also shown along with the three-layered electron density profile as a function of depth obtained from the fitting of specular reflectivity data. Reprinted with permission from Sanyal *et al. J. Phys. Chem. C*, 2008, **112**, 1739. © 2008 American Chemical Society.

the equation relating the model electron density profile ($\rho_m(z)$) with actual electron density profile ($\rho_e(z)$) as

$$\rho_e(z) = FT^{-1}\left[\sqrt{\frac{R_e}{R_m}}\int \rho'_m(z)\exp(iq_z z)dz\right] \tag{6}$$

Here, R_e is the experimental reflectivity curve and R_m is the model reflectivity curve calculated from the model profile $\rho_m(z)$ by the standard slicing technique. In each iteration, $\rho_m(z)$ is replaced by $\rho_e(z)$ obtained from the above equation and the process was continued until R_e and R_m become indistinguishable. The water–toluene profile (with a small increase in toluene density due to presence of $Au(PPh_3)Cl$ to match the critical angle) have been used (inset of Fig. 11(a)) as the initial $\rho_m(z)$. The extracted final electron density profiles, for these six reflectivity data, which always take the value of water density toward the end of the film, are shown in Fig. 11(b). The dip in reflectivity data around 0.123 Å^{-1} and the subsequent modulation require a strong peak in electron density around 70 Å above the water surface. The width of this peak and the simple model shown in Fig. 11(b) suggest the size of the uppermost gold particle to be around 12 Å. There is a broad composite structure around the middle of the film (marked as 40 Å) and a small hump just before reaching the water density. These three layers have a vertical separation of about 30 Å from each other. The electron density between these layers takes the value ($\sim$0.32 electrons/Å^3) of a typical organic material. The electron density of the uppermost layers decreases, and the peak electron densities of all of the peaks tend to become equal as time progresses, but the separation of the layers remain nearly constant. The profiles obtained in Fig. 11(b) with continuous reduction of electron density of the uppermost layer have been explained assuming coexistence of a monolayer of individual nanoparticles having around a 12 Å gold core, with an 11 Å organic shell, and a layer of a 13-member "magic cluster" of these nanoparticles. With the progress of reaction, the number of individual nanoparticles is reduced and only the monolayer of the "magic cluster" remains at the toluene–water interface.

The authors have concluded by an *in-situ* X-ray reflectivity study that a monolayer of clusters having 13 gold nanoparticles of 12 Å diameter, with large (180 Å) in-plane cluster–cluster separation, forms at the toluene–water interface (Fig. 11(c)). The electron density of such aligned clusters exhibited three layers along the depth with the central electron density values of 0.33, 0.37, and 0.33 electrons/Å.^{3} Reflectivity measurements confirmed the presence of three layers with the lower two layers showing slightly less than the calculated values probably due to partial coverage. The presence of low electron densities in between these peak values confirms the monodispersity of these aligned clusters. The higher electron density of the top peak indicates the presence of individual gold nanoparticles with these clusters (Fig. 11(d)).

X-ray reflectivity has also been used to study the formation and growth of single crystalline CuS film at organic aqueous interface.[57] Formation of CuS crystals at the organic-aqueous interface occurs through self-assembly of nanocrystallites, which were formed at the interface within a few minutes of reaction. *In situ* x-ray reflectivity measurements showed that the thickness of the composite nanofilm was around 27 nm over the entire reaction time, involving several hours, and the thickness of CuS rich layer was around 6 nm.

X-ray photoelectron spectroscopy

X-ray photoelectron spectroscopy (XPS) is a quantitative technique that measures the elemental composition, empirical formula, chemical state and electronic state of the elements that exist within a nanomaterial.[58] XPS spectra were obtained by irradiating a given material with a beam of Al or Mg X-rays while simultaneously measuring the kinetic energy and number of electrons that escape from the top 1 to 10 nm of the material. XPS requires ultra-high vacuum. For example, XPS analysis of B-C-N nanotubes indicates the presence of B and N in the carbon nanotubes and gives the composition to be $BC_{28}N$.[59] Photoelectron spectroscopy is subdivided in two parts, depending on the exciting radiation. XPS using soft X-rays (200–2000 eV) to examine core levels (inner-shell electrons) while UV photoelectron spectroscopy (UPS) employs UV radiation to investigate valance levels (outer-shell electrons). Thus, the valence bands and changes in the Fermi level of materials are effectively studied by UPS.

XPS has been used extensively to determine the doping element and content in the nanostructured materials. For example, nitrogen and boron doping in graphene have been identified by XPS.[60] Doping amount were 1.4% and 3.4% for nitrogen and boron in graphene respectively. X-ray photoelectron spectra of nanocrystalline $AgBiS_2$ and $AgBiS_{1.92}Se_{0.08}$ confirms the presence of Se in solid solution nanocrystals (Fig. 12). Two strong peaks at 367.8 and 373.7 eV (Fig. 12(a)) with a peak splitting of 5.9 eV were observed due to Ag $3d_{5/2}$ and Ag $3d_{3/2}$ levels, which is consistent with the standard Ag(I). The peaks at 158.3 and 163.7 eV correspond to Bi $4f_{7/2}$ and Bi $4f_{5/2}$ (Fig. 12(b)). The binding energy of S 2s was located at 225.3 eV (Fig. 12(c)). Peaks at 54.1 and 57.1 eV assigned to Se $3d_{5/2}$ and Se $3d_{3/2}$ levels (Fig. 12(d)), respectively, suggest the incorporation of Se in the S sublattice of $AgBiS_2$.Variable photon energy XPS have been used to understand the internal hetero structure of luminescent ZnS/CdSe/ZnS quantum-dot-quantum-well structures.[61] X-ray photoelectron spectroscopy can provide layer-resolved information on the internal structure of highly complex heterostructured nanocrystals because of the technique's extremely short depth of probing that matches the dimensions of such nanocrystals. This short depth, characterized by the inelastic mean free path of photoejected electrons, can be varied by changing the photon energy over the wide range available from a

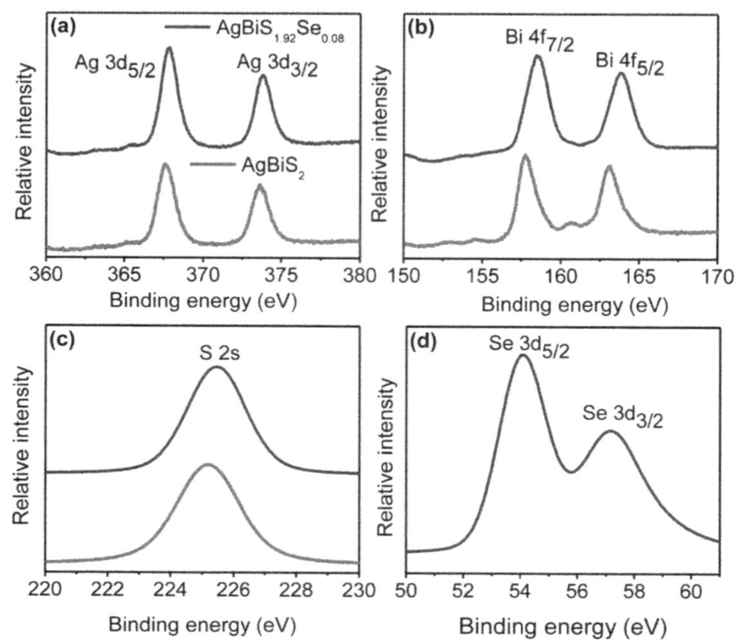

Fig. 12 XPS spectra of as prepared $AgBiS_2$ and $AgBiS_{1.92}Se_{0.8}$ nanocrystals. (a) Ag 3d, (b) Bi 4f, (c) S 2s, and (d) Se 3d spectra. Reprinted with permission from Guin *et al. Chem. Mater.*, 2013, **25**, 3225 © 2013 American Chemical Society.

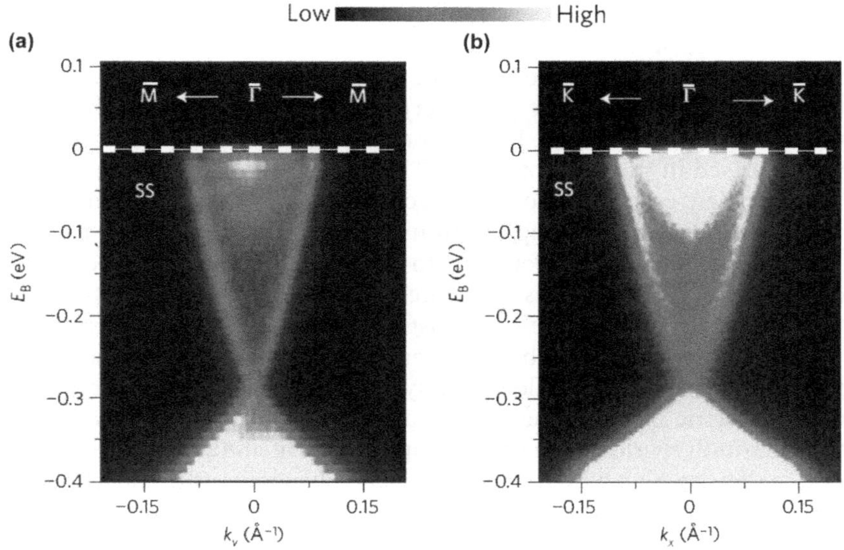

Fig. 13 High-resolution ARPES measurements of surface electronic band dispersion on Bi_2Se_3(111). Electron dispersion data measured with an incident photon energy of 22 eV near the Γ-point along the (a) Γ–M and (b) Γ–K momentum-space cuts. Reprinted with permission from Xia *et al. Nature Phys.*, 2009, **5**, 398. © 2009 Nature Publishing Group.

high-brilliance synchrotron source, thereby making quantitative analysis possible. The internal structure of the composite nanocrystals was shown to consist of a graded alloy core whose composition gradually changes from ZnS at the very center, to CdSe at the onset of a CdSe layer. The outer shell is ZnS with a sharp interface.

Topological insulators are spin Hall quantum phases with metallic surface states, embedded within an insulating band gap. They resist electron back-scattering even at defects, offering low dissipation charge transport at the surface.[62,63] These exotic surface states were experimentally realized in Bi_2Se_3 and Bi_2Te_3 in the form of a gapless single Dirac cone.[64,65] The first predicted material was a Bi_xSb_{1-x} alloy, whose surface electronic band structure was directly confirmed by angle-resolved photoemission spectroscopy (ARPES). In ARPES, electrons in the crystal are ejected by high-energy photons, whose energy and momentum are measured and used to obtain the electronic band structure. A key advance was the prediction and experimental confirmation of topological insulators in layered binary chalcogenides, including Bi_2Se_3, Bi_2Te_3 and Sb_2Te_3 (Fig. 13); their helical spin texture was later verified by spin-resolved ARPES. In this figure, Γ-point is the centre of the Brillouin zone and M, K are high symmetry points on the zone boundary.

Energy dispersive X-ray analysis

Energy dispersive X-ray spectroscopy (EDS or EDAX) is used mainly for elemental analysis or chemical characterization of samples.[66] The technique is based on the fact that every element has a unique atomic structure and that X-rays characteristic of different atomic structures are readily distinguished from one another. The incident electron beam excites an electron in an inner shell, causing its ejection and in the formation of an electron hole in the electronic structure of atom. An electron from a higher-energy (outer) shell fills the hole, and the difference in energy between the higher-energy shell and the lower energy shell gets released in the form of X-rays. The X-rays so released is analyzed by means of an energy dispersive spectrometer. EDAX systems are commonly found with scanning electron microscope (SEM) as well as TEM. For example, EDAX has been used to obtain elemental analysis of Au–Ag alloy nanocrystalline films generated at the liquid–liquid interface.[67] Endotaxially arranged SrTe nanocrystals at only 2 mol% concentration in the PbTe matrix doped with Na_2Te very effectively inhibit the heat flow in this system while the hole mobility is immune, allowing for a large thermoelectric figure of merit to be achieved.[21] Although it was difficult to quantitatively determine the compositions of individual precipitates due to their overlap with the matrix, energy dispersion x-ray spectroscopy (EDAX) indicated an increase in the Sr signal from the precipitates (dark areas in the scanning transmission electron microscopy (STEM) image) compared to the matrix regions (Fig. 14). Similarly, presence of MgTe,

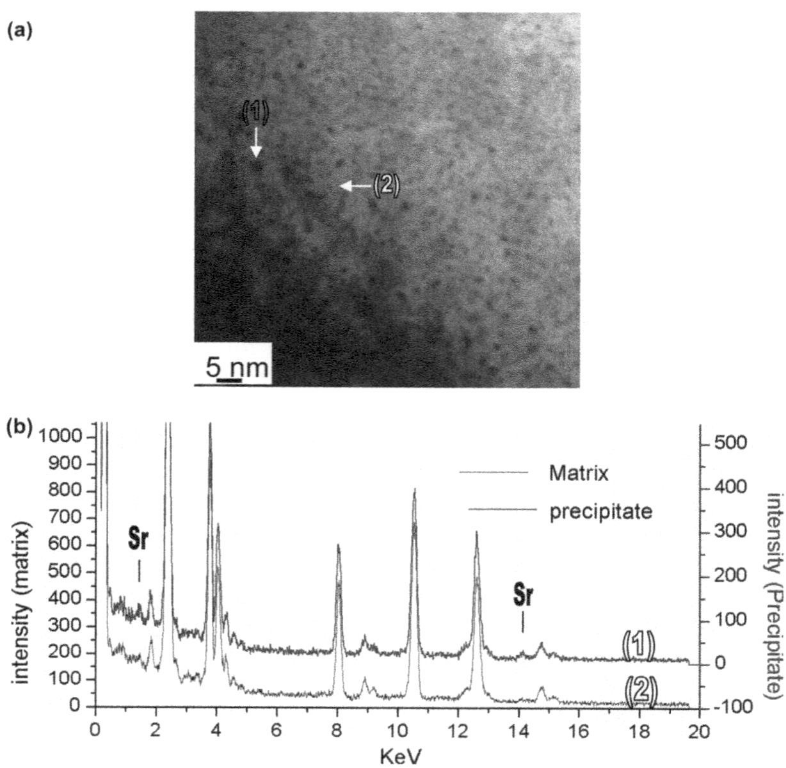

Fig. 14 (a) High resolution scanning transmission electron microscopy (STEM) images of PbTe-SrTe (2%) doped with 1% Na_2Te sample. (b) Comparison of EDAX between the precipitate and matrix. Reprinted with permission from Biswas *et al. Nature Chem.*, 2011, **3**, 160. © 2011 Nature Publishing Group.

CaTe, BaTe, CdTe and HgTe nanoprecipitates were confirmed by EDAX analysis.[68–71]

Conclusions and future outlook

X-ray related techniques have the versatility to determine not only the crystal structure but also local bonding environment and electronic structure of nanomaterials. PXRD is the essential tool to study the average crystal structure and phase transition behaviour of inorganic nanomaterials. Detailed local structural information of nanomaterials can be obtained by PDF analysis. Mention must be made that PDF analysis can be used not only on crystalline nanomaterials but also on amorphous materails such as chalcogenide glasses, gels and fibers. SAXS enables the evaluation of the size and shape of nanoparticles of both crystalline and amorphous inorganic materials. Long range particle-particle correlation can also be obtained by SAXS investigation. X-ray reflectivity technique can determine the thickness and electron density profile of nanocrystalline films. XPS is really a powerful technique to determine the elemental composition, empirical formula, chemical state and electronic state of the elements that exist within a nanomaterial. ARPES have been used extensively to determine the electronic band structure of inorganic materials. There are several challenges in developing better X-ray techniques for the characterization of nanomaterials. Improved resolution of X-ray spectroscopic techniques can provide valuable insights into the properties and phenomena in nanomaterials. Aspects of nanomateials such as self-assembly and reactions at nanoscale can be studied by improved X-ray techniques. *In-situ* studies are essential to understand the dynamics of chemical reaction in nanoscale. Furthermore, combined and logical used of various techniques are necessary to understand crystal and electronic structure and dynamical behaviour in nanoscale regime.

Acknowledgement

KB greatly appreciates the support of a DST Ramanujan Fellowship, the New Chemistry Unit and Sheikh Saqr Laboratory in JNCASR. I apologize in advance to all the investigators whose research could not be cited owing to the requirement of brevity.

References

1 E. N. Kaufmann ed. *Characterization of materials*, Vol. 1, 2 . New Jersey: Wiley-Interscience, 2003.
2 M. Hosokawa, K. Nogi, N. Naito and T. Yokoyama, eds. *Nanocrystal Technology Handbook*, 1st edn, Amsterdam: Elsevier. 2007. pp. 269–312.
3 H. M. Rietveld, *J. Appl. Crystallogr.*, 1969, **2**, 65.
4 C. N. R. Rao and B. Raveau, *Transition Metal Oxides*, Weinheim: Wiley-VCH, 2nd edn, 1995.
5 K. Biswas and C. N. R. Rao, *J. Phys. Chem. B*, 2006, **110**, 842.
6 W. S. Seo, J. H. Shim, S. J. Oh, E. K. Lee, N. H. Hur and J. T. Park, *J. Am. Chem. Soc.*, 2005, **127**, 6188.

7 J. Park, K. An, Y. Hwang, J. G. Park, H. J. Noh, J. Y. Kim, J. H. Park, N. M. Hwang and T. Hyeon, *Nature Mater.*, 2004, **3**, 891.
8 R. B. Soriano, C. D. Malliakas, J. Wu and M. G. Kanatzidis, *J. Am. Chem. Soc.*, 2012, **134**, 3228.
9 R. B. Soriano, I. U. Arachchinge, C. D. Malliakas, J. Wu and M. G. Kanatzidis, *J. Am. Chem. Soc.*, 2013, **135**, 768.
10 S. N. Guin and K. Biswas, *Chem. Mater.*, 2013, **25**, 3225.
11 I. T. Sines, R. Misra, P. Schiffer and R. E. Schaak, *Angew. Chem. Int. Ed.*, 2010, **49**, 4638.
12 M. E. Norako, M. J. Greaney and R. L. Brutchey, *J. Am. Chem. Soc.*, 2012, **134**, 23.
13 D. Pan, L. An, Z. Sun, W. Hou, Y. Yang, Z. Yang and Y. Lu, *J.Am. Chem. Soc.*, 2008, **130**, 5620.
14 K. Senevirathne, R. Tackett, P. R. Kharel, G. Lawes, K. Somaskandan and S. L. Brock, *ACS Nano*, 2009, **3**, 1129.
15 M. Ghosh, K. Biswas, A. Sundaresan and C. N. R. Rao, *J. Mater. Chem.*, 2006, **16**, 106.
16 M. Ghosh, E. V. Sampathkumaran and C. N. R. Rao, *Chem. Mater.*, 2005, **17**, 2348.
17 K. Biswas and C. N. R. Rao, *Chem. Eur. J.*, 2007, **13**, 6123.
18 K. Biswas, K. Sardar and C. N. R. Rao, *Appl. Phys. Lett.*, 2006, **89**, 132503.
19 H. S. S. R. Matte, A. Gomathi, A. K. Manna, D. J. Late, R. Datta, S. K. Pati and C. N. R. Rao, *Angew. Chem. Int. Ed.*, 2010, **49**, 4059.
20 M. K. Jana, K. Biswas and C. N. R. Rao, *Chem. Eur. J.*, 2013, **19**, 9110.
21 K. Biswas, J. He, Q. Zhang, G. Wang, C. Uher, V. P. Dravid and M. G. Kanatzidis, *Nature Chem.*, 2011, **3**, 160.
22 K. Biswas, J. He, I. D. Blum, C. I. Wu, T. P. Hogan, D. N. Seidman, V. P. Dravid and M. G. Kanatzidis, *Nature*, 2012, **489**, 414.
23 J. Androulakis, C. H. Lin, H. J. Kong, C. Uher, C. I. Wu, T. Hogan, B. A. Cook, T. Caillat, M. Paraskevopoulos and M. G. Kanatzidis, *J. Am. Chem. Soc.*, 2007, **129**, 9780.
24 S. N. Girard, J. He, C. Li, S. Moses, G. Wang, C. Uher, V. P. Dravid and M. G. Kanatzidis, *Nano Lett.*, 2010, **10**, 2825.
25 S. N. Girard, K. S. Rohr, T. C. Chasapis, E. Hatzikraniotis, B. Njegic, E. M. Levine, A. Rawal, K. M. Paraskevopoulos and M. G. Kanatzidis, *Adv. Func. Mater*, 2013, **23**, 747.
26 C. Xiao, X. Qin, J. Zhang, R. An, J. Xu, K. Li, B. Cao, J. Yang, B. Ye and Y. Xie, *J. Am. Chem. Soc.*, 2012, **134**, 18460.
27 C. N. R. Rao, *Acc. Chem. Res.*, 1984, **17**, 83.
28 C. Xiao, J. Xu, K. Li, J. Peng, J. L. Yang and Y. Xie, *J. Am. Chem. Soc.*, 2012, **134**, 4287.
29 S. H. Tolbert and A. P. Alivisatos, *Science*, 1994, **265**, 373.
30 K. Biswas, D. V. S. Muthu, A. K. Sood, M. B. Kruger, B. Chen and C. N. R. Rao, *J. Phys.: Condens. Matter*, 2007, **19**, 436214.
31 J. S. Pedersen, *Adv. Colloid Interface Sci.*, 1997, **70**, 171.
32 J. S. Pedersen, In *Neutrons, X-rays and Light: Scattering Methods Applied to Soft Condensed Matter*, ed. P. Lindner, Th. Zemb, 2002. pp. 391–420. North-Holland: New York.
33 K. Biswas, N. Varghese and C. N. R. Rao, *Small*, 2008, **4**, 649.
34 K. Biswas, N. Varghese and C. N. R. Rao, *J. Mater. Sci. Technol.*, 2008, **24**, 615.
35 L. Rayleigh, *Proc. Roy. Soc. London, Ser. A*, 1911, **84**, 25.
36 N. Varghese, K. Biswas and C. N. R. Rao, *Chem. Asian J.*, 2008, **3**, 1435.
37 R. Viswanatha, S Sapra, H. Amenitsch, B Sartori and D. D. Sarma, *J. Nanosci. Nanotech.*, 2007, **7**, 1726.

38 R. Viswanatha, H. Amenitsch and D. D. Sarma, *J. Am. Chem. Soc.*, 2007, **129**, 4470–4475.
39 R. Viswanatha, P. K. Santra, C. Dasgupta and D. D. Sarma, *Phys. Rev. Lett.*, 2007, **98**, 255501.
40 K. Biswas, B. Das and C. N. R. Rao, *J. Phys. Chem. B*, 2008, **112**, 2404.
41 G. Fournet, *Bull. Soc. Franc. Minéral. Crist.*, 1951, **74**, 39.
42 J. S. Pedersen, *J. Appl. Crystallogr.*, 2000, **33**, 637.
43 M. K. Bera, M. K. Sanyal, L. Yang, K. Biswas, A. Gibaud and C. N. R. Rao, *Phys. Rev. B*, 2010, **81**, 115415.
44 R. J. Macfarlane, B. Lee, M. R. Jones, N. Harris, G. C. Schatz and C. A. Mirkin, *Science*, 2011, **334**, 204.
45 C. Zhang, R. J. Macfarlane, K. L. Young, C. H. J. Choi, L. Hao, E. Auyeung, G. Liu, X. Zhou and C. A. Mirkin, *Nature Mater.*, 2013, **12**, 741.
46 S. J. L. Billinge and M. G. Kanatzidis, *Chem. Commun*, 2004, 749.
47 K. Page, Th. Proffen, T. Humberto, M. Terrones, L. Lee, Y. Yang, S. Stemmer, R. Seshadri and A. K. Cheetham, *Chem. Phys. Lett.*, 2004, **393**, 385.
48 V. Petkov, D. Cozzoli, R. Buonsanti, R. Cingolani and Y. Ren, *J. Am. Chem. Soc.*, 2009, **131**, 14264.
49 A. S. Masadeh, E. S. Bozin, C. L. Farrow, G. Paglia, P. Juhas, S. J. L. Billinge, A. Karkamkar and M. G. Kanatzidis, *Phys. Rev. B*, 2007, **76**, 115413.
50 V. Petkov, S. J. L. Billinge, P. Larson, S. D. Mahanti, T. Vogt, K. K. Rangan and M. G. Kanatzidis, *Phys. Rev. B*, 2002, **65**, 92105.
51 S. Bag, P. N Trikalitis, P. J. Chupas, G. S. Armatas and M. G. Kanatzidis, *Science*, 2007, **317**, 490.
52 Y. Oh, S. Bag, C. D. Malliakas and M. G. Kanatzidis, *Chem. Mater.*, 2011, **23**, 2447.
53 C. N. R. Rao, G. U. Kulkarni, P. J. Thomas, V. V. Agrawal and P. Saravanan, *J. Phys. Chem. B*, 2003, **107**, 7391.
54 C. N. R. Rao and K. P. Kalyanikutty, *Acc. Chem. Res.*, 2008, **41**, 489.
55 P. J. Thomas, E. Mbufu and P. O'Brien, *Chem. Commun.*, 2013, **49**, 118.
56 M. K. Sanyal, V. V. Agrawal, M. K. Bera, K. P. Kalyanikutty, J. Daillant, C. Blot, S. Kubowicz, O. Konovalov and C. N. R. Rao, *J. Phys. Chem. C*, 2008, **112**, 1739.
57 S. Maiti, M. K. Sanyal, N. Varghese, B. Satpati, D. Dasgupta, J. Daillant, D. Carriere, O. Konovolov and C. N. R. Rao, *J. Phys.: Condens. Matter.*, 2013, **25**, 395401.
58 C. N. R. Rao and K. Biswas, *Annu. Rev. Anal. Chem.*, 2009, **2**, 435.
59 R. Sen, B. C. Satishkumar, A. Govindaraj, K. R. Harikumar, G. Raina, J.-P. Zhang, A. K. Cheetham and C. N. R. Rao, *Chem. Phys. Lett.*, 1998, **287**, 671.
60 L. S. Panchakarla, K. S. Subrahmanyam, S. K. Saha, Achutharao Govindaraj, H. R. Krishnamurthy, U. V. Waghmare and C. N. R. Rao, *Adv. Mater.*, 2009, **21**, 4726.
61 P. K. Santra, R. Viswanatha, S. M. Daniels, N. L. Pickett, J. M. Smith, P. O'Brien and D. D. Sarma, *J. Am. Chem. Soc.*, 2009, **131**, 470.
62 D. Kong and Y. Cui, *Nature Chem.*, 2011, **3**, 845–849.
63 D. Hsieh, D. Qian, L. Wray, Y. Xia, Y. S. Hor, R. J. Cava and M. Z. Hasan, *Nature*, 2008, **452**, 970.
64 Y. Xia, D. Qian, D. Hsieh, L. Wray, A. Pal, H. Lin, A. Bansil, D. Grauer, Y. S. Hor, R. J. Cava and M. Z. Hasan, *Nature Phys.*, 2009, **5**, 398.
65 Y. L. Chen, J. G. Analytis, J.-H. Chu, Z. K. Liu, S.-K. Mo, X. L. Qi, H. J. Zhang, D. H. Lu, X. Dai, Z. Fang, S. C. Zhang, I. R. Fisher, Z. Hussain and Z.-X. Shen, *Science*, 2009, **325**, 178.
66 J. Goldstein, D. Newbury, D. Joy, C. Lyman, P. Echlin, Lifshin, L. Sawyer and J. Michael, *Scanning Electron Microscopy and X-ray Microanalysis*, 3rd edn, New York: Springer, 2003.

67 V. V. Agrawal, P. Mahalakshmi, G. U. Kulkarni and C. N. R. Rao, *Langmuir*, 2006, **22**, 1846.
68 S. H. Lo, J. Q. He, K. Biswas, M. G. Kanatzidis and V. P. Dravid, *Adv. Func. Mater.*, 2012, **22**, 5175.
69 K. Biswas, J. He, G. Wang, C. Uher, V. P. Dravid and M. G. Kanatzidis, *Energy Environ. Sci.*, 2011, **4**, 4675.
70 M. Ohta, K. Biswas, S.-H. Lo, J. He, D.-Y. Chung, V. P. Dravid and M. G. Kanatzidis, *Adv. Energy Mater*, 2012, **2**, 1117.
71 K. Ahn, K. Biswas, J. He, I. Chung, V. P. Dravid and M. G. Kanatzidis, *Energy Environ. Sci.*, 2013, **6**, 1529.

Raman measurements of stress in films and coatings

Dong Liu[a] and Peter E. J. Flewitt*[b]

DOI: 10.1039/9781782621485-00141

Films and coatings have become widely used in structures and components to protect the underlying material from mechanical degradation, corrosion, oxidation and high temperatures or improve surface properties. Residual stress generated in these multi-layer systems is one of the main causes of coating delamination and eventual failure. Systematic measurement and monitoring of the residual stresses are a vital basis for integrity evaluation and remaining life prediction. Raman spectroscopy has been recognised as one of the most important approaches to measure the stress in films and coatings. This review considers the measurement of stresses in films and coatings using Raman spectroscopy. It addresses the following questions: what is Raman spectroscopy, why is stress important for films and coatings, how is strain/stress derived from Raman spectra and what confidence do we have in this technique and the limitations. To elucidate specific issues related to the application of the Raman technique for stress measurement, despite the wide range of coatings available, important films and coating are chosen as representative examples.

1 Introduction

Raman spectroscopy is a powerful non-destructive analytical technique with applications that involve either gas, liquid and solid phase materials. It is widely used for the characterisation of stresses in bulk systems as well as thin films and coatings, single crystals and polycrystalline materials such as poly-diamond and poly-Si. It is based on the Raman effect discovered by C. V. Raman in 1928 where there is interaction between monochromatic light and the chemical bonds within the material. The spectra produced may be used to identify the characteristic energies of chemical bonds to distinguish between the phases within the same material *e.g.* diamond and graphite. However, it is also possible to obtain information about the strain from the shift of frequency in the spectra caused by a change in chemical bonding under either internal stress or externally applied load. It has been extensively used in solid state physics for the detection of stresses in semi-conductor materials such as silicon where stress state can limit the working condition of the devices. Advantages of the method are that it is non-destructive and can reach a nanometre scale resolution, 100 to 600 nm for example.[1–3] To achieve realistic measurements generally little or no preparation of the samples is necessary. In section 2, we will explore more fully the derivation of strains and stresses within films and coatings on substrates.

[a]*Interface Analysis Centre, School of Physics, University of Bristol, Bristol, BS8 1TL, UK*

[b]*HH Wills Physics Laboratory, School of Physics, University of Bristol, Bristol, BS8 1TL, UK. E-mail: Peter.Flewitt@bristol.ac.uk*

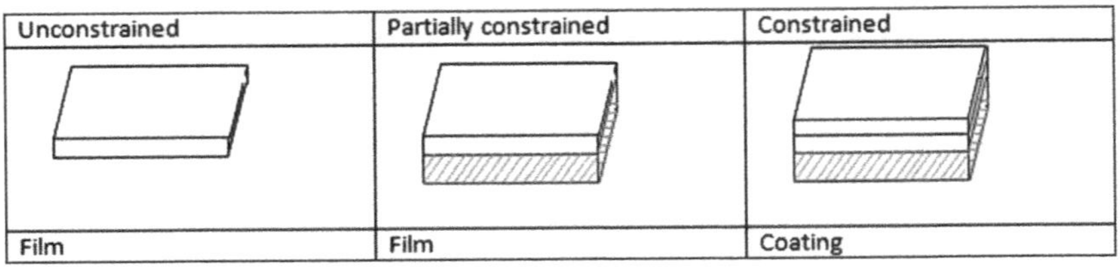

Fig. 1 Definition of films and coatings in terms of geometry and amount of constraint.

It is important to have a classification to describe the difference between a solid thin film and a coating. This is best achieved by considering the geometrical configuration and the constraint on deformation imposed by the surroundings.[4] In this context as shown in Fig. 1, a film or coating will be small in one dimension compared with the other two. However, the degree of constraint on deformation is unconfined if the boundaries associated with the thin dimension are free to displace. Full constraint is present when the thin dimension is confined against deformation. This is usually a result of the presence of another material which shares a boundary at a common interface and is usually the case for a coating. Intermediate between these conditions is when the boundaries are not constrained in all these dimensions; partial constraint. Within this chapter we consider mainly partially and fully constrained conditions. For consistency we have associated coatings with multi-layer systems on a substrate. This compares to films which are either free standing or a single layer on a substrate.

Residual stress is the stress that remains in a material or body without any external influence such as a force, a displacement or a thermal gradient, and it is in equilibrium with its surroundings.[5] It is this residual stress that usually has been measured in films and coatings. The presence of a residual stress can be very detrimental to the performance of a material or the life of a component. However, there are cases where beneficial residual stresses can be introduced deliberately to enhance performance. In general, residual stresses are more difficult to predict than externally applied stresses on which they superimpose.[6] With respect to the scale of residual stress, this has been divided into three categories.[7] Type I, where σ_{I} refers to macro-scale stress which varies within the material over a length-scale much larger than the grain size, Fig. 2. Type II and Type III are meso- and micro-scale residual stress. Type II are where σ_{II} changes on the scale of a grain, Fig. 2, which is normally caused by the anisotropy of each grain within a single phase material or by the difference in properties of a multi-phase body. Type III stresses are where σ_{III} describes the variation of stress within a grain usually as a result of dislocations and other crystalline defects. Techniques used for residual stress measurement in films and coatings include material removal (hole drilling, layer removal), mechanical methods (curvature, displacement measurements), diffraction (neutron, X-ray) and spectroscopy (Raman and photo-stimulated luminescence spectroscopy (PLPS)) and so on. The measurement resolution, requirements for specimen preparation and types of residual stresses acquired by each

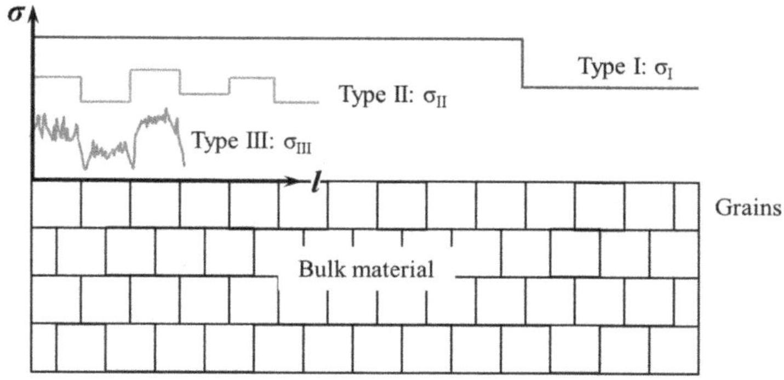

Fig. 2 Three types of residual stresses, Type I, II and III with respect to the length scale of variation, *l*. The squares provide a simple representation of grains in a polycrystalline material.

technique are set out in Appendix I. Among these Raman spectroscopy has the best spatial resolution, is straightforward to make a measurement and requires little preparation of the specimens.

Residual stresses usually accompany the manufacture of films and coatings due to thermal mismatch differences between the materials and the substrate, especially when there is cooling to room temperature from a higher coating temperature involved (*e.g.* 700–900 °C for deposition used for conventional diamond films[8]). A basic linear elastic approximation of the stress, $\sigma_{f/c}$, can be expressed by:

$$\sigma_{f/c} = E(T - T_0)(\alpha_s - \alpha_{f/c}) = E\Delta T\Delta\alpha \tag{1}$$

where $\alpha_{f/c}$ and E are the coefficient of thermal expansion (CTE) and elastic modulus of the film or coating; T is the final temperature and T_0 is the initial temperature; α_s is the coefficient of expansion of the substrate. This equation applies only to simple elastic conditions for thin layers. For thicker layers stress gradients and relaxation in the vicinity of the coating/substrate interface result in an over-estimation of the stress levels if this equation is used.[9,10] Moreover, taking into account the dependence of the CTE and E on temperature makes the equation more complex. Hence a better way of calculating the stress would be finite element (FE) modelling. In practice, when the film or coating is on a curved substrate the stress is more complex especially when non-linearity occurs.[11,12]

It has been recognised that the manufacturing process introduces additional 'intrinsic' stress into a structure. These residual stresses can be compressive, for example in the case of chemical vapour deposited (CVD) diamond films onto a silicon substrate,[2] or tensile, such as the case of air plasma sprayed yttria stabilised zirconia (APS-YSZ) thermal barrier coatings (TBC) applied to Ni-based superalloys.[13,14] The other sources of the residual stress, including the evolution of residual stress in the films and coatings during service, arise from the changes of microstructure or the formation of new oxide layers. For example sintering in thermal barrier coatings leads to stiffening in the ceramic layer and the formation of thermally grown oxide (TGO) threatens the stability of the interface.[15] Other examples are where lattice mismatch, bonding at grain boundaries and bonding changes during growth in CVD films cause segregation, relaxation and generation of residual stresses.[16]

When the residual stress is small, it can sometimes improve the mechanical behaviour of the film/coatings, for example, a 1 GPa compressive stress has been found to improve the effective fracture toughness of diamond films[17] and if strain is applied in Si micro-electronic devices the electron mobility is increased.[18,19] Stress modifies photons in a crystal, *i.e.* a tensile stress usually causes mode softening whilst a compressive stress results in the opposite.[19] High compressive residual stresses, especially in brittle coatings on a ductile substrate, can lead to failure by buckling; if the residual stresses are tensile through-thickness cracking of, and segregation within the film or coating tends to occur.[12,20–22] In addition, interfacial adhesion plays an important role in the integrity of the entire coating system, but residual stresses are still important in controlling degradation.[14,20,23,24] Therefore, measurement of the stress in films or a coating system is key to the integrity evaluation. Raman spectroscopy offers a non-destructive option for evaluating stresses as part of the quality control during manufacture or monitoring during the service life.

In this review, relationships between Raman spectroscopy and strain/stress are discussed in section 2. Representative examples of the application of Raman spectroscopy are given in section 3 to elucidate the particular issues that occur when making practical measurements.

2 Raman spectroscopy: stress measurement

2.1 Relationship between strain/stress and Raman shift

The vibrations of a crystal are described not in terms of the vibrations of individual atoms but in terms of collective motions in the form of waves, called lattice vibrations.[25] It is these lattice vibrations with certain types of symmetry that give rise to Raman scattering; such vibrations are said to be Raman active.[26–28] The measurement of the Raman spectrum of a crystal is one of the main methods for obtaining information about its lattice vibration frequencies. Incident light of angular frequency ω_{in} can interact with the crystal to create or destroy one of more lattice vibration quanta (photons) and the energy ΔE gained or lost by the lattice is compensated by a decrease or increase in the frequency ω_s of the scattered light ($\omega_s = \omega_{in} + \Delta E/hc$; h is the Planck constant).[28] With the presence of a strain/stress in a crystal, the lattice vibration changes and this results in a shift of the acquired Raman spectrum.

Prior to the description of the relationship between strain/stress and Raman shift, we provide a brief introduction to the general relationship between stress and strain. When an external load is applied to a body, the response can be presented as a stress-strain curve. On loading a body to a stress, σ_A, in Fig. 3 the material responds elastically so that on removing the load the strain, ε_A, reduces to zero. However, if loaded to a stress of σ_B, the material responds non-linearly above σ_A. Usually this non-linear response is associated with the on-set of plasticity and therefore on

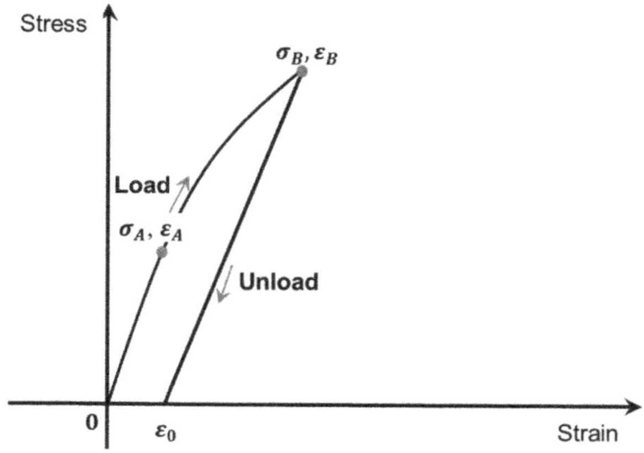

Fig. 3 A typical stress-strain curve showing the elastic and plastic response of a material.

removal of the applied stress a permanent strain, ε_0, remains. There can be other reasons for the non-linear response of a material above the σ_A applied stress. For example the materials have quasi-brittle characteristic.[29] In general, as a stress is applied in the range zero to σ_A there is a corresponding change in the lattice spacing within a crystalline body. Beyond σ_A, no further change of lattice spacing occurs but the lattice vibration still responds with further loading. Hence, for diffraction techniques such as X-ray diffraction and neutron diffraction, which are based on the measure of the change of lattice spacing in crystalline materials, they can only measure strains in the elastic region. For Raman spectroscopy, however, elastic strain as well as plastic strain can be measured *via* the change of lattice vibration.[30]

Loudon[28] summaries the Raman tensors for each of the 32 crystal classes (symmetry point groups). In the case of silicon for example, there are three Raman tensors and these three modes are optionally visible for different polarization directions of optical phonons.[25] In the absence of strain, the corresponding three optical Raman modes, $\omega_i, \omega_j, \omega_k$, in silicon have the same frequency of about 520 cm^{-1}. As described earlier in this section, local strain changes the frequencies of Raman modes. One of the early papers addressing the relationships between strain (linear strain) and the Raman modes is by Ganesan *et al.*[31] It was proposed that the frequencies of the three optical modes with the presence of strain can be derived by solving:[3,31]

$$\begin{vmatrix} p\varepsilon_{11} + q(\varepsilon_{22} + \varepsilon_{33}) - \lambda & 2r\varepsilon_{12} & 2r\varepsilon_{13} \\ 2r\varepsilon_{12} & p\varepsilon_{22} + q(\varepsilon_{33} + \varepsilon_{11}) - \lambda & 2r\varepsilon_{23} \\ 2r\varepsilon_{13} & 2r\varepsilon_{23} & p\varepsilon_{33} + q(\varepsilon_{11} + \varepsilon_{22}) - \lambda \end{vmatrix} = 0 \tag{2}$$

where p, q, r are the photon deformation potential constants; $\varepsilon_{ij}(i, j = 1, 2, 3)$ are the strain components. Hence the change of frequency of each mode, $\Delta\omega_j(j = 1, 2, 3)$, from strain-free condition, ω_{j0}, to strained condition, ω_j, can be calculated from the eigenvalues:

$$\Delta\omega_j = \omega_j - \omega_{j0} \approx \frac{\lambda_j}{2\omega_{j0}} \quad (3)$$

where

$$\lambda_1 = p\varepsilon_{11} + q(\varepsilon_{22} + \varepsilon_{33}) \quad (4.a)$$

$$\lambda_2 = p\varepsilon_{22} + q(\varepsilon_{33} + \varepsilon_{11}) \quad (4.b)$$

$$\lambda_3 = p\varepsilon_{33} + q(\varepsilon_{11} + \varepsilon_{22}) \quad (4.c)$$

Therefore,

$$\Delta\omega_1 = [p\varepsilon_{11} + q(\varepsilon_{22} + \varepsilon_{33})]/2\omega_{10} \quad (5.a)$$

$$\Delta\omega_2 = [p\varepsilon_{22} + q(\varepsilon_{33} + \varepsilon_{11})]/2\omega_{20} \quad (5.b)$$

$$\Delta\omega_3 = [p\varepsilon_{33} + q(\varepsilon_{11} + \varepsilon_{22})]/2\omega_{30} \quad (5.c)$$

For a given linear-elastic material, to convert the strain components into stress, according to Hook's law the elastic compliance tensor elements, $S_{ij}(i, j = 1, 2, 3)$, have to be determined and applied. For example when a uniaxial strain is applied to a body, $\varepsilon_{11} = S_{11}\sigma$, $\varepsilon_{22} = S_{22}\sigma$ and $\varepsilon_{33} = S_{33}$-$\sigma$; For silicon, using $S_{11} = 7.68 \times 10^{-12} Pa^{-1}$, $S_{22} = 2.14 \times 10^{-12} Pa^{-1}$, $S_{33} = 12.7 \times 10^{-12} Pa^{-1}$ and $p = -1.43\omega_0^2$, $q = -1.89\omega_0^2$ and $r = -0.59\omega_0^2$, equation 5.c gives[3,25]

$$\Delta\omega_3 = -2 \times 10^{-9}\sigma \quad (6)$$

This complies with the rule that a compressive stress results in an increase in Raman frequency whereas a tensile stress leads to a decrease in this frequency.

However, this analysis applies to simplified conditions where linear-elastic strain/stress is applied in a known direction with known material constants. For those materials with crystal structures that are more complex than silicon, the photon deformation potential constants may not be available and the strain distribution is normally non-uniform in the probed volume. Especially for the residual stresses in coatings and films, the distribution of stress is normally very difficult to determine. Most importantly, for a multi-phase material, the conversion factor from strain to stress is not straightforward. Hence, workers recently suggested that for engineering materials a calibration has to be undertaken to provide the required confidence in the conversion from Raman peak shift to stress.[13,32]

2.2 The conversion factor

Initially, workers applied hydrostatic/uniaxial stresses to crystalline materials such as diamond,[31,33] silicon[3] and quartz[34] to measure the evolution of Raman peaks, with the aim of improving the understanding of the origin of the optical spectra. These researches addressed both the electronic energy band structure and the lattice vibrational levels.[3,33,35] For example, in stressed silicon, splitting and shifts of the peaks in the silicon Raman spectra were identified. The effects of strain and stress on the Raman frequency of the optical phonons of silicon are well known and extensively documented.[3,36,37] The first experimental work on determination of the effects of static uniaxial stress on the frequency of the $q \approx 0$ optical phonons in a diamond-type crystal (Ge, GaAs, GaSb, InAs) was carried out by Anastassakis *et al.* in 1970.[3] In their work the phonon deformation potentials which describe the changes in the 'force constant' of these phonons with strain were determined by applying strain in the $\langle 001 \rangle$ and $\langle 111 \rangle$ crystallographic directions.

In addition, stress measurements using Raman spectroscopy on a material with known or unknown crystallographic orientations have been undertaken. Indeed, this is a reverse of the approach based on the established photon deformation potential constants. To gather strain/stress information a factor converting the measured Raman spectra, typically shifts of wavenumber and line width, has to be determined. There is not a common fixed parameter/relationship between the strain/stress and the Raman characteristics; it has to be calibrated case by case. The photon deformation potential constants of crystalline diamond have been reported by Grimsditch *et al.*[38] and are well documented for Si.[36,39,40] The precision in the derivation of lattice constants, uncertainty in the mode Grüneisen parameters[19,41–43] and the fact that Raman modes depend significantly on the polarization configurations are all factors that can hinder the precise measure of strain/stress. Therefore, it is necessary to obtain the full set of quantitative photon deformation potential values before the strain/stress tensors can be resolved from the Raman spectra.

For silicon, for example, ways of deriving the stress tensors from the Raman spectra have been proposed for both single crystals with known orientations and polycrystalline material with arbitrary orientations.[1,44] For simple cases, such as a uniaxial stress applied along the $\langle 110 \rangle$ direction of a silicon crystal, Wolf *et al.*[37] polarised the incident laser light in the crystal $\langle 110 \rangle$ direction and observed the Raman signal in the back-scattering geometry in the $\langle 001 \rangle$ direction. This gives

$$\Delta\nu(\mathrm{cm}^{-1}) = -1.93\times10^{-9}\left(\frac{\mathrm{cm}^{-1}}{Pa}\right)\sigma(\mathrm{Pa}) \qquad (7)$$

Similar relationships can be derived for hydrostatically and biaxially applied stresses which are characterised by one stress value. For polycrystalline materials where a large number of grains are variously oriented, an additional analysis of the Raman peak intensity is required. This process is considered to be able to provide information on the presence of any preferred orientation of the grains because the intensity of the spectrum depends on the polarization direction of the incident and scattered laser light. In general, a Raman frequency larger than the stress free frequency indicates the presence of a compressive stress in the specimen, while a Raman frequency smaller than for the stress-free condition indicates a tensile stress.[37,44–46]

When there is more than one unknown stress tensor, σ_{11},σ_{22},σ_{12}, a secular equation has been proposed which can be solved by recording spectra with polarised light in the back-scattering geometry and with the incident beam tilted by a known angle.[42,47] However, in the general case it is not possible to deduce all stress components from the Raman spectra. Atkinson *et al.*[44] and Wolf *et al.*[37] both proposed use of numerical modelling to predict the expected spectrum from an assumed or calculated stress distribution and then compare these spectra with those measured. This approach involves the transformation of crystal axes to the macro-scale specimen axes which introduce over simple assumptions. Alternatively, combining Raman spectroscopy with other techniques such as X-ray diffraction can provide more information about the unknown stress states in the materials.[48,49]

Most of the important fundamental work on photon deformation potential constant measurement and calculations for stress evaluation and the analysis of Raman spectra on diamond films were undertaken by pioneers in the 20th century.[50,51] Approaches such as X-ray diffraction (normal or low angle X-ray diffraction), neutron diffraction and substrate curvature are normally employed to assist the interpretation and validation of the stress measurements derived from Raman spectroscopy. In the present decade, CVD films doped/sintered with other elements such as B, Co *etc.* have become more popular and the focus for stress measurement has shifted to application. A typical example being the determination of stresses within polycrystalline diamond (PCD) coatings on cutting tools.[52] These applications usually adopt the photon deformation potential values previously established.[53] Combined with other techniques such as digital image correlation (DIC),[54] focus ion beam (FIB)[54] and atomic force microscopy (AFM) and so on the stress measurement using Raman spectroscopy has become more versatile in application.[53]

2.3 Limitations and confidence in measurements

In this section, we now consider how strain and stress can be measured within films and coatings on substrates. Derivation of strain/stress in practical components is further complicated by the unknown stress state and varying properties of the materials. Also, it is important to understand the confidence with which measurements can be made and the sources of potential errors which can arise from difficulties in determining the sampling volume and influence of temperature on the spectra.

There are various Raman instruments developed according to different needs. That most commonly used to undertake stress measurements is micro-Raman. In conventional applications of micro-Raman spectroscopy the incident laser beam is a few micrometres in diameter. Since the laser is focussed through a set of lens onto the surface of the test material, identified by an optical microscope, the upper limit of the resolution (spot size of the laser beam on the specimen surface) is $\sim$1 to 2 μm due to the diffraction limit of the geometrical optical system.[55,56]

The resolution of the technique can be improved by up to λ/20, where λ is the incident radiation wavelength, for example a resolution of 25 nm can be reached using a 488 nm wavelength radiation. This can be achieved by combining near-field optical technology with Raman spectroscopy.[56–59] Nevertheless, this technique requires the specimen to have a high-quality surface finish because the distance between the lens and the surface is normally of nanometre scale or, at best, up to several micrometres.[56] Mapping stresses on polished cross-sections has been achieved using a step size of 200 nm over an rectangular area of 2.5 μm × 5 μm.[57]

It has to be noted that prior to using the Raman spectrometer for stress measurement, calibration of the system and the stress conversion factor to relate the strain/stress and frequency change for a particular material needs to be undertaken. Local heating is another potential factor that could contribute to spectrum shift. Hence when the acquisition time is long a sufficient cooling system has to be adopted.[60] Apart from the physical resolution of the system, the actual sampled volume is influenced significantly by the scattering of the laser beam within translucent materials. These factors and their influence on the strain/stress derived from Raman spectra will be discussed in this section.

2.3.1 Sampled volume. As discussed in section 2, Raman spectroscopy is becoming more widely used to study stress distributions in films and coatings because of its potential for high spatial resolution. Typically the diameter of a incident focused laser beam is 1 μm but the penetration depth can vary from tens of nanometres to several millimetres and thus the incident beam samples a potentially large volume of the strained material.[44] Each point within this volume scatters light with a Raman wavenumber characteristic of the local stress. Because the beam is usually focused, the distribution of light intensity in this volume and the interpretation of the resulting spectrum are complicated. Atkinson *et al.*[44] proposed that the shift of Raman spectra can be caused by strains/stresses in different directions and individual stress tensors for a given stress state. Hence, one resultant Raman shift has the potential to be interpreted in various ways: A shift towards lower frequency from the stress-free state can be caused by a uniaxial compressive stress or a hydrostatic compressive stress or a combined larger compressive stress acting in one direction and a lower tensile stress in another.[37,61] Therefore, as pointed out by Wolf *et al.*[37] a preliminary knowledge of the stress state in the measured material assists the interpretation of the Raman data. If there is a stress variation in the sampled volume, the Raman shift measured is a combination of all contributions. It is possible to deconvolute the stress tensors in films with known or assumed random crystallographic orientation, such as SiC, Si and diamond, however, it is less likely for more complex materials and coatings.[44]

The sampling depth for Raman spectroscopy depends upon the specific microstructure and optical properties of the material. Higher resolution measurements can be expected for more opaque materials such as silicon[62,63] or when using ultraviolet laser radiation.[63] Even though the depth resolution can be improved by the use of a confocal Raman system combined with oil-immersed objective lenses,[64,65] the depth resolution in translucent materials can be reduced significantly due to refraction[66] and spherical aberration[67] within the out-of-focus region of the laser beam.[68] The lateral spreading of the laser beam is directly related to the spatial resolution for measurements. In bulk material, the size of the cone-like excitation volume depends on the numerical aperture of the objective lens and the difference in refractive index between the medium in which the objective is immersed and the material being investigated.[66,69,70] For materials such as air plasma sprayed yttria stabilized tetragonal zirconia (YSZ), which is typically applied on turbine blades as a thermal barrier coating, the lateral resolution can be further reduced by the scattering of the laser beam due to the presence of pores and unmelted particles.[71] In addition, it has been shown that ray guiding[72] can also occur when the incident beam strikes features such as the internal splat boundaries at an angle greater than the critical refraction angle.[73,74] This dispersion of the beam has to be established experimentally because there are too many unquantified variables within the overall microstructure[75] to allow realistic, quantitative theoretical calculations to be made. Although not exactly within the scope this current review, it is worth mentioning that the penetration/sampling depth for Raman spectroscopy is significant when measuring the thickness of multi-layered polymer laminates.[65,76] Confocal Raman spectroscopy is normally adopted in such cases, but due to the absorption of the laser beam by the material, the interfaces between two different materials are usually blurred and shifted so that it is essential to consider the sampled volume. To summarise, knowledge of spatial resolution is crucial when either undertaking stress measurements or characterising the microstructure of a material using the Raman spectroscopy.

2.3.2 Local heating. When an incident laser beam interacts with a material there is the potential for a local rise in temperature. Silicon is the standard frequently adopted to calibrate the peaks in Raman spectroscopy. It is well-documented[25,60] that tensile stresses in silicon result in lower frequency (due to a softening of the phonon frequency) and line broadening (due to a reduction in phonon decay time); whereas compressive stresses result in higher frequency (due to a stiffening of the phonon frequency) but do not produce line broadening.[60,77] When a highly focused laser beam causes local heating of a material this results in Raman peak shifts towards a lower frequency[78] that can be confused with a stress-induced shift. The Raman signal originates mostly in the wings of the beam profile where the temperature is substantially lower (by a factor of ten) than at the centre.[60] Combined with a local compressive stress imposed by the cooler surrounding substrate material this shifts the Raman peak to a higher frequency.[79] Obviously, there is a balance between the opposing residual stresses generated by these two sources. Local heating effects result in a correspondingly greater compressive stress as laser spot size is decreased.[37,60,80] When using a 632.80 nm wavelength incident laser, the position of the Raman peak, for stress-free silicon is approximately 520 cm^{-1} and it is possible to detect shifts of $\sim 0.05\,cm^{-1}$ ($\sim$25 MPa for silicon) if the spectrometer is sensitive and stable.[37] Therefore, in a series of consistent measurements, it is appropriate to employ the same lens setting and the same acquisition time. It is important to optimise the measurement conditions to minimise and in many cases eliminate local heating contributions.

3 Applications

Raman spectroscopy has been, and is applied to the measurement of strain and stress in a wide range of films and coatings. These span thermally and chemically grown oxides, sulphides *etc.* through to dedicated surface layers to improve the engineering properties of electronic devices and large machine components. In this section, we select two specific examples to highlight the benefits of Raman spectroscopy for making stress and strain measurements in films and coatings. The first relates ubiquitous carbon including diamond films and graphene, the second to an engineering application on which many of us depend, thermal barrier coatings used for aero engines and land based gas turbine components.

Films and coatings such as chemical vapour deposition (CVD) diamond films, graphite coatings, the more recently discovered 2D graphene, amorphous carbon and diamond-like carbon (DLC) coatings are now used extensively. The various carbon bonding configurations give rise to specific structural arrangements, so that sp-bonding is associated with chain structures (*e.g.*, carbynes), sp^2-bonding with planar structures (*e.g.*, graphite), and sp^3-bonding with tetrahedral structures (*e.g.*, diamond).[81] Application of Raman spectroscopy in these materials becomes vital because it provides an ability to distinguish each allotrope of carbon. Indeed, most of the published papers on these films and coatings invoke the use of Raman spectroscopy.[18]

The group IV element, Si, is a fundamental conventional semiconductor material widely used for microelectronic devices. Silicon has a single sharp Raman peak at $\sim 520\,cm^{-1}$, and this peak is usually used to calibrate a Raman system prior to undertaking stress measurements.[60] Raman spectroscopy has been studied extensively for the measurement of stress on silicon films because of the potential to achieve high spatial resolution. A knowledge of the stresses in the silicon films is a vital part of the non-destructive inspection to establish the integrity of films and coating applied to micro-electronic devices. The non-destructive measurement of these stresses by Raman spectroscopy provides a powerful tool for this industry. But since silicon and diamond both have crystals with cubic symmetry and/or polycrystalline structures, the stress derivation procedure is broadly the same. In this review, to avoid repeatability between diamond films and silicon films, we have selected diamond as a representative example to discuss the application of Raman spectroscopy for stress measurement.

In addition, zirconias have been extensively investigated due to their superior properties of low thermal conductivity, high dielectric constant, high optical refractive index, chemical durability, mechanical strength and fracture toughness.[82] One of the most important applications for zirconia is as a thermal barrier coating (TBC) for components working at high temperatures. Typically, a thin layer (~200 to ~300 μm) of yttria stabilized tetragonal zirconia (YSZ) is applied to the blades used in the high pressure stage turbines for aero engines and land-based power generators. The thickness varies with position on the blade surface. The low thermal conductivity of the layer allows coated components to operate at higher temperatures than the underlying parent metal can normally withstand. Therefore it is crucial for the layers to remain attached to the components for an appreciable length of time and to be tolerant to residual stresses and thermal cycling encountered during service.[15,83] These thermal barrier coatings are multi-layer consisting of a top layer of YSZ, a metallic bond coat (BC) that provides bonding between the metallic superalloy substrate and YSZ. During service a thermally grown oxide (TGO) layer can form between the TBC and BC as a result of the thermal exposure.[84] The growth of this oxide[15] and the sintering[85] of the YSZ layer during thermal aging are the main sources of residual stresses that can lead to the eventual failure of the coating system. The extensive use of the YSZ layer as a thermal barrier coating (TBC) has given rise to much research focused on how these coating systems degrade and what are the roles of the spraying/deposition methods during manufacturing,[86] the microstructure of the coatings,[87,88] and the residual stress distributions in the YSZ and TGO layers[23,89,90] on the final failure of these protective coatings. Hence, in this review, YSZ applied as a TBC is selected since it is a representative of many complex coating materials.

3.1 Thin films

3.1.1 Chemical vapour deposited diamond films and Raman spectra. Diamond is the archetypal covalent material. Each atom in a sp^3 configuration is bonded to four nearest neighbours. Many technological and scientific applications of diamond arise from its unique properties including optical transparency in the ultraviolet to infrared range, electrical insulation, thermodynamic stability, and unsurpassed strength and hardness.[91] It has been recognised as the hardest material which is very useful for cutting device tools, loading and other mechanical devices. But the high hardness makes diamond difficult to machine into a desired shape and the rareness of the natural material limits the maximum size and the application. Through-out the 20th century, several methods of growing diamond/diamond films synthetically were developed. One of the most important films is produced by chemical vapour deposition (CVD). Depending on the conditions selected, the films are single crystal (if grown epitaxially on a diamond substrate) or consist of micro- or nano- diamond crystallites bound together by amorphous carbon grain boundaries. These films maintain the merits of high hardness, high elastic modulus, and high thermal conductivity (1800 W/mK, because heat transfer is *via* lattice vibrations instead of electrons), a low coefficient of friction and broad-band optical transparency. CVD films are commonly used in cutting tools (cutting, grinding and polishing) and micro-fabrication of semiconductor devices.

The optimization of diamond films as valuable engineering materials for a wide variety of applications has required the development of robust methods for their characterization. Amongst the many methods used, Raman spectroscopy is perhaps the most extensively adopted to investigate the phonon modes of the different allotropes and non-crystalline phases of carbon because it provides readily distinguishable signatures of each carbon allotrope[92] (*e.g.* single crystal diamond at 1332 cm^{-1}; G peak at 1575 cm^{-1} for single crystal graphite and D peak at 1355 cm^{-1} for disordered graphite; severely broadened G and D peaks between 1500 and 1550 cm^{-1} in amorphous carbon). In addition to spatially resolved maps of the different forms of carbon, stresses can be mapped within a specimen.[64–66]

Natural diamond shows a first-order symmetric peak (it is triply degenerate TO(X) phonon of F_{2g} symmetry) at 1332 cm^{-1} in Fig. 4 (full-width-half-maximum is 1.2 cm^{-1}). In an isotropic diamond crystal they consist of one longitudinal mode (singlet) relative to the direction of propagation and two degenerate transverse (doublet) modes. In CVD films, however, non-diamond components, commonly in the form of amorphous carbon, often exist and produce a broad peak at 1500 to 1550 cm^{-1}. The shift of the first order peak of diamond has been extensively studied and the change of this peak with applied stress and temperature is well known.[43,81,93] Second-order and third-order peaks are shown in Fig. 4, but the stress measurements use the first-order peak shifts. The peaks are often fitted with a combined Gaussian and Lorentzian function and a sloping background. High compressive stresses can be generated in the CVD film especially when deposited on a stiff substrate. Ager *et al.*[94] found that a residual in-plane compressive stress

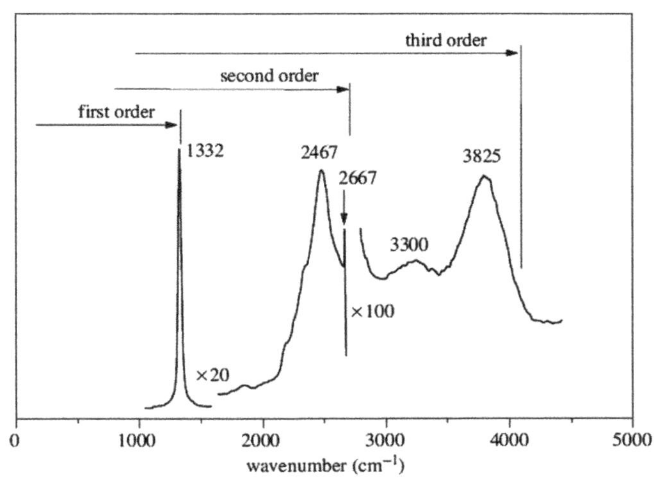

Fig. 4 A general Raman spectrum of gem-quality diamond excited at room temperature at a wavelength of 228.9 nm. The first-, second- and third-order Raman peaks are shown (Bormett *et al.*,[50] Zaitsev[43] and Prawer *et al.*[2])

of ~ 7.1 GPa is produced by the deposition of a 1 μm thick weakly oriented polycrystalline diamond film to an common aircraft alloy, Ti-6Al-4V, substrate. The magnitude of stress can reach 17 GPa which can then lead to buckling failure.

3.1.2 The relationship between stress and Raman spectrum.

Under hydrostatic pressure, Occelli *et al.*[95] revealed that in single crystal diamond the 1332 cm^{-1} peak shifts to 1650 cm^{-1} at 140 GPa with the covalent bond only slightly stretched so that the Raman peak width increased by 2 cm^{-1}. The change of peak position with hydrostatic pressure, P (Pa), is expressed by:

$$\nu = \nu_0 + 2.83P - 3.65P^2 \tag{8}$$

where $\nu(cm^{-1})$ is the shift caused by hydrostatic pressure, and $\nu_0(cm^{-1})$ is the stress-free frequency of the diamond peak, ~1332 cm^{-1}. The factor 2.83 in equation (8) has the unit of $\left(\frac{cm^{-1}}{Pa}\right)$ and −3.65 has the unit of $\left(\frac{cm^{-1}}{Pa^2}\right)$. Equation (8) suggests that the ratio of the force constants for bond 'stretching' and 'bending' vary with applied pressure and there is an increased bond strength with increase in pressure which is not commonly encountered in other materials.[95,96] Diamond is very stable under hydrostatic pressure. But large deformations and phase transformations of diamond have been observed under a non-hydrostatic pressure of around 150 GPa.[91] Indeed Gogotsi *et al.*[97] demonstrated that a transition to disordered graphite is possible under uni-axial compression applied by Vickers pyramidal indentation. The changes of 1332 cm^{-1} peak position in single crystal diamond with applied low hydrostatic pressure (less than 40 GPa) have been measured, and a conversion factor of 2.58 $GPa^{-1}\,cm^{-1}$ derived.[98] It is worth noting that the constant 2.83 in front of P in Equation (8) is consistent with the 2.58 $GPa^{-1}\,cm^{-1}$ derived at lower pressures.

For thin CVD films, several micrometres or even nanometers[100] thick compared with the millimetre scale substrate, the use of the hydrostatic factor is inappropriate because an in-plane stress state exists and the out-of-plane stress component is zero. This conclusion has been drawn by comparing the stress measured by Raman spectroscopy with other approaches, such as the substrate curvature method.[101,102] An in-plane equi-biaxial stress can be considered to comprise a 2/3 hydrostatic stress and a 1/3 shear stress. Therefore, a 2/3 correction for the gauge factor has been proposed

$$\begin{pmatrix} \sigma & 0 & 0 \\ 0 & \sigma & 0 \\ 0 & 0 & 0 \end{pmatrix} = \frac{2}{3}\begin{pmatrix} \sigma & 0 & 0 \\ 0 & \sigma & 0 \\ 0 & 0 & \sigma \end{pmatrix} + \frac{1}{3}\begin{pmatrix} \sigma & 0 & 0 \\ 0 & \sigma & 0 \\ 0 & 0 & -2\sigma \end{pmatrix} \tag{9}$$

Thus, the stress calculated from the hydrostatic factor has to be multiplied by 3/2. Windischman *et al.*[101] showed an example of in-plane stress in a CVD film deposited onto a molybdenum substrate measured to be 2078 MPa using the hydrostatic factor multiplied by 3/2. A very similar result of 2098 MPa was given by the substrate curvature method.

In general, one Raman peak is observed under zero or hydrostatic stress. But peak shifts and peak splitting depend upon the relative direction of the applied loading and the crystallographic orientation of the material. Under a biaxial or a uniaxial stress, the diamond Raman peak splits into two to give a singlet and a doublet which are distinguishable by their polarisation characteristics, Fig. 5. These split peaks add complexity when resolving the stress tensors. However, this deconvolution can be undertaken by controlling the polarisation of the incident laser light and the orientation of the specimen. Grimsditch *et al.*[38] first determined the photon deformation potential constants by applying an uniaxial stress up to 1 GPa along the $\langle 001 \rangle$ and $\langle 111 \rangle$ directions on single crystal diamond respectively, Fig. 6a and b. Polarisation configurations have been adopted to separate the different components and the stress sensitivity has been evaluated in the two directions. The peak shift, $\Delta\nu$, under hydrostatic pressure is[38]

$$\Delta\nu = 3.2 \pm 0.2 \text{GPa}^{-1}\ \text{cm}^{-1} \tag{10.a}$$

The shifts along $\langle 001 \rangle$ and $\langle 111 \rangle$ under the 1 GPa hydrostatic pressure are[38]

$$\Delta\nu_{\langle 001 \rangle} = 0.73 \pm 0.1 \text{GPa}^{-1}\ \text{cm}^{-1} \tag{10.b}$$

$$\Delta\nu_{\langle 111 \rangle} = 2.2 \pm 0.2 \text{GPa}^{-1}\ \text{cm}^{-1} \tag{10.c}$$

Ager *et al.*[94] experimentally and numerically developed for a polycrystalline diamond film a more general solution for the three degenerate components, based on the two peaks at 1338 cm^{-1} and 1352 cm^{-1} deconvoluted from the broadened and split diamond peak under a

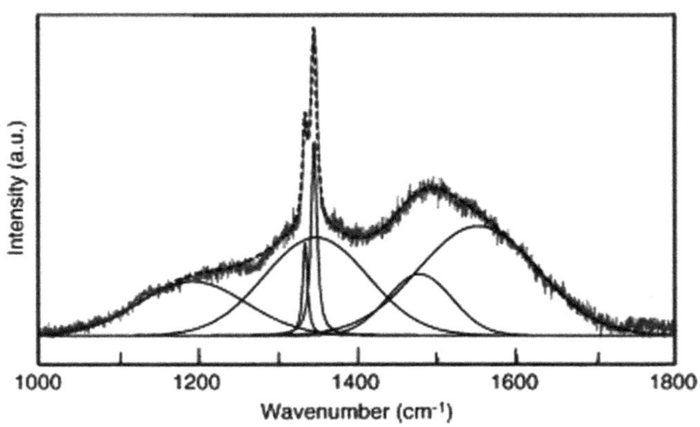

Fig. 5 A typical spectrum acquired by micro-Raman on a polycrystalline diamond coating deposited on Ti-6Al-4V showing the shift and splitting of the 1332 cm^{-1} diamond peak, the other broad peaks at 1350, 1480 and 1560 cm^{-1} and the band at about 1180 cm^{-1} correspond to co-existence with other carbon phases.[99]

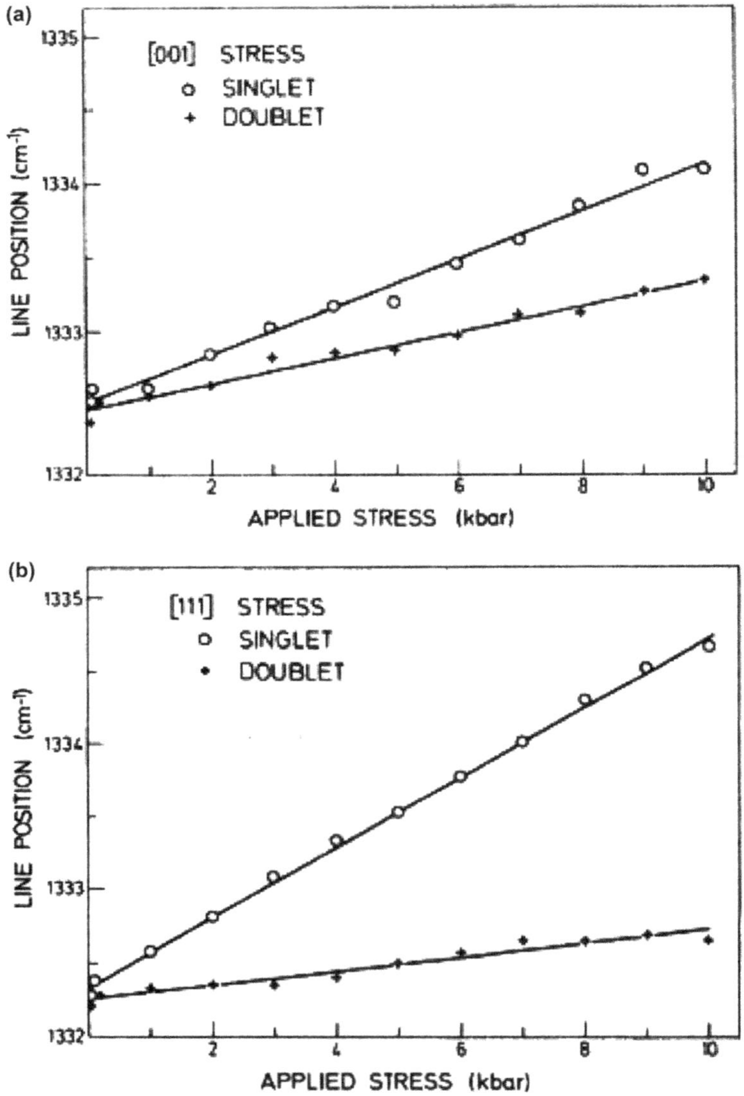

Fig. 6 The stress dependence of the frequencies of the Raman phonons of diamond (a) under a compressive force along (a) the ⟨001⟩ direction and (b) the ⟨111⟩ direction at room temperature. The solid lines are least-square fits used to extract the parameters, Equations (11.a) and (11.b).[38]

compressive load for an in-plane stress state as a function of the crystallographic directions ⟨110⟩, ⟨111⟩, ⟨110⟩ and ⟨112⟩:

$$\sigma_1 = -\frac{1.08}{\nu - 1338}\left(\frac{\text{GPa}}{\text{cm}^{-1}}\right) \tag{11.a}$$

$$\sigma_2 = -\frac{0.384}{\nu - 1352}\left(\frac{\text{GPa}}{\text{cm}^{-1}}\right) \tag{11.b}$$

where σ_1 and σ_2 are the two in-plane stress tensors, ν is the position of the two peak shift corresponding to 1338 cm^{-1} and 1352 cm^{-1} respectively. Equation (11.a) is for the single peak and Equation (11.b) is for the doublet.

However, strain is not the only reason why peaks split, since both twins and stacking faults cause peak splitting.[50,100,103] In addition, temperature change, domain size and crystallographic variations in the stress sensitivity of the peak orientation in sampled volumes all introduce peak shifts.[41] The increase of temperature causes the peak to shift to lower values, for example a −0.2 cm^{-1}/10 °C change in wavenumber was reported and this is equivalent to a 80 MPa error.[101] In addition, Prawer *et al.*[104] pointed out that the intensity ratio between the non-diamond part and the diamond part of the Raman spectrum in a CVD film can change by a factor of 2.5 by rotating the crystal with respect to a fixed polarizer and analyser, and this can lead to a peak shift up to 3 cm^{-1}. Windischmann *et al.*[98] measured compressive thermal stress introduced during microwave plasma manufacturing of diamond films together with a tensile intrinsic micro-scale stress up to 0.84 GPa due to grain boundary relaxation. It is noted that the stress caused by manufacturing is not exclusively compressive, but depends on the specific process[2,53] and the stiffness of the substrate, *e.g.* a tensile stress was found in those films applied to SiAlON substrates[105] whereas a compressive stress was present in diamond deposited on WC-Co substrates.

3.1.3 Advances in the technique. Different wavelength lasers were found to enhance or supress certain active features in Raman spectra obtained from CVD diamond films, so that when applying the technique to stress measurement the parameters could be tailored and optimised according to the target material.[92,106,107] The elimination of the photoluminescence (PL) background when using 244 or 229 nm UV excitation allows detailed measurements of the shape and width of the first-order Raman peak even for nanocrystalline CVD films.[100,108] The diamond Raman spectrum obtained by using a UV laser as the excitation source is dependent on grain size as demonstrated by Sun *et al.*[109] They used a 244 nm wavelength UV laser on nanocrystalline diamond films where the diamond peak at 1332 cm^{-1} was enhanced, while the D and G peaks of graphite as well as photoluminescence were suppressed compared with visible Raman (514.5 nm). When decreasing the particle size from 120 to 28 nm, the diamond peak shifts from 1332.8 to 1329.6 cm^{-1}, the width of the peak becomes broader and the intensity ratio of diamond and G peak decreases.

For CVD diamond films with nano-scale thickness, Raman spectra are not sensitive to the surface structure of the films even when the confocal Raman technique is adopt due to the large skin penetration depth of white light (normally several hundreds of nanometres). In these cases, a thin layer of metal particles, normally Ag, is deposited onto the surface of the diamond film, which enhances the near surface signal by a factor of between 10^3 to 10^8.[2,106] Knight *et al.*[110] were the first to adopt this treatment and the enhancement is shown in Fig. 7 where a 5 nm Ag coating is applied to a 20 to 40 nm thick CVD film on a silicon substrate. Without the Ag layer, Fig. 7a, the spectrum failed to reveal a diamond

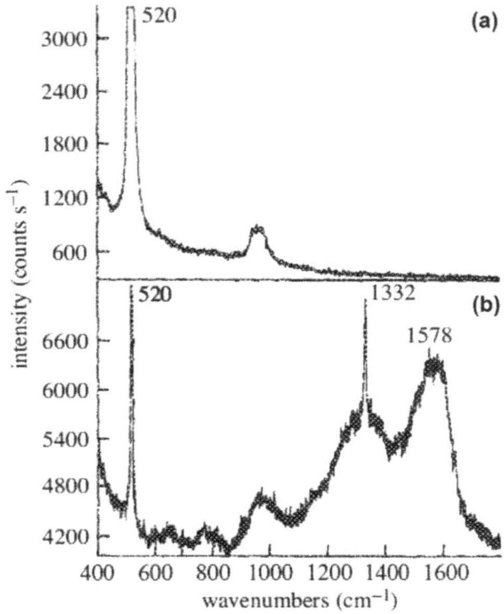

Fig. 7 Raman spectra of a 20 to 40 nm CVD diamond film deposited on silicon substrate when (a) without Ag coating and (b) with 5 nm Ag layer.[110]

signal and it is indistinguishable from that of a bare silicon substrate; whereas the Raman spectrum from the Ag coated film revealed the signature of a typical CVD film superimposed with signals from amorphous carbon, Fig. 7b. Subsequently, workers have validated the detected peaks and measured the stresses within these films.[111,112]

To summarise, the stress measurements using Raman spectroscopy in thin films are not straightforward. These are highlighted by consideration of the application to CVD diamond films where complications arise due to the following points:

(i) The single Raman peak of natural stress-free diamond splits into two or even three not necessarily well resolved peaks, when non-hydrostatic pressure/stress are applied or are present. This is often complicated by peak asymmetry and broadening leading to a range of different interpretations. A pre-understanding of the likely stress state of the film can help to narrow down the deconvoluting options and interpretations.

(ii) Four-rank tensors, both for the elastic constants and photon deformation potential constants in most computation methods are assumed to average all the components representing the random orientation of the crystals. For diamonds, and diamond-like cubic polycrystals the photon deformation potential constants are determined in terms of their atomic arrangements. For polycrystals, it is always assumed that there is no texture in the analysed volume because the laser penetration depth and the dispersion of the beam within the material samples a large number of crystals.[39] In reality the growth of the crystals is influenced by the substrate leading to preferred orientations and mixed multi-directions and phases which complicate the stress evaluation.

(iii) Certainly, the influence on peak shift and asymmetry from confinement and temperature has to be taken into account when evaluating strain/stress in these films. Similar to bulk diamond, the downshift and line broadening of Raman spectra in nanocrystalline particles can be caused by local laser heating. However, this change of shift and peak width varies with the particle size in the film, a typical example is provided by Chaigneau *et al.*[113] where it was proposed that this is due to the photon confinement effect. In addition, the sensitivity of the Raman spectra towards the temperature, the defects in diamond and the photon confinement provides useful information on the processing and growth of diamond crystals, the detection of point defects and the crystal size.[2]

In general, Raman spectroscopy provides a relatively accurate measurement of stresses in CVD thin films. For the measure of more components of the stress tensors a complex general methodology should be invoked such as that developed by Anastassakis[36,39,114] for calculating photon deformation potential constants and elastic constants in cubic polycrystalline materials.

3.2 Ultra-thin films

There a few examples such as graphene and ZnO of ultra-thin films. Graphene is recognised to provide an ultimate in thin films since it is only one-atom thick.[115–117] This monolayer of atoms is tightly packed into a 2D honeycomb atomic lattice and is produced by either micromechanical cleavage of graphite[115,118] or epitaxial growth on a silicon carbide substrate.[119,120] It is deemed to be a 2D building brick for carbon allotropes of other dimensions *i.e.* it can build 3D graphite, rolled into 1D nanotubes and wrapped into 0D fullerenes.[121] Raman spectroscopy has been identified as the most important tool for investigating the electron-photon interactions in graphene and establishing the strain/stress characteristics.[18,122] Mohiuddin *et al.*[19] applied uniaxial strains, from zero to 0.8%, to graphene on a substrate in bending, Fig. 8a, and measured the change of Raman shift with strain. The two peaks, G peak at $\sim$1575 cm^{-1} for single crystal graphite split into G$^-$ and G$^+$ with the increase of strain[123] and second-order D peak, named as '2D', at $\sim$2675 cm^{-1}, Fig. 8b and c. The shifts of these peaks with increasing strain are shown in Fig. 8b for the G peak and Fig. 8c for the 2D peak.

If there is more than one layer of graphene present, the 2D single peak for monolayer splits into four for bilayer graphene as a result of the evolution of the peaks,[122] Fig. 9. Details of the assignment of peaks and the characteristic Raman peak lie beyond the scope of this work and can be found in literatures.[115,117,122] The shifts of the G$^-$, G$^+$ and 2D peaks derived from Fig. 8b and c can be fitted with linear equations:

$$\frac{\partial \nu_{G^-}}{\partial \varepsilon} = -31.7\ \mathrm{cm}^{-1}/\% \tag{12.a}$$

$$\frac{\partial \nu_{G^+}}{\partial \varepsilon} = -10.8\ \mathrm{cm}^{-1}/\% \tag{12.b}$$

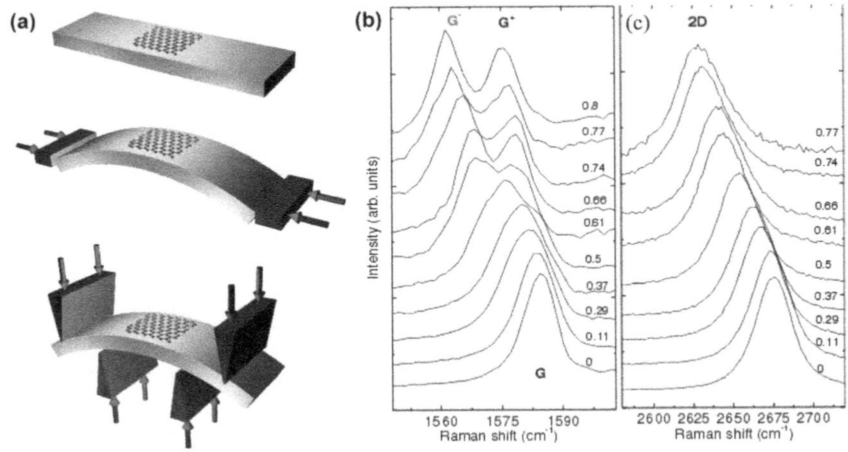

Fig. 8 (a) Schematic of the loading system for a monolayer of graphene: two point and four point loading; (b) shifts of G peaks and (c) 2D peak with applied strain.[19]

$$\frac{\partial \nu_{2D}}{\partial \varepsilon} = -64\ \text{cm}^{-1}/\% \qquad (12.\text{c})$$

These conversion factors have been validated against first-principle calculations for free-standing graphene. Based on the previously reported values for elastic moduli,[124,125] Mohiuddin *et al.*[19] predicted that under a biaxial strain the constant is $\sim$144 cm^{-1}/% for the peak 2D, which is nearly twice the magnitude for uniaxial conditions as in Equation (8c). However, this only applies for a single layer of graphene. When several layers stack together, broadening and shifts of the 2D band occur as shown in Figs. 9a and b, and the relative intensity of the G and 2D band increases with the number of layers.[121] Wang *et al.*[126] investigated the influence of substrate on the Raman spectrum for monolayer of graphene. It was concluded that G band for an epitaxial, monolayer of graphene on a SiC substrate showed a significant shift towards the blue end of the spectrum. This is caused by the interaction *via* the covalent bonding between epitaxial graphene and the SiC substrate which resulted in the change of the lattice constant and hence a nominal compressive stress.[121,126] Such features have to be taken into account when evaluating the relationship between peak shift and strain.

3.3 Coatings

3.3.1 Yttria stabilized tegragonal zirconia thermal barrier coating. Zirconia based materials have been extensively investigated due to their numerous applications and superior properties. Zirconia has low thermal conductivity, 2.3 W · m^{-1} · K^{-1} at 1000 °C for a fully dense material,[127,128] this is due to the presence of high concentrations of point defects such as oxygen vacancies, and substitutional solute atoms which can scatter heat-conducting phonons in terms of lattice waves.[129] The high thermal expansion (10×10^{-6} deg C^{-1} to 11×10^{-6} deg C^{-1}) is close to that of a metal such as a ductile superalloy which is often used as a substrate for

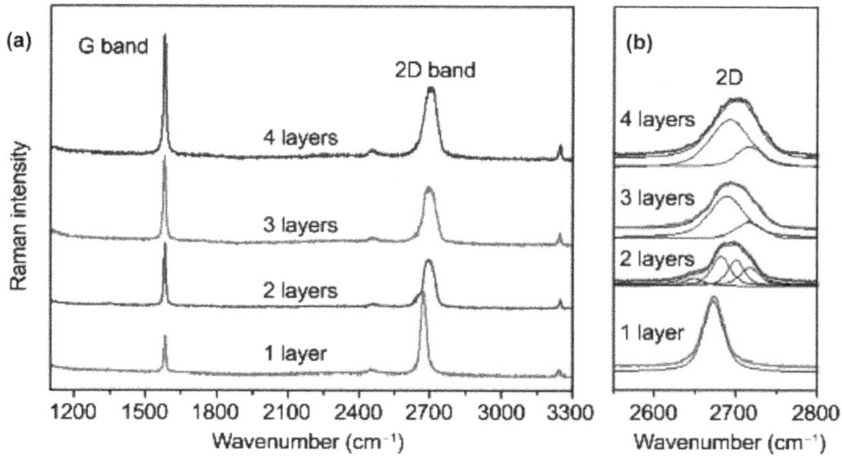

Fig. 9 (a) Raman spectra for graphene with 1, 2, 3, and 4 layers, and the 2D peak was fitted in (b) that broadening and shifting are observed.[121,126]

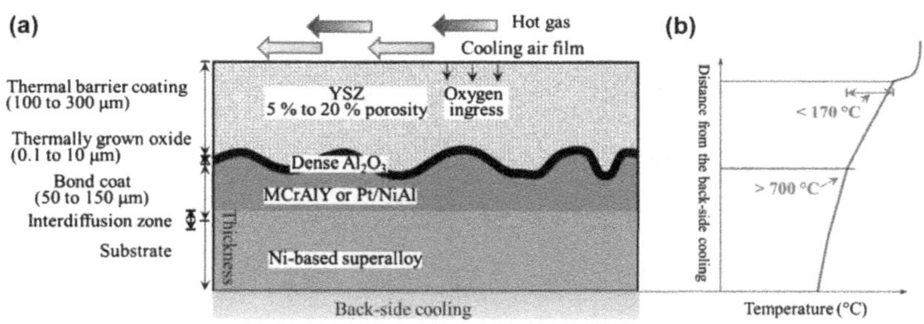

Fig. 10 Schematic of (a) a thermal barrier coating applied on superalloy substrate; the Al_2O_3 is developed during service and (b) the associated temperature distribution.

mechanical components (14×10^{-6} deg C^{-1} to 17×10^{-6} deg C^{-1}). This alleviates mismatch stresses when applied as a coating,[127] It also has high optical refractive index, chemical durability, high mechanical strength and fracture toughness[82] and good bonding to Ni-based metals. A disadvantage is that zirconia undergoes phase transformations at high temperatures. Therefore, rare earth element oxides such as Y_2O_3, MgO, CeO_2, Sc_2O_3, In_2O_3, CaO and GdO_2 are often used to stabilize ZrO_2 to retain the tetragonal phase.[130–132] In this context, 6 to 8 wt.% Y_2O_3 is typically used for commercially applications.[15,133,134]

One of the most important applications for yttria stabilized tetragonal zirconia (YSZ) is as a thermal barrier coating (TBC) on components working at high temperatures, Fig. 10. Typically, a layer of ~100 to ~300 μm thick YSZ is applied on the blades in the high pressure turbine of land-based power generators and aero-engines. The low thermal conductivity of the layer allows coated components to operate at higher temperatures than the underlying parent metal can withstand. Therefore it is crucial for the layers to remain attached to the components for an appreciable length of time and to be tolerant to the residual stresses and thermal cycling encountered during service.[15,83] These TBCs are multi-layer systems consisting of a top layer of YSZ, a metallic bond coat (BC) that provides bonding between the metallic substrate and YSZ. During thermal exposure a thermally grown oxide (TGO) layer, predominantly Al_2O_3, forms between the YSZ and BC,[84] Fig. 10a. Combined with the internal cooling system the TBC layer can potentially sustain a temperature difference of up to 170 °C, Fig. 10b.

In terms of manufacturing routes, there are mainly two types of TBCs: (i) air-plasma-spraying (APS-TBC) and (ii) electron beam physical vapour deposition (EBPVD-TBC), Fig. 11. APS-TBC has a splat structure, Fig. 11a, which gives a lower thermal conductivity but greater lateral stiffness. The EBPVD-TBC has a columnar structure, Fig. 11b, which is more flexible in the lateral direction and has a higher thermal conductivity. EBPVD-TBC is more costly to produce compared with APS-TBC, and normally APS-TBC has less limitations with respect to thickness. The EBPVD-TBC systems tend to be applied only to critical rotary components exposed to severe environments such as the blades and vanes in turbines and aero-engines. For the stress measurements in these two coatings the

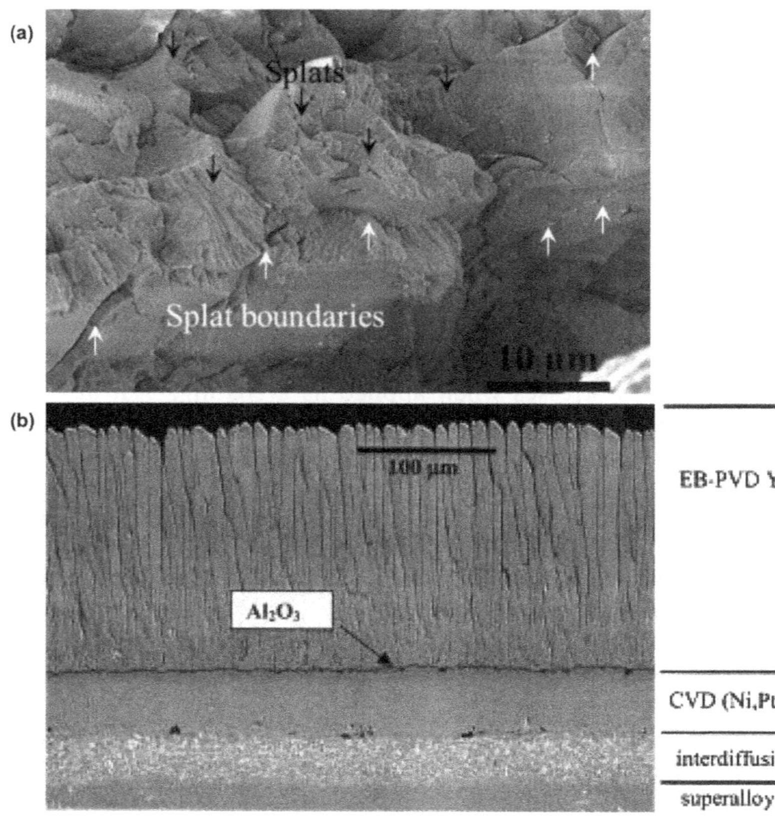

Fig. 11 Scanning electron microscopic images of (a) fracture APS-TBC revealing the splat structures with pores and defects and (b) polished EBPVD-TBC with the columnar structure deposited on PtAl-BC and superalloy substrate.

requirements and procedures are largely similar, but this review focuses on APS-TBC.

The failure of TBC systems in-service is complex and the processes are not fully understood, but normally occur in the form of local detachment and spallation of the TBC layer.[15,135,136] The stored strain energy in the system is recognised to be the main driving force for the failure.[137,138] Stress measurement of the individual layers provides important information with respect to the evaluation and prediction of service life, degradation and failure modes of the coating system. As a non-destructive approach, Raman spectroscopy has been utilized widely to measure the stress in the TBC layer whereas photo-stimulated luminescence piezo-spectrosocopy (PLPS) is applied for the measurement of stress in the TGO by collecting the ruby fluorescence from Cr^{3+} embedded in the Al_2O_3 lattice.[72,139–141] The extensive use of the YSZ layer as a thermal barrier coating (TBC) has given rise to much research on the residual stress distributions in the YSZ and TGO layers,[23,89,90] the microstructure[87,88] and the degradation of this coating system.[86]

3.3.2 Origin of residual stresses. Residual stresses in these YSZ-TBC coatings generally arise from three main contributions:[15]

(i) **The manufacturing process.** In APS, a high-temperature plasma jet is generated by an arc discharge with typical temperatures higher than 15000 K (12000 to 20000 K), which makes it possible to spray refractory materials such as ZrO_2.[127] The high velocity impinges the ceramic particles in the form of splats onto the surface of the substrate to form a coating with a lamellar structure interspersed with pores and unmelted particles, Fig. 12. For the APS-TBC, upon spraying the residual stress in the first layer of splats is tensile as the substrate constrains the splat during cooling and shrinking[14,142] as shown schematically in Fig. 12. If the tensile stress is large enough, through thickness cracks are generated within each splat.[143] With more splats being laid down during spraying, overlapping occurs with associated complex stress relaxation and redistribution.[144,145]

(ii) **Mismatch between the TBC and the substrate.** Mismatch is still a very important source of compressive stress although the selected TBC material has the closest thermal expansion coefficient with superalloy substrate compared with other ceramics (section 3.3.1). This mismatch stress puts the entire coating into compression which leads to typical

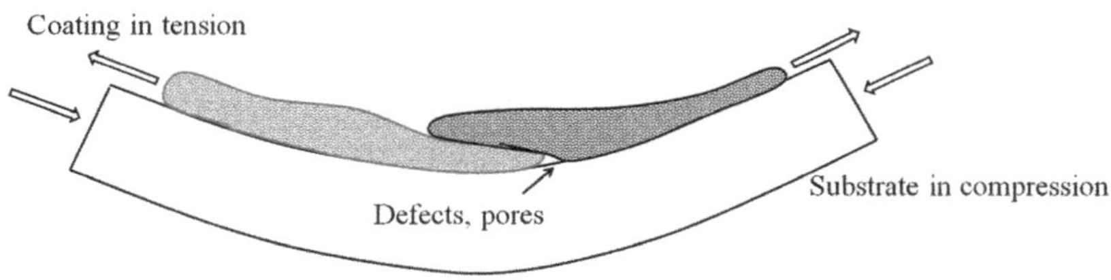

Fig. 12 Schematic showing the tension in splats upon spraying. Defects and pores are introduced within and between splats.[14]

buckling failures in the case for brittle coatings on ductile substrates.[11,146] In addition, the elastic modulus of the TBC increases significantly with thermal exposure and the porosity decreases by sintering.[85,147]

(iii) **The growth of the TGO.** During service oxide growth introduces an extra volume between the TBC and BC[15] to generate significant residual stresses in the TGO. Moreover, it modifies the residual stress distribution in the TBC. Liu *et al.*[14] measured compressive stress at the TBC/TGO interface and a decrease in this stress with increasing distance from the interface. This is consistent with significant constraint within TBC at the interface in the presence of TGO.

Recently, Liu[145,148] measured the residual stresses systematically in typical APS-TBCs exposed to a range of temperatures between 850 °C to 1050 °C and for times up to 30000 hrs in laboratory air. It was proposed that there is a progressive change from the initial tensile stress in as-sprayed YSZ coatings to compressive stresses with extended thermal cycles. The compressive stress introduced into the TBCs is proposed to be an effect of all the above three contributions. For practical applications such as the evaluation of stresses during service life, it is important that non-destructive measurements of residual stress by Raman spectroscopy are carried out at fixed intervals. The integrity and lifetime evaluations of the coating systems then become possible from an understanding of the progressive evolution of residual stresses.

3.3.3 Stress measurement. Raman spectra collected from a typical APS-TBC are shown in Fig. 13[14] when using an incident laser with the wavelength of 514 nm. Tetragonal ZrO_2 has six peaks at ~150 cm^{-1} (E_g), ~265 cm^{-1}(E_g), ~320 cm^{-1}((B_{1g}), ~460 cm^{-1}(E_g), ~602 cm^{-1}(A_{1g}), ~640 cm^{-1}(B_{1g}), Fig. 13, according to the symmetry assignment by Bouvier *et al.*[149] Any of these peaks has the potential to be used to measure stress. However peak 6 (640 cm^{-1}), Fig. 13, which has a pronounced intensity and high signal-to-noise ratio, is usually selected and the profile fitted using a mixed Gaussian and Lorentzian method to extract parameters such as the peak position, full-width-half-maximum (FWHM) and intensity. The other peaks located within the lower frequency range, as shown in Fig. 13, have significantly reduced intensity at many locations measured in a practical APS-YSZ TBCs. This scenario limits, for example the application of the peak centred at 465 cm^{-1} for calibration of APS-YSZ, as proposed by Limarga *et al.*[150] based on bulk, dense, pure YSZ material.

More specifically, the peaks in the Raman spectra associated with the YSZ have been found by many workers to shift as a linear function of local strain. Typically this is observed when the strain is less than 8 GPa to 10 GPa, which is normally well above the maximum strain measured in APS-YSZ):[13,149,151,152] This shift in wavenumber is given by:

$$\sigma_{\text{triaxial}} = \Delta\nu/\Pi \tag{13}$$

Here $\Delta\mathbf{v}$ is the Raman shift of the peak (cm^{-1}) from the stress-free state, and $\mathbf{\Pi}$ is the conversion factor (cm^{-1}/Pa). It is important to recognize that

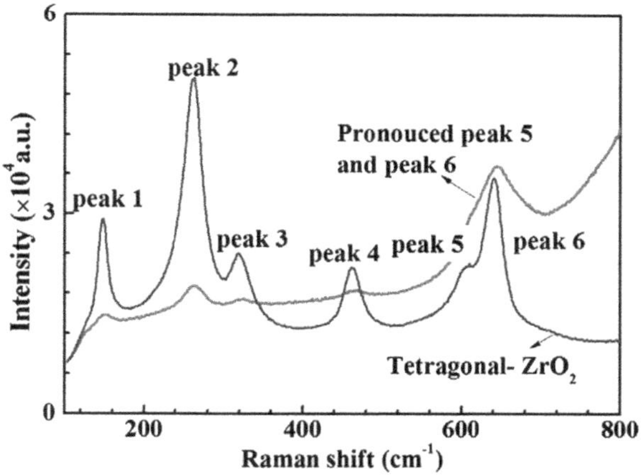

Fig. 13 Typical Raman spectra for tetragonal ZrO_2 with six peaks obtained using a 514 nm Ar^+ incident laser beam in a Renishaw Ramascope, model 2000.[14]

the stress, σ_{triaxial}, is triaxial as proposed by Liu *et al.*[13,74] Raman spectroscopy does not measure the near-surface stress, because the penetration depth is typically 30 μm to 40 μm and the cone shaped laser beam expands from 1.5 μm to 160 μm after being scattered by a 180 μm thick APS-YSZ.[74] A shift to higher frequency indicates the presence of a compressive stress. The conversion factor, Π, in this equation has to be calibrated so that Raman spectra can be used to calculate the stress in these ceramic coating.[13,153,154] There are a range of values for the conversion factor, Π, that have been reported previously for tetragonal zirconia material under different conditions. Teixeira *et al.*[151] proposed a linear relationship between the applied tensile stress and the Raman shift where each shift represents 220 MPa in stress[32] which indicates a conversion factor of 4.54 $GPa^{-1} \cdot cm^{-1}$. This value is fairly close to the value of 5.55±0.1 $GPa^{-1} \cdot cm^{-1}$ obtained for the 640 cm^{-1} peak by Liu *et al.*[13] But Teixeira did not provide detail for the loading and measurement system used. Other factors reported vary with the specific material tested and the loading applied. For example, Tanaka *et al.*[32] proposed a uniaxial factor of 25 $GPa^{-1} \cdot cm^{-1}$ for a free-standing 8 wt.% YSZ manufactured by APS, whereas Cai *et al.*[155] obtained a uniaxial value of 1.1 $GPa^{-1} \cdot cm^{-1}$ for dense yttria stabilized cubic polycrystalline zirconia. For bulk dense materials under hydrostatic loading conditions, the variation in the conversion factors diminishes such that Alzyab *et al.*[156] obtained a value of 3.2 $GPa^{-1} \cdot cm^{-1}$ for 5 wt.% YSZ and Bouvier and Lucazeau[149] reported 2.79 $GPa^{-1} \cdot cm^{-1}$ for nanometric tetragonal zirconia. It is obvious that these values vary significantly depending on the material (single crystal or powder or APS-YSZ) tested.

Recently, Liu *et al.*[13] calibrated the conversion factor for a series of micro-scale APS-YSZ specimens sampled from a practical coating in both the as-coated and heat treated (at 1000 °C for 1050 hrs) conditions. The test was undertaken using a standard diamond anvil cell (DAC) with ruby grains as an indicator of stress in a methanol-ethanol filled chamber.

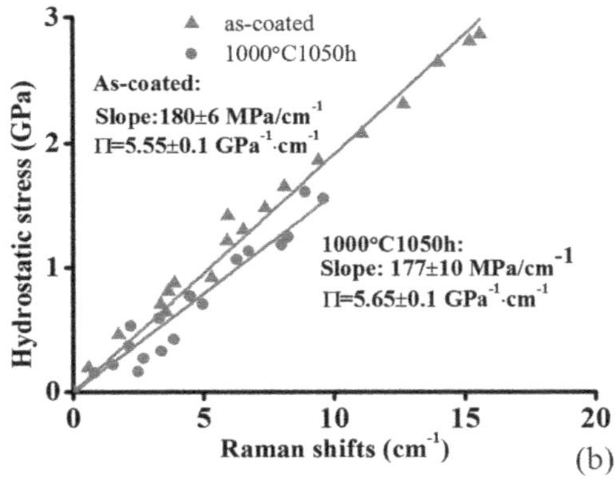

Fig. 14 The shifts of the 640 cm^{-1} peak for as-coated and thermally heat treated TBC specimens subject to hydrostatic stress.[13]

Raman spectra were acquired from the micro-scale specimen at each applied hydrostatic stress; specifically the peaks at 608 cm^{-1} and 640 cm^{-1} were studied. Linear relationships were found between the applied hydrostatic stress and the shifts of these two peaks. These factors obtained from the 640 cm^{-1} peak are 5.65 $GPa^{-1} \cdot cm^{-1}$ for the as-coated condition, and 5.55 $GPa^{-1} \cdot cm^{-1}$ when heat treated at 1000 °C for 1050 hrs, Fig. 14. The corresponding values for the peaks at 608 cm^{-1} are 5.07 $GPa^{-1} \cdot cm^{-1}$ and 5.05 $GPa^{-1} \cdot cm^{-1}$ respectively. An important conclusion reached was that the heat treatment history of this practical APS-YSZ coating will not influence the conversion factor within the range of experimental errors so that a consistent value can be adopt for stress measurement. This provided further confidence for stress measurements in practical APS-TBCs over service life. Combined with focus ion beam milling and secondary electron beam imaging, it was argued by Liu *et al.*[13] that considering the thickness and surface roughness of the TBC, the stress triaxiality cannot be neglected in these type of coatings.[13] However, the conversion factor is material specific so that calibration of the conversion factor must be carried out case by case.

Using the determined conversion factor, the residual stress could be derived from the Raman spectra. One example shown in Fig. 15 indicates that residual stress in the TBC changes progressively from tensile to compressive with prolonged thermal exposure. It is during the thermal exposure that TGO develops at the TBC and BC interface. The residual stress in TGO relaxed after a maximum compressive value was reached. It is worth noting that this maximum ~ -1.5 GPa is at the point where the TBC residual stress changes sign from tensile to compressive. Ultimately the residual stress in TGO reduces to a similar value of stress as that in TBC.[145] Liu *et al.*[148] considered that this evolution of residual stress could be used to build a failure prediction diagram, Fig. 15. The failure-prediction diagram shows that if the residual stress in TBC and TGO can be monitored regularly during shutdown of components, degradation of the TBC system is predictable. The shape of the failure diagram could be either a triangle or a polygon, but the key message that emerges from this

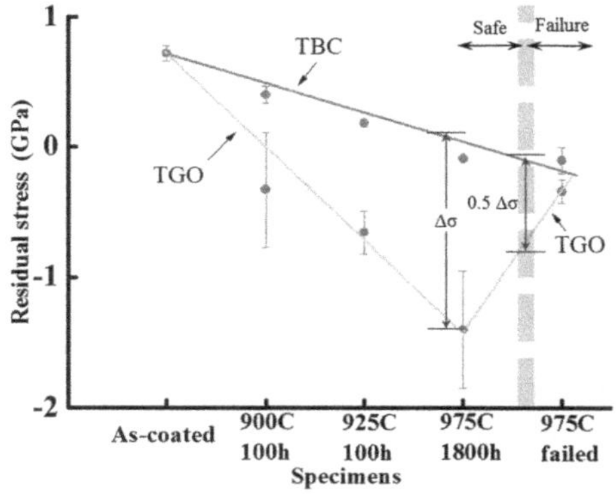

Fig. 15 The failure-prediction diagram formed by residual stresses in TBC and TGO at the same position on five selected specimens showing the stresses changing progressively with exposure (time and temperature).[145]

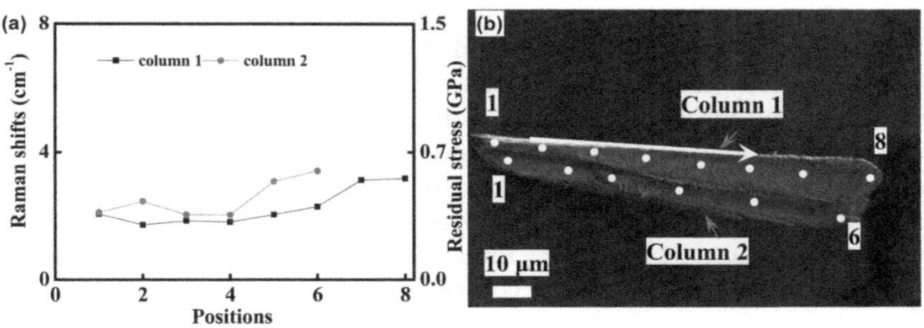

Fig. 16 The Raman shift measured in two free-standing columns: (a) Raman shifts indicating tensile stress in the two columns in (b). (b) Secondary electron images showing eight positions measured along column 1 and six in column 2. The white spots are the positions where the laser beam focused during acquisition. Position 1 is at the base and the numbers increase with increasing distance from the base.[145]

diagram is the relaxation of the stress in TGO and the reduced difference between the residual TBC and TGO stresses. This is an indication of failure initiation or proof for occurrence of failure in the system. An appropriate margin between the safe and a 'failure conceded' condition would have to be set. It is proposed that if the largest difference in the residual stresses in TBC and TGO is $\Delta\sigma = \sigma_{TBC} - \sigma_{TGO}$ prior to the relaxation of the TGO stress, Fig. 15, potential failure is expected when the difference is reduced to $0.5\Delta\sigma$.

Additionally, due to the higher resolution of Raman spectroscopy and the in-situ view of the surface of the measured material, normally *via* an optical microscope, allows the measurement and mapping of the stress at small features. Here we show an example for mapping of stress along free-standing columns sampled from an as-deposited EBPVD-TBC coating using Raman spectroscopy.[145] Figure 16a is the Raman shift and residual stress measured from two columns (column 1 and 2 shown in Fig. 16b). The white spots on the columns in Fig. 16b are the measured positions from 1 to 8 for column 1 and 1 to 6 for column 2. Note that there are significant tensile stresses present in these as-deposited free-standing EBPVD-TBC columns.

3.3.4 Confidence in measurement. In Raman spectroscopy a laser beam is focused on the specimen surface and the information collected is from the illuminated volume. The zirconia based ceramic coatings are translucent because the optical band gap for these materials is ~12 eV, which is well above the energy of the normal laboratory lasers such as argon-ion and He-Ne lasers ($\lambda = 514$ nm and $\lambda = 632.8$ nm).[157] The complex microstructure of the translucent material reduces the resolution of the measurements dramatically due to incident laser beam dispersion.[74,158,159] The assumption generally made for the stress measurements on the APS-YSZ is that the Raman signal acquired is limited to the near surface but the authors do not specify the range of depth sampled.[152] Recent experimental studies show that the sampling depth can reach 20 μm to 50 μm even in dense YSZ.[70] The only published work on the calibration of sampled depth in APS-YSZ TBC is from Liu *et al.*,[74] and these workers specifically calibrated this parameter using a wedge APS-YSZ sample by measuring the change of the Raman spectra intensity when moving the laser beam from thinner to thicker end of the wedge, *e.g.* position 1 to 4 in Fig. 17a. The intensity of the Raman spectra increases with the increased thickness of the sampled materials, Fig. 17b, and this intensity change is plotted as a function of the thickness of the wedge specimen in Fig. 17c. Effective sampling depth is defined as the point when the intensity becomes a constant value. It was found that the Raman signal collected from the surface extends to about 40 μm (about one fifth of the YSZ layer thickness).[74] The illuminated volume below the focus point of the incident laser beam is cone shaped,[76] and as sampling depth increases the resolution decreases due to beam dispersion. This dispersion is a consequence of scattering from a range of microstructural features within the APS-YSZ TBC such as internal boundaries, pores and other imperfections.

The evaluated surface stress derived from a Raman spectrum shift is obtained from the volume over which the stress is averaged. Although most of the Raman scattering and the collected signal is from the focus or near-focus region of the incident laser beam, scattering within the out-of-focus region will still contribute to the intensity.[160] There is a gradient of the effective signal across the point of focus on the surface to a depth of ~40 μm as proposed by Presser *et al.*[70] and 90% of the signal comes from half of the sampled depth.

A reliable measure of beam spreading within the TBC is a pre-requisite for mapping of stresses. Two types of non-destructive mapping of residual stresses are often required in TBCs:[24,161,162] mapping of residual stress in the top layer of YSZ[152] and in the underlying TGO.[139] Liu *et al.*[74] suggested that for the mapping of the residual stress in APS-YSZ a step interval of ~35 μm should be considered in order to avoid beam overlap and thereby mis-interpretation of the evaluated stresses. For stress mapping in the TGO the steps should be set as ≥160 μm for APS-YSZ

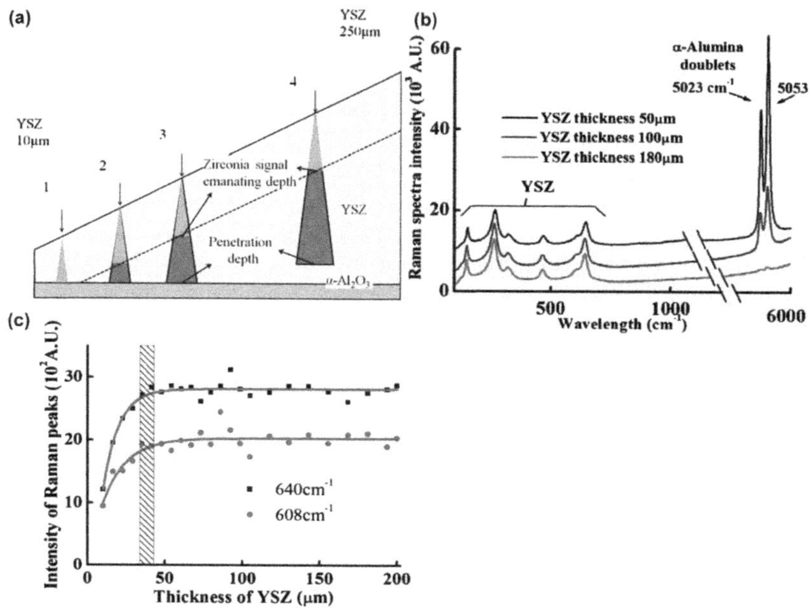

Fig. 17 (a) The schematic of the setup for the measurements obtained from intensity obtained from the Raman spectra of a wedged YSZ/α-alumina sample; (b) Raman spectra for YSZ and α-alumina, three spectra of the YSZ and α-alumina collected at thicknesses of 50, 100 and 180 μm; these are vertically displaced to show the significant drop in intensity for the α-alumina peaks (5023 cm^{-1} and 5053 cm^{-1}) with the increasing thickness of the YSZ layer; (c) the exponential increase of the intensities of two Raman peaks with YSZ thickness. Stable platform is reached when the thickness is larger than 40 μm, with an error in thickness of about 5 μm.[74]

and ≥80 μm for EBPVD-YSZ. It is important to make specific measurements of beam dispersion for the particular YSZ coating being addressed. For other applications of Raman spectroscopy such as the phase detection in materials, the sampling volume is crucial for determining the location and distribution of the detected phases.[65,66]

4 Concluding comments

The stress measured in the films and coatings by Raman spectroscopy arises from the effect on atomic bonding at different length-scales when a strain is applied.[101,145,163] It is noted that comparing the stress derived by Raman spectroscopy with other techniques (see Appendix I), *e.g.* substrate curvature, may not be a reliable approach to validate the conversion factor. This is because different approaches measure stresses from different sources and at various length-scales,[106] *i.e.* substrate curvature evaluates the general macro-scale strain energy stored in the thin film or coating, but Raman spectroscopy acquires all the information at the micro-scale. The change of the length of the bonding atoms can be affected by applied macro-scale stress as well as micro-scale intrinsic stress which may be omitted by the macro-scale stress measurement approaches. In other words, the measured nominal stress derived from the shift of the Raman peak and that from the change of curvature represents two distinct ways of considering stress contributions.[47,51,54,105] Therefore, unless the stress components accommodated in a specific volume of material have been completely and appropriately deconvoluted, the measured stresses obtained by the various techniques may not be directly comparable.

Raman spectroscopy measurement of stress in films and coatings provides unique information related to the atomic bond vibration. These spectra contain indirect characterisation of the change of the bonding with applied strain. As a consequence, deconvolution of the spectra to resolve the strain components often requires pre-knowledge of the stress state in the measured layer. In such cases, this can be assisted by other methods such as curvature, X-ray diffraction and digital image correlation *etc.*, see Appendix I.

In films and coatings the intrinsic stress introduced by the manufacturing process and the evolution during subsequent service or use is unique. This stress and contribution from factors such as the macro-scale thermal expansion mismatch are convoluted and reflected in the shift and shape of spectra collected by Raman spectroscopy. However, the interpretation of the exact strain/stress state from the Raman spectra is not always straightforward since many parameters contribute to one Raman feature (peak shift, width and intensity *etc.*). For example, apart of strain/stress, the shift of a Raman peak can additionally be caused by temperature[37,44,60,61,100] and inhomogeneous local distribution of stress. Therefore, the stress measured by Raman spectroscopy is an average of the volume of sampled material plus the influence imposed by external factors (temperature). Interpretation has to take account of the condition of the films and coatings. For example if the film is opaque it is the near-surface signal that is collected by the detector, hence the stress can be assumed as plane-stress. The shift from Raman peak is the main approach to derive averaged stress values. In addition, other parameters of the spectrum have been explored by workers to assist the deconvolution of all stress components: the variation of strain field is found to relate to the peak broadening, and the intensity ratios between peaks are sometimes an indication of crystal orientation.

One ultimate goal of non-destructive stress measurement is to provide a reliable basis for the integrity evaluation of the films and coatings. A single independent value obtained by Raman spectroscopy may be insufficient, but the progressive measurements of stresses at fixed intervals or selected key stages in a film or coating from manufacturing to the end of service life allows potentially reliable integrity criteria to be established. In addition, combining the stresses measured with other information in the spectra such as the broadening of the peaks, the peak intensity ratios and the phase transformations, with thermal-mechanical history provides a precise understanding about the condition of films and coatings.

Mapping of stresses using a motorised stage to systematically shift specimens to pre-set coordinates has become popular since it provides contours of stress over relatively large areas (from square micrometres to millimetres). However, as discussed in section 3, the sampling volume and dispersion of the laser within the structure of the coating becomes crucial for stress mapping to a high degree of confidence.

In summary, Raman spectroscopy is a versatile technique for stress characterisation in films and coatings. This review has covered the application of Raman spectroscopy to the films and coatings used in commercial applications. For crystals with simple structures, such as cubic diamond and silicon, the deconvolution of the stress tensors from the spectra is possible when manipulating the polarisation of incident lasers. However, for complex coating materials, such as YSZ, this becomes experimentally and theoretically difficult. With many fundamental mechanisms unknown, Raman spectroscopy still provides a relatively reliable non-destructive method for measuring average stresses developed in films and coatings. Calibration of conversion factors combined with a knowledge of sampling volumes and heating effects are key factors that need to be considered for making measurements of the required reliability.

Appendix I

Common techniques for residual stress measurement in films and coatings and their associated advantages and limitations.[6,70,164–166]

Techniques	Resolution	Specimen preparation requirement	The type of stress measured
Material removal; Curvature method	Average values of a layer, $\leq$0.2 mm	Thin, flat substrate; Can solve finely graded coatings	Measure macro stress Type I in films and coatings. Determine stress profiles across coating thickness. Destructive.
Hole drilling	~50 μm in depth, 2 mm dia.[167]	Thicker (>100 μm) and ductile coatings; Relatively flat surface	In-plane macro-stress in coatings; Gradient of stress across the depth of the coating. Destructive.
Neutron diffraction	Phase distinctive, millimetre scale, *e.g.* 0.2 mm to 250 mm in depth in Al	Thicker coatings, no limit on size or shape	Measure macro-stress and determines stress profiles in thick coatings; 3-dimensional strain. Non-destructive.
X-ray diffraction	Phase distinctive, 20 μm to 30 μm from the surface layer	No requirement on coating thickness; A reasonable flat area is desirable	Measures macro-stress to high spatial resolution; If applied to depth profiles of stress, layer removal is required; Peak shift measure both Type I and average Type II stresses, while Type III stresses give peak broadening. Non-destructive.

Techniques	Resolution	Specimen preparation requirement	The type of stress measured
Piezo-spectroscopic (Raman/PLPS)	Laterally can reach $\sim 1\,\mu m$ or less in dia.; Depth depends on the absorption and scattering of the material (sub-micrometre to $\sim 50\,\mu m$[44])	No special requirement in specimen geometries	A combined stress in the sampled volume; Deconvolution of stress tensors from the spectrum is complex.[44,168] Peak shift samples both Type I and average Type II stresses, while Type III stresses give peak broadening. Non-destructive.
Electron diffraction	Convergent electron beam is used for highest resolution (~ 10 nm)	Only very thin films (<100 nm) can be examined	Type II and Type III stresses can be measured in very small electronic devices; Results are vulnerable to surface relaxation effects and strain values represent integral through the thin foil thickness. Non-destructive.

Acknowledgement

We would like to acknowledge the support of The Energy Programme, which is a Research Councils UK cross council initiative led by EPSRC and contributed to by ESRC, NERC, BBSRC and STFC, and specifically the Supergen initiative (Grants GR/S86334/01 and EP/F029748) and the following companies; Alstom Power Ltd., Doosan Power, E.ON, National Physical Laboratory, Praxair Surface Technologies Ltd, QinetiQ, Rolls-Royce plc, RWE npower, Siemens Industrial Turbomachinery Ltd. and Tata Steel, for their valuable contributions to the project. Also to Prof. David Rickerby (Rolls Royce plc.), Dr. Maud Seraffon (NPL), Prof. Martyn Pavier (University of Bristol) and Dr. Keith Hallam (University of Bristol) for useful discussions.

References

1 M. Becker, H. Scheel, S. Christiansen and H. P. Strunk, *J. Appl. Phys.*, 2007, **101**, 063531–063510.
2 S. Prawer and R. J. Nemanich, *Phil. Trans. R. Soc. Lond. A*, 2004, **362**, 2537–2565.
3 E. Anastassakis, A. Pinczuk, E. Burstein, F. H. Pollak and M. Cardona, *Solid. State. Commun.*, 1970, **8**, 133–138.
4 L. B. Freund and S. Suresh, Cambridge University Press, Cambridge, UK, 2003.
5 P. J. Withers and H. K. D. H. Bhadeshia, *Mater. Sci. Tech.*, 2001, **17**, 366–375.
6 P. J. Withers and H. K. D. H. Bhadeshia, *Mater. Sci. Tech.*, 2001, **17**, 355–365.

7 I. C. Noyan and J. B. Cohen, *Residual stress: Measurement by Diffraction and Interpretation*, Springer-Verlag, New York, 1987.
8 F. Long, Q. Wei, Z. M. Yu, J. Q. Luo, X. W. Zhang, H. Y. Long and X. Z. Wu, *MRS proceedings, Symposium F - Adhesions in Solids*, 2011, **119**, 185–203.
9 P. Y. Hou, J. Ager, J. Mougin and A. Galerie, *Oxid. Met.*, 2011, **75**, 229–245.
10 J. Stringer and G. C. Wood, MRS Proceedings, Symposium F-Adhesions in Solids, 1988
11 S. Faulhaber, C. Mercer, M. W. Moon, J. W. Hutchinson and A. G. Evans, *J. Mech. Phys. Solids.*, 2006, **54**, 1004–1028.
12 J. W. Hutchinson, *J. Mech. Phys. Solids.*, 2001, **49**, 1847–1864.
13 D. Liu, O. Lord and P. E. J. Flewitt, *Appl. Spectrosc.*, 2012, **66**, 1204–1209.
14 D. Liu, M. Seraffon, P. E. J. Flewitt, N. Simms, J. Nicholls and D. Rickerby, *J. Eur. Ceram. Soc.*, 2013, **33**, 3345–3357.
15 A. G. Evans, D. R. Mumm, J. W. Hutchinson, G. H. Meier and F. S. Pettit, *Prog. Mater. Sci.*, 2001, **46**, 505–553.
16 H. Windischmann and K. J. Gray, *Diamond Relat. Mater.*, 1995, **4**, 837–842.
17 J. P. Sullivan, T. A. Friedmann and K. Hjort, *MRS Bulletin*, 2001, **26**, 309–311.
18 A. C. Ferrari, *Solid. State. Commun.*, 2007, **143**, 47–57.
19 T. M. G. Mohiuddin, A. Lombardo, R. R. Nair, A. Bonetti, G. Savini, R. Jalil, N. Bonini, D. M. Basko, C. Galiotis, N. Marzari, K. S. Novoselov, A. K. Geim and A. C. Ferrari, *Phys. Rev. B*, 2009, **79**, 205433.
20 F. Ahmed, K. Bayerlein, S. M. Rosiwal, M. Göken and K. Durst, *Acta. Mater.*, 2011, **59**, 5422–5433.
21 M. D. Thouless, *J. Am. Ceram. Soc*, 1993, **76**, 2936–2938.
22 G. Giola and M. Ortiz, *Adv. Appl. Mech.*, 1997, **33**, 531–559.
23 V. K. Tolpygo, D. R. Clarke and K. S. Murphy, *Surf. Coat. Tech.*, 2004, **188–189**, 62–70.
24 X. Wang, A. Atkinson, L. Chirivì and J. R. Nicholls, *Surf. Coat. Tech.*, 2010, **204**, 3851–3857.
25 W. Ingrid De, *Semicond. Sci. Technol.*, 1996, **11**, 139–154.
26 P. E. J. Flewitt and P. K. Wild, *Physical Methods for Materials Characterisation*, 2 edn., Institute of Physics Publishing, 2003.
27 J. B. Hopkins, L. A. Farrow and G. J. Fisanick, *Appl. Phys. Lett.*, 1984, **44**, 535–537.
28 R. Loudon, *Adv. Phys.*, 1964, **13**, 423–482.
29 P. J. Heard, M. R. Wootton, R. Moskovic and P. E. J. Flewitt, *J. Nucl. Mater.*, 2011, **418**, 223–232.
30 B. Champagnon, C. Martinet, M. Boudeulle, D. Vouagner, C. Coussa, T. Deschamps and L. Grosvalet, *J. Non-Cryst. Solids.*, 2008, **354**, 569–573.
31 S. Ganesan, A. A. Maradudin and J. Oitmaa, *Ann. Phys.*, 1970, **56**, 556–594.
32 M. Tanaka, M. Hasegawa, A. F. Dericioglu and Y. Kagawa, *Mater. Sci. Eng., A*, 2006, **419**, 262–268.
33 S. S. Mitra, O. Brafman, W. B. Daniels and R. K. Crawford, *Phys. Rev.*, 1969, **186**, 942–944.
34 Y. D. Harker, C. Y. She and D. F. Edwards, *Appl. Phys. Lett.*, 1969, **15**, 272–275.
35 F. Cerdeira, C. J. Buchenauer, F. H. Pollak and M. Cardona, *Phys. Rev. B.*, 1972, **5**, 580–593.
36 E. Anastassakis, A. Cantarero and M. Cardona, *Phys. Rev. B.*, 1990, **41**, 7529–7535.
37 I. De Wolf, H. E. Maes and S. K. Jones, *J. Appl. Phys.*, 1996, **79**, 7148–7156.
38 M. H. Grimsditch, E. Anastassakis and M. Cardona, *Phys. Rev. B.*, 1978, **18**, 901–904.

39 E. Anastassakis, *J. Appl. Phys.*, 1999, **86**, 249–258.
40 W. Zhu, A. Leto, K.-y. Hashimoto and G. Pezzotti, *J. Appl. Phys.*, 2012, **112**, 103526–103525.
41 D. S. Knight and W. B. White, *J. Mater. Res.*, 1989, **4**, 385–393.
42 S. Narayanan, S. R. Kalidindi and L. S. Schadler, *J. Appl. Phys.*, 1997, **82**, 2595–2602.
43 A. M. Zaitsev, *Optical Properties of Diamond: A Data Handbook*, Springer, 2001.
44 A. Atkinson and S. C. Jain, *J. Raman. Spectrosc.*, 1999, **30**, 885–891.
45 L. A. Falkovsky, J. M. Bluet and J. Camassel, *Phys. Rev. B.*, 1997, **55**, 14697–14700.
46 K. S. Wan, S. Tochino, W. L. Zhu, S. Ohtsuka and G. Pezzotti, *J. Phys. D: Appl. Phys.*, 2010, **43**, 205501–205508.
47 G. H. Loechelt, N. G. Cave and J. Menendez, *Appl. Phys. Lett.*, 1995, **66**, 3639–3641.
48 N. G. Ferreira, E. Abramof, E. J. Corat and V. J. Trava-Airoldi, *Carbon*, 2003, **41**, 1301–1308.
49 T. S. Perova, J. Wasyluk, K. Lyutovich, E. Kasper, M. Oehme, K. Rode and A. Waldron, *J. Appl. Phys.*, 2011, **109**, 033502.
50 L. Bergman and R. J. Nemanich, *J. Appl. Phys.*, 1995, **78**, 6709–6719.
51 S. A. Catledge, Y. K. Vohra, R. Ladi and G. Rai, *Diam. Relat. Mater.*, 1996, **5**, 1159–1165.
52 H. Li, T. Zhang, L. Li, X. Lü, B. Li, Z. Jin and G. Zou, *J. Cryst. Growth.*, 2010, **312**, 1986–1991.
53 R. M. Erasmus, J. D. Comins, V. Mofokeng and Z. Martin, *Diam. Relat. Mater.*, 2011, **20**, 907–911.
54 F. Ahmed, M. Krottenthaler, C. Schmid and K. Durst, in *Surf. Coat. Technol.*, 2013, http://dx.doi.org/10.1016/j.surfcoat.2013.1007.1025.
55 M. Abbe, *Arch. Mikrosc. Anat. Entwichlungsmech.*, 1873, **9**, 413.
56 T. Tomimatsu, *J. Appl. Phys.*, 2007, **101**, 123527.
57 T. Tomimatsu, S. J. Zhu and Y. Kagawa, *Scripta. Mater.*, 2004, **50**, 137–141.
58 D. W. Pohl, W. Denk and M. Lanz, *Appl. Phys. Lett.*, 1984, **44**, 651–653.
59 S.-K. Eah, W. Jhe and Y. Arakawa, *Appl. Phys. Lett.*, 2002, **80**, 2779–2781.
60 S. Kouteva-Arguirova, T. Arguirov, D. Wolfframm and J. Reif, *J. Appl. Phys.*, 2003, **94**, 4946–4949.
61 I. De Wolf, H. Norstrom and H. E. Maes, *J. Appl. Phys.*, 1993, **74**, 4490–4500.
62 S. Nakashima and H. Harima, *Phys. Status. Solidi.*, 1997, **162**, 39–64.
63 H. Hiroshi, *Microelectron. Eng.*, 2006, **83**, 126–129.
64 H. J. Van Manen, R. Bloemenkamp and O. F. V. d. Brink, *Appl. Spectrosc.*, 2009, **63**, 378–380.
65 C. A. Froud, I. P. Hayward and J. Laven, *Appl. Spectrosc.*, 2003, **57**, 1468–1474.
66 N. J. Everall, *Appl. Spectrosc.*, 2000, **54**, 1515–1520.
67 S. Michielsen, *J. Appl. Polym. Sci.*, 2001, **81**, 1662–1669.
68 N. Everall, *Appl. Spectrosc.*, 2008, **62**, 591–598.
69 K. J. Baldwin and D. N. Batchelder, *Appl. Spectrosc.*, 2001, **55**, 517–524.
70 V. Presser, M. Keuper, C. Berthold and K. G. Nickel, *Appl. Spectrosc.*, 2009, **63**, 1288–1292.
71 P. S. Anderson, X. Wang and P. Xiao, *Surf. Coat. Technol.*, 2004, **185**, 106–119.
72 G. Lee, A. Atkinson and A. Selcuk, *Surf. Coat. Technol.*, 2006, **201**, 3931–3936.
73 W. Zhang, J. Gan, Z. Hu, W. Yu, Q. Li, J. Sun, N. Xu, J. Wu and Z. Ying, *Appl. Spectrosc.*, 2011, **65**, 522–527.

74 D. Liu, O. Lord, O. Stevens and P. E. J. Flewitt, *Acta. Mater.*, 2013, **61**, 12–21.
75 D. M. Lipkin and D. R. Clark, *J. Appl. Phys.*, 1995, **77**, 1855–1863.
76 A. M. Macdonald, A. S. Vaughan and P. Wyeth, *J. Raman. Spectrosc.*, 2005, **36**, 185–191.
77 T. R. Hart, R. L. Aggarwal and B. Lax, *Phys. Rev. B.*, 1970, **1**, 638–642.
78 M. M. Gentleman, V. Lughi, J. A. Nychka and D. R. Clarke, *Int. J. Appl. Ceram. Technol.*, 2006, **3**, 105–112.
79 H. W. Lo and A. Compaan, *J. Appl. Phys.*, 1980, **51**, 1565–1568.
80 P. H. Duong, H. K. Phan, N. T. T. Tam, D. X. Thanh, N. X. Nghia and P. Lavallard, *J. Lumin.*, 2000, **87–89**, 353–356.
81 M. S. Dresselhaus, in *Encyclopedia of Materials: Science and Technology* (Second Edition), eds. K. H. J. B. Editors-in-Chief, W. C. Robert, C. F. Merton, I. Bernard, J. K. Edward, M. Subhash and V. Patrick, Elsevier, Oxford, 2001, pp. 995–998.
82 K. E. Tae and Y. S. Gil, *Thin. Solid. Films.*, 1993, **227**, 7–12.
83 P. Fauchais, G. Montavon and G. Bertrand, *J. Therm. Spray. Technol.*, 2010, **19**, 56–80.
84 R. A. Miller, *Surf. Coat. Techol.*, 1987, **30**, 1–11.
85 M. Ahrens, R. Vaßen, D. Stöver and S. Lampenscherf, *J. Therm. Spray. Technol.*, 2004, **13**, 432–442.
86 D. S. Rickerby and M. R. Winstone, *Mater. Manuf. Process.*, 1992, **7**, 495–526.
87 D. S. Rickerby, G. Eckold, K. T. Scott and I. M. Buckley-Golder, *Thin. Solid. Films.*, 1987, **154**, 125–141.
88 R. T. Wu, X. Wang and A. Atkinson, *Acta. Mater.*, 2010, **58**, 5578–5585.
89 R. J. Christensen, D. M. Lipkin, D. R. Clarke and K. Murphy, *Appl. Phys. Lett.*, 1996, **69**, 3754–3756.
90 T. Tomimatsu, S. Zhu and Y. Kagawa, *Acta. Mater.*, 2003, **51**, 2397–2405.
91 H. K. Mao and R. J. Hemley, *Nature*, 1991, **351**, 721–724.
92 A. C. Ferrari and J. Robertson, *Philos. Trans. Roy. Soc. London Ser. A.*, 2004, **362**, 2477–2512.
93 M. S. Liu, L. A. Bursill, S. Prawer and R. Beserman, *Phys. Rev. B.*, 2000, **61**, 3391–3395.
94 J. W. Ager III and M. D. Drory, *Phys. Rev. B.*, 1993, **48**, 2601–2607.
95 F. Occelli, P. Loubeyre and R. LeToullec, *Nat. Mater.*, 2003, **2**, 151–154.
96 K. Kunc, I. Loa and K. Syassen, *Phys. Rev. B.*, 2003, **68**, 094107.
97 Y. G. Gogotsi, A. Kailer and K. G. Nickel, *Nature*, 1999, **401**, 663–664.
98 H. Windischmann, G. F. Epps, Y. Cong and R. W. Collins, *J. Appl. Phys.*, 1991, **69**, 2231–2237.
99 T. Gries, L. Vandenbulcke, P. Simon and A. Canizares, *Surf. Coat. Technol.*, 2008, **202**, 2263–2267.
100 E. Perevedentseva, A. Karmenyan, P. H. Chung and C. L. Cheng, *J. Vac. Sci. Technol. B. Nanotechnol. Microelectron.*, 2005, **23**, 1980–1983.
101 H. Windischmann and K. J. Gray, *Diam. Relat. Mater.*, 1995, **4**, 837–842.
102 J. Yu, J. G. Kim, J. O. Chung and D. H. Cho, *J. Appl. Phys.*, 2000, **88**, 1688–1694.
103 S. A. Stuart, S. Prawer and P. S. Weiser, *Appl. Phys. Lett.*, 1993, **62**, 1227–1229.
104 S. Prawer, K. W. Nugent and P. S. Weiser, *Appl. Phys. Lett.*, 1994, **65**, 2248–2250.
105 H. Mohrbacher, K. V. Acker, B. Blanpain, P. V. Houtte and J. P. Celis, *J. Mater. Sci.*, 1996, **11**, 1776–1782.
106 X. Ma, H. Yin, Z. Fu, H. Zhang, X. Zhang, J. Yan, C. Zhao, D. Chen and T. Ye, *MRS Proceedings*, 2012, **1427**.
107 P. W. May, J. A. Smith and K. N. Rosser, *MRS Proceedings*, 2007, **1039**.

108 R. W. Bormett, S. A. Asher, R. E. Witowski, W. D. Partlow, R. Lizewski and F. Pettit, *J. Appl. Phys.*, 1995, **77**, 5916–5923.
109 Z. Sun, J. R. Shi, B. K. Tay and S. P. Lau, *Diam. Relat. Mater.*, 2000, **9**, 1979–1983.
110 D. S. Knight, R. Weimer, L. Pilione and W. B. White, *Appl. Phys. Lett.*, 1990, **56**, 1320–1322.
111 D. Roy, Z. H. Barber and T. W. Clyne, *J. Appl. Phys.*, 2002, **91**, 6085–6088.
112 G. Gouadec and P. Colomban, *Prog. Cryst. Growth. Ch.*, 2007, **53**, 1–56.
113 M. Chaigneau, G. Picardi, H. Girard, J.-C. Arnault and R. Ossikovski, *J. Nanopart. Res.*, 2012, **14**, 1–8.
114 E. Anastassakis and M. Siakavellas, *J. Appl. Phys.*, 2001, **90**, 144–152.
115 K. S. Novoselov, A. K. Geim, S. V. Morozov, D. Jiang, Y. Zhang, S. V. Dubonos, I. V. Grigorieva and A. A. Firsov, *Science*, 2004, **306**, 666–669.
116 K. S. Novoselov, D. Jiang, F. Schedin, T. J. Booth, V. V. Khotkevich, S. V. Morozov and A. K. Geim, *Proceedings of the National Academy of Sciences of the United States of America*, 2005, **102**, 10451–10453.
117 A. K. Geim and K. S. Novoselov, *Nat. Mater.*, 2007, **6**, 183–191.
118 K. S. Novoselov, A. K. Geim, S. V. Morozov, D. Jiang, M. I. Katsnelson, I. V. Grigorieva, S. V. Dubonos and A. A. Firsov, *Nature*, 2005, **438**, 197–200.
119 C. Berger, Z. Song, X. Li, X. Wu, N. Brown, C. Naud, D. Mayou, T. Li, J. Hass, A. N. Marchenkov, E. H. Conrad, P. N. First and W. A. de Heer, *Science*, 2006, **312**, 1191–1196.
120 C. Berger, Z. Song, T. Li, X. Li, A. Y. Ogbazghi, R. Feng, Z. Dai, A. N. Marchenkov, E. H. Conrad, P. N. First and W. A. de Heer, *J. Phys. Chem. B*, 2004, **108**, 19912–19916.
121 Z. Ni, Y. Wang, T. Yu and Z. Shen, *Nano. Res.*, 2008, **1**, 273–291.
122 A. C. Ferrari, J. C. Meyer, V. Scardaci, C. Casiraghi, M. Lazzeri, F. Mauri, S. Piscanec, D. Jiang, K. S. Novoselov, S. Roth and A. K. Geim, *Phys. Rev. Lett.*, 2006, **97**, 187401.
123 S. Piscanec, M. Lazzeri, J. Robertson, A. C. Ferrari and F. Mauri, *Phys. Rev. B.*, 2007, **75**, 035427.
124 A. Bosak, M. Krisch, M. Mohr, J. Maultzsch and C. Thomsen, *Phys. Rev. B.*, 2007, **75**, 153408.
125 C. Lee, X. Wei, J. W. Kysar and J. Hone, *Science*, 2008, **321**, 385–388.
126 Y. Y. Wang, Z. H. Ni, T. Yu, Z. X. Shen, H. M. Wang, Y. H. Wu, W. Chen and A. T. Shen Wee, *J. Phys. Chem. C.*, 2008, **112**, 10637–10640.
127 N. P. Padture, M. Gell and E. H. Jordan, *Science*, 2002, **296**, 280–284.
128 D. P. H. Hasselman, L. F. Johnson, L. D. Bentsen, R. Syed, H. L. Lee and M. V. Swain, *Am. Ceram. Soc. Bull.*, 1987, **66**, 799–806.
129 P. G. Klemens, *Thermal Conductivity*, Technomics, Lancaster, PA, 1993.
130 K. Nomura, Y. Mizutani, M. Kawai, Y. Nakamura and O. Yamamoto, *Solid. State. Ionics.*, 2000, **132**, 235–239.
131 J. I. Eldridge and C. M. Spuckler, *J. Am. Ceram. Soc.*, 2008, **91**, 1603–1611.
132 K. I. M. Byung Kook, H. Jin Woo and H. A. N. Kyoung R, *J. Mater. Sci. Lett.*, 1997, **16**, 669–671.
133 B. Liang, G. Zhang, H. Liao, C. Coddet and C. Ding, *J. Therm. Spray. Tech.*, 2010, **19**, 1163–1170.
134 A. Eichler, *Phys. Rev. B.*, 2001, **64**, 174103.
135 V. K. Tolpygo and D. R. Clarke, *Surf. Coat. Technol.*, 2003, **163–164**, 81–86.
136 E. A. G. Shillington and D. R. Clarke, *Acta. Mater.*, 1999, **47**, 1297–1305.
137 D. R. Clarke, *Acta. Mater.*, 2003, **51**, 1393–1407.
138 V. Lughi, V. K. Tolpygo and D. R. Clarke, *Mater. Sci. Eng. A-Struct.*, 2004, **368**, 212–221.

139 X. Wang, R. T. Wu and A. Atkinson, *Surf. Coat. Technol.*, 2010, **204**, 2472–2482.
140 X. Wang, G. Lee and A. Atkinson, *Acta. Mater.*, 2009, **57**, 182–195.
141 A. M. Limarga, J. Iveland, M. Gentleman, D. M. Lipkin and D. R. Clarke, *Acta. Mater.*, 2011, **59**, 1162–1167.
142 R. Elsing, O. Knotek and U. Balting, *Surf. Coat. Technol.*, 1990, **43–44**, 416–425.
143 M. Karger, R. Vaßen and D. Stöver, *Surf. Coat. Technol.*, 2011, **206**, 16–23.
144 J. Matejicek and S. Sampath, *Acta. Mater.*, 2003, **51**, 863–872.
145 D. Liu, Ph.D. Thesis. University of Bristol, 2012
146 A. Strawbridge and H. E. Evans, *Eng. Fail. Anal.*, 1995, **2**, 85–103.
147 A. C. F. Cocks and N. A. Fleck, *Acta. Mater.*, 2010, **58**, 4233–4244.
148 D. Liu, P. E. J. Flewitt and K. R. Hallam, *Mat. High. Temp.*, 2012, **29**, 181–186.
149 P. Bouvier and G. Lucazeau, *J. Phys. Chem. Solids.*, 2000, **61**, 569–578.
150 A. M. Limarga, R. Vassen and D. R. Clarke, *J. Appl. Mech-T. ASME.*, 78 , 3–11
151 V. Teixeira, M. Andritschky, W. Fischer, H. P. Buchkremer and D. Stover, *J. Mater. Process. Technol.*, 1999, **93**, 209–216.
152 M. Tanaka, M. Hasegawa, A. F. Dericioglu and Y. Kagawa, *Mater. Sci. Eng., A*, 2006, **419**, 262–268.
153 P. Bouvier, E. Djurado, G. Lucazeau and T. L. Bihan, *Phys. Rev. B.*, 2000, **62**, 8731–8737.
154 A. M. Limarga and D. R. Clarke, *J. Am. Ceram. Soc.*, 2007, **90**, 1272–1275.
155 J. G. Cai, Y. S. Raptis and E. Anastassakis, *Appl. Phys. Lett.*, 1993, **62**, 2781–2783.
156 B. Alzyab, C. H. Perry and R. P. Ingel, *J. Am. Ceram. Soc.*, 1987, **70**, 760–765.
157 J. He and D. R. Clarke, *J. Am. Ceram. Soc.*, 1995, **78**, 1347–1353.
158 P. Barberis, T. Merle-Mejean and P. Quintard, *J. Nucl. Mater.*, 1997, **246**, 232–243.
159 S. Guo and R. I. Todd, *Acta. Mater.*, 2011, **59**, 2637–2647.
160 V. Presser and C. Glotzbach, *J. Raman. Spectrosc.*, 2009, **40**, 499–508.
161 C. Rinaldi, L. De Maria and M. Mandelli, *J. Eng. Gas. Turb. Power.*, 2010, **132**, 114501–114504.
162 X. Wang and A. Atkinson, *Mat. Sci. Eng. A-Struct.*, 2007, **465**, 49–58.
163 E. Smith and G. Dent, *Modern Raman Spectroscopy: A Practical Approach*, John Wiley & Sons, Ltd., West Sussex, UK, 2005.
164 O. Kesler, J. Matejicek, S. Sampath, S. Suresh, T. Gnaeupel-Herold, P. C. Brand and H. J. Prask, *Mater. Sci. Eng. A-Struct.*, 1998, **257**, 215–224.
165 T. Clyne and S. Gill, *J. Therm. Spray. Tech.*, 1996, **5**, 401–418.
166 O. Kesler, M. Finot, S. Suresh and S. Sampath, *Acta. Mater.*, 1997, **45**, 3123–3134.
167 G. Montay, A. Cherouat, J. Lu, N. Baradel and L. Bianchi, *Surf. Coat. Technol.*, 2002, **155**, 152–160.
168 A. P. Mirgorodsky, M. B. Smirnov, P. E. Quintard and T. Merle-Mejean, *Phys. Rev. B.*, 1995, **52**, 9111–9114.

Use of Raman spectroscopy and scanning electron microscopy for the detection and analysis of road transport pollution

Jose Antonio Carrero,* Gorka Arana and Juan Manuel Madariaga

DOI: 10.1039/9781782621485-00178

In this chapter we review the applications of Raman spectroscopy and scanning electron microscopy for analysis of the road transport pollution. Raman spectroscopy has been applied on roadside soils and plants and on buildings facades in order to detect traffic emitted compounds. The emission of particles by diesel engines is another important field of study regarding to the air pollution in urban areas. In this sense, apart from Raman spectroscopy, the use of scanning electron microscopy coupled to Energy Dispersive X-ray Spectroscopy (SEM-EDS) plays an important role. In the literature appear several works which focus on the analysis of road dust; particles emitted by brake and tire wear are characterised by SEM images together with X-ray absorption spectra (EDS).

1 Introduction

Road transport emissions are one of the most important environmental problems in many cities and the main source of pollution of urban areas.[1,2] The dispersion of those pollutants is affected by the climatic conditions and they are deposited in surrounding areas affecting soil, water and/or building's facades.

CO_x, NO_x and SO_x emitted in the exhaust fumes, together with organic compounds coming from the combustion of fossil fuels (aromatic hydrocarbons) have been the most problematic pollutants originated by road traffic, leading to the photochemical smog in big cities with elevated traffic density.[3,4] But in recent decades, an increasing concern with inorganic pollutants has grown. The road traffic involves several potential sources of metals, such as combustion products of petrol and oil, products of tire and brake wear, building materials of the road and for traffic safety, and re-suspension of soil and road dust.[1,2,5–8] Apart from the well-known case of lead pollution, due to its use as anti-knocking additive in gasoline, the accumulation of several other metals in roadside and urban soils, as well as sediments, has been detected, mainly Pb, Ba, Zn, Cd and Cu.[9–11] Brake linings emit principally Cu and Sb,[12–14] whereas tire wear leads to Ba and Zn particles.[6,15–17]

There are several studies showing the contamination of roadside and urban soils as a consequence of traffic activity.[9–11,18–20] But in these, the most common approach carried out is to make an acid digestion of the sample and subsequent elemental analysis by means of an atomic spectroscopy technique (ICP-OES or ICP-MS in most of the cases) in order

Department of Analytical Chemistry, University of the Basque Country (UPV/EHU), P.O. Box 644, E-48080 Bilabo, Spain. E-mail: joseantonio.carrero@ehu.es

to obtain the metal concentration. However, the state in which the metals are found in the soil is unknown in this manner, and they can be more or less mobile in the soil, in relation to their geochemical form, which affects their solubility and thus their bioavailability.[21] Therefore, the total concentration is not very representative of the potential toxicity and it is essential to know the chemical form of the metals. Moreover, metal pollution is confined to a narrow area alongside the road, but not limited to the soil surface. Contaminated roadside soils act as secondary pollution sources and may pose a risk to ecosystems and human health, as well as to the built heritage if they are transferred to other reservoirs.[15,19,22] For instance, these pollutants can be transported to the aerial parts of the vegetation, bioaccumulating in them.[18,23] Consequently, metal speciation by molecular spectroscopy techniques is an important task in order to evaluate metal mobility and potential bioavailability of hazardous compounds. In this sense, Raman spectroscopy results are very useful for this purpose since they provide structural and molecular information about the compounds originating from traffic emissions and give an idea of the potential risk of the pollutants in soil. They also help to understand the reaction paths that pollutants undergo once in the soil, leading to the formation of other compounds that could be more mobile than the original pollutant.

1.1 Raman spectroscopy

The main spectroscopic techniques employed to detect vibrations in molecules are based on the processes of infrared absorption and Raman scattering.[24] They are widely used to provide information on chemical structures and physical forms, to identify substances from the characteristic spectral patterns ('fingerprinting') and to determine quantitatively or semi-quantitatively the amount of a substance in a sample. Samples can be examined in a whole range of physical states; for example, as solids, liquids or vapours, in hot or cold states, in bulk, as microscopic particles, or as surface layers. Spectroscopic techniques show a very wide application range and provide solutions to a host of interesting and challenging analytical problems.

Raman spectroscopy is a spectroscopic technique based on inelastic scattering of monochromatic light, usually from a laser source in the visible, near infrared, or near ultraviolet range, called Raman scattering or Raman effect.[25] It was discovered in 1928 by the Indian physicist C.V. Raman, but wide application was delayed until the development of the laser. The importance of this discovery was recognized with the Nobel Prize in Physics in 1930. When light interacts with matter, the photons which make up the light can be absorbed or scattered, or may not interact with the material and may pass straight through it. The process of absorption requires that the energy of the incident photon corresponds to the energy gap between the ground state of a molecule and an excited state. It is this change which is measured in absorption spectroscopy by the detection of the loss of that energy of radiation from the light and is the basic process used in a wide range of spectroscopic techniques. However, when a photon, considered as a propagating oscillating dipole, passes over a molecule, it can interact and distort the cloud of electrons around the nuclei. This energy is released in the form of scattered radiation. In this case there is no need for the photon to have an energy which matches the difference between two energy levels of the molecule. The scattered photons can be observed by collecting light at an angle to the incident light beam.

Consider a molecule in its ground state; when absorbing a laser photon, the molecule is excited towards a virtual energy level. This level is called 'virtual state' since, according to the laws of quantum mechanics, this is a forbidden level. Therefore, the new situation is an unstable state (on the order of 10^{-14} s life-time) and the molecule quickly relaxes towards the ground state, emitting a photon of the same wavelength as the incident photon (Rayleigh scattering). This is an elastic effect which means that the light does not gain or lose energy during the scattering. Raman scattering is different in that it is inelastic. The light photons lose or gain energy during the scattering process, and therefore increase or decrease in wavelength respectively. If the molecule is promoted from a ground to a virtual state and then drops back down to a higher energy vibrational state then the scattered photon has less energy than the incident photon, and therefore a longer wavelength. This is called Stokes scattering. If the molecule is in a higher energy vibrational state to begin with and after scattering is in its ground state then the scattered photon has more energy, and therefore a shorter wavelength. This is called anti-Stokes scattering.[26] At room temperature, most molecules, but not all, are present in the lowest energy vibrational level. The Rayleigh process will be the most intense process since the majority of light undergoes Rayleigh scattering. Raman scattering is inherently a weak process, in that only one in every 10^6–10^8 scattering photons is Raman scattered.[24] In itself this does not make the process insensitive since with modern lasers and microscopes, very high power densities can be delivered to very small samples but it follows that other processes such as sample degradation and fluorescence can readily occur.

Figure 1 shows above mentioned basic processes which occur for one vibration. The relative intensities of the Stokes and anti-Stokes scattering processes depend on the population of the various states of the molecule, but might also be affected by other factors (e.g. fluorescence, surface enhancement, *etc.*). The population ratio for a vibrational state can be worked out from the Boltzmann equation (equation 1) but at room temperature, the number of molecules expected to be in an excited vibrational state other than any really low-energy ones will be small.[24] Note that the line width of the diagram of Fig. 1 is in accordance with the Raman signal intensity.

$$\frac{N_\mathrm{n}}{N_\mathrm{m}} = \frac{g_\mathrm{n}}{g_\mathrm{m}} \exp\left[\frac{-(E_\mathrm{n} - E_\mathrm{m})}{kT}\right] \qquad (1)$$

N_n is the number of molecules in the excited vibrational energy level (n),
N_m is the number of molecules in the ground vibrational energy level (m),
g is the degeneracy of the levels n and m,
$E_\mathrm{n} - E_\mathrm{m}$ is the difference in energy between the vibrational energy levels,
k is Boltzmann's constant (1.3807×10^{-23} JK^{-1}).

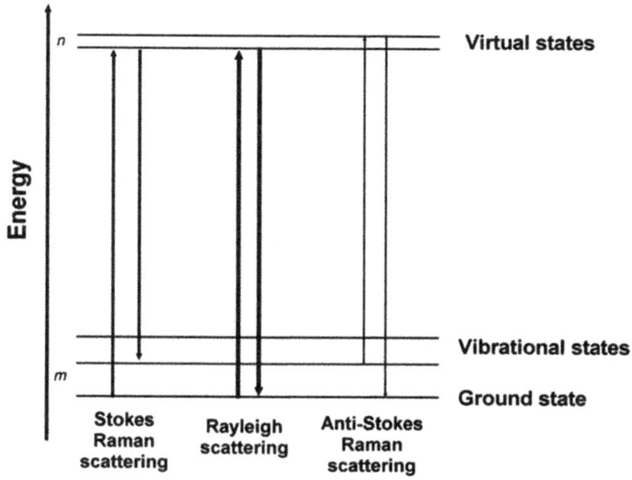

Fig. 1 Energy diagram of Raman and Rayleigh scattering.

Raman spectroscopy is a non-destructive spectroscopic technique, requiring in general only minimal or no sample preparation. Solid (crystalline or amorphous) and liquid samples can be measured as well as transparent or non-transparent samples or samples with different surface textures. Raman scattering usually is measured by irradiating a sample with a narrow spectral line from a continuous laser, but time-resolved measurements also can be made by using a pulsed laser as the light source. Light scattered at 90° or another convenient angle from the axis of incidence is collected through a monochromator and the intensity of the signal is plotted as a function of the difference in frequency or wavenumber between the excitation light and the scattered photons.[27] Strictly speaking, Raman scattering should be expressed as a shift in energy from that of the exciting radiation and should be referred to as $\Delta\,cm^{-1}$ but it is nowadays expressed simply as cm^{-1}. Although different energy ranges are possible, the information of interest to most users is in the 50 to 4000 cm^{-1} range since this includes most modes characteristic of a molecule, including those of metal oxides and oxy-hydroxides in minerals. At wavenumbers between 50 and 1200 cm^{-1}, there are frequently inorganic bands which are valuable for describing the mineral composition of the soils. In addition, to make easier the interpretation of Raman spectra, the use of elemental techniques such as X-ray fluorescence (XRF) is highly recommended.[28,29]

Recent advances in instrument technology have simplified the equipment and reduced the problems substantially.[25] These advances, together with the ability of Raman spectroscopy to examine samples without any preparation, have led to a rapid growth in the application of the technique. Raman microscopy couples a Raman spectrometer to a standard optical microscope, allowing high magnification visualisation of a sample and Raman analysis with a microscopic laser spot. Traditionally Raman microscopy has been used to measure the Raman spectrum of a point on a sample (few micrometres). The high spatial resolution, and the use of automated stages, enables mapping and imaging experiments to be carried out relatively easily. However there are disadvantages. For example, obtaining a representative spectrum from a sample, which may be inhomogeneous at the microscopic level, is difficult. From the optical engineer's point of view, the use of a microscope to detect the scattering has some advantages. A relatively low-powered laser can be used since it will be focussed to give a very small spot giving a high power density at the sample and also a large collection angle. The microscope can be set up as a simple microscope or can be set up confocally. The advantages of using a microscope have so far focussed only on the X–Y plane. The microscope can also be used to advantage by changing the focus in the Z direction. This configuration allows the possibility to carry out also depth profile analysis.[24]

1.2 Scanning electron microscopy

The scanning electron microscope (SEM) is one of the most popular and user-friendly imaging tools that reveal the surface topography of a sample. It is also widely used for structural characterization of materials and devices. A typical SEM consists of an electron gun, an electron lens system, various electron beam deflection coils, electron detectors, and display and recording devices.[30] The electron beam is focused into a fine probe, which is scanned point-by-point across the surface from a region of interest in the sample. The electron beam is emitted from an electron gun in high vacuum and is accelerated towards the specimen (0.5–40 keV) while is confined and focused using metal apertures and magnetic lenses into a thin, focused, monochromatic beam. The resulting signals from the interaction of the electron beam with the sample, which include among others, secondary, and backscattered electrons along with characteristic X-rays, can be collected with suitable detectors and transformed into an image providing highly detailed spatial and compositional information.[31]

Signal generation in SEM is a result of the interaction between the incident electron beam and a thin surface layer of the sample, which depends on the beam energy. These interactions can be divided into two major categories: elastic interactions and inelastic interactions. The interaction is said to be inelastic if some of the energy of the primary electron is lost during the interaction. If no energy is lost the interaction is said to be elastic.[32] Elastic scattering results from the deflection of the primary electrons (PE) by the specimen atomic nucleus or by outer shell electrons of similar energy. This kind of interaction is characterized by negligible energy loss during the collision and by a wide-angle directional change of the scattered electron. Primary electrons that are elastically scattered through an angle of more than 90° are called backscattered electrons (BSE). Inelastic scattering occurs when the primary electrons strikes the sample surface causing the ionization of specimen atoms. If the energy of the incident electron is high enough, the valence electrons of the surface atoms can easily be released from the atoms (Fig. 2).

These electrons are called secondary electrons (SE). The secondary electron signal yields an image with a three-dimensional perspective,

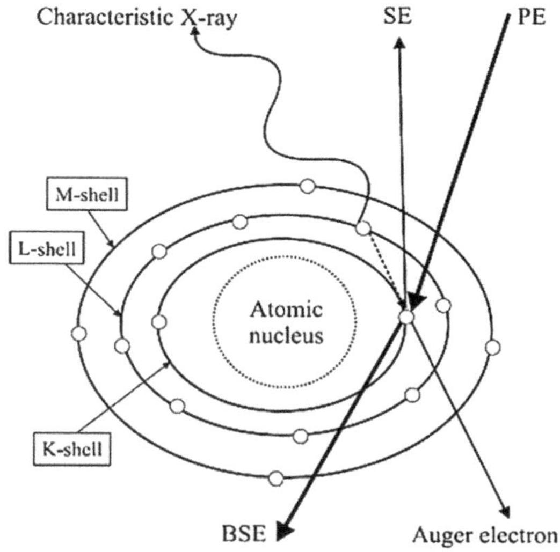

Fig. 2 Inelastic interaction between primary electrons and electrons in atomic shells. Taken from reference 32 with permission of Springer.

high depth-of-field and the appearance of overhead illumination. The backscattered electron signal yields an image containing compositional information because the signal is dependent on the atomic number of the particle being examined. Higher atomic number elements generate more backscattered electrons than lower atomic number elements. This results in intensity variations in which features consisting of higher atomic number elements appear brighter than those consisting of low atomic number elements. With this attribute, the backscattered mode offers the ability to discriminate between phases within a particle.[31]

When the sample is bombarded by the electron beam of the SEM, electrons are rejected from the atoms on the specimens' surface. A resulting electron vacancy is filled by an electron from a higher shell, and an X-ray is emitted to balance the energy difference between the two electrons. The Energy Dispersive Spectroscopy (EDS) X-ray detector (also called XEDS or EDX) measures the number of emitted X-rays *versus* their energy. The energy of the X-ray is characteristic of the element from which the X-ray was emitted. In practice, EDS is most often used for qualitative elemental analysis, simply to determine which elements are present and their relative abundance.[33]

Since the SEM is operated under high vacuum the specimens that can be studied must be compatible with high vacuum ($\sim 10^{-5}$ mbar). This means that liquids and materials containing water and other volatile components cannot be studied directly. Also fine powder samples need to be fixed firmly to a specimen holder substrate so that they will not contaminate the SEM specimen chamber. Non-conductive materials need to be attached to a conductive specimen holder and coated with a thin conductive film by sputtering or evaporation. Typical coating materials are Au, Pt, Pd, and their alloys, as well as carbon.[34]

The high spatial resolution of a SEM makes it a powerful tool to characterise a wide range of specimens at the nanometre to micrometre length scales. The ability to provide detailed information on the morphological and chemical characteristics associated with individual particles has made the SEM the method of choice in many studies where bulk analytical methods have insufficient resolution to identify the source(s) effecting ambient air quality.

2 Applications of Raman spectroscopy on traffic pollution

2.1 Detecting traffic impact on soils

Raman spectroscopy is a very good tool for the examination of inorganic materials or those materials containing inorganic components. In fact, the technique was initially used to study inorganic materials, as early work carried out in Raman spectroscopy showed its strength with those compounds. Particulates in urban dust such as anhydrite, calcite, dolomite and quartz have been identified and characterized. Nowadays the applications of Raman spectroscopy are multiple and its use is spreading to different areas, such as, pharmaceutical, forensics or polymers.[24] The potential of the application of Raman spectroscopy to soils has been shown. Metal speciation by Raman spectroscopy has made possible the evaluation of metal mobility and potential bioavailability of hazardous compounds in urban soils and sediments.[29,35] In this chapter we review the recent application of the technique for the analysis of different samples, in order to detect the impact of traffic pollution in urban areas.

Raman spectroscopy has been performed on several sample types but little work is available where it is applied on soils. Carrero *et al.*[29] have recently published a study where they use Raman spectroscopy to assess the impact of traffic pollutants and guardrails on roadside soil. Soil has a complex matrix and obtaining good Raman spectra is a difficult task. The clay matrix of the soil gives a high level of fluorescence, which could saturate the Raman signal. In addition to this, the presence of organic matter can enhance fluorescence phenomena, and give rise to interferences in the observation of the characteristic Raman spectra of the desired compounds present in the soil matrix. Both problems make the observation and interpretation of the spectra difficult. Therefore, different sample pre-treatments must be carried out to minimize the fluorescence phenomena and make spectral interpretation easier.[29] On the one hand, short acquisition times at low intensity power for spectra acquisition are recommended to be used in order to avoid saturation (exposure times from 5 to 30 s and 1–10 accumulations at 1 or 10% power). On the other hand, soil samples must be first submitted to a solid–liquid extraction with acetone during 30 minutes in an ultrasonic bath to eliminate the possible organic matter, according to a previous study.[36] Nevertheless, in some samples high fluorescence can still be obtained and the resulting spectra can be too noisy making the application of baseline correction and smoothing necessary. Using this methodology, Carrero *et al.*[29] could detect the degradation compound of the galvanized zinc layer coating the surface of the guardrail; hydrozincite ($Zn_5(CO_3)_2(OH)_6$). This compound can be easily dissolved by the rain around the road environment that contains carbonic and nitric acids, as the most concentrated ones. The attack of both acids on hydrozincite leads to Zn^{2+}, HCO_3^- and NO_3^- as the main dissolved ions in the rain-wash of the guardrail. This degradation process can be considered to be responsible for the high concentration of Zn found in soils under the guardrail. The subsequent Raman analysis of the soil samples under guardrails showed the presence of $Zn(NO_3)_2$ which proves in terms of mobility of metals in soil a major risk due to its higher solubility. In the same study the authors were also able to detect the presence of $BaCO_3$ deposited on the guardrail, as well as in soil. The use of Raman spectroscopy enabled them to explain the presence of traffic related elements in elevated concentrations in roadside soils, and the reaction paths that they underwent once emitted by the traffic.

In another paper published by the same authors,[37] a number of different compounds were described to be found in roadside soil. Once again, clay matricies of the soil represented a drawback for the detection of pollutants. Soil samples gave a high background Raman signal and the less intense bands (weak) were covered by this fluorescence emission. In most cases, only the principal bands (very strong) were clearly observed in the spectra. It is common to obtain spectra of some of the major components of the soil, such as quartz (SiO_2), calcite ($CaCO_3$), silicates and different crystalline forms of titanium and iron oxides. The power of the technique goes further than to be able to identify the molecular form of the pollutants in soil. Not only is it possible to determine that the metal is as oxide, carbonate, sulphate, nitrate or whatever other possible molecular form, but also to distinguish among different structural forms of the same molecular compound. That work identified three different crystalline forms of the titanium oxide: anatase, rutile and brookite, all three of them with the same molecular form TiO_2 (Fig. 3). In the Raman spectrum of the Fig. 3a, a weak band can be seen at 300 cm^{-1} together with those corresponding to the anatase compound. This band matches with the Raman spectrum of the standard of litharge (PbO) and therefore was attributed to it. Leaded gasoline was banned in Spain in 2001, but high levels of Pb concentration still remain in roadside soils nowadays. Analysis by means of elemental techniques, such as ICP-MS or XRF, gives information about the total concentration of Pb in soil. But the presence of lead as an oxide form detected by Raman spectroscopy gives a more valuable information. This form of lead is not very mobilizable and accumulates in top soil,[9,38] which explains the high Pb concentration in roadside soil ten years after the phase-out of its use as an additive in gasoline.

According to the authors, the spectrum containing anatase and litharge bands in Fig. 3a have some other not fully attributed bands. They seem to correspond to other oxides and, possibly, to an organic complex form of Cu, because some of the peaks match well with the Raman spectra of the Cu–O bands. They also stated that they could identify a

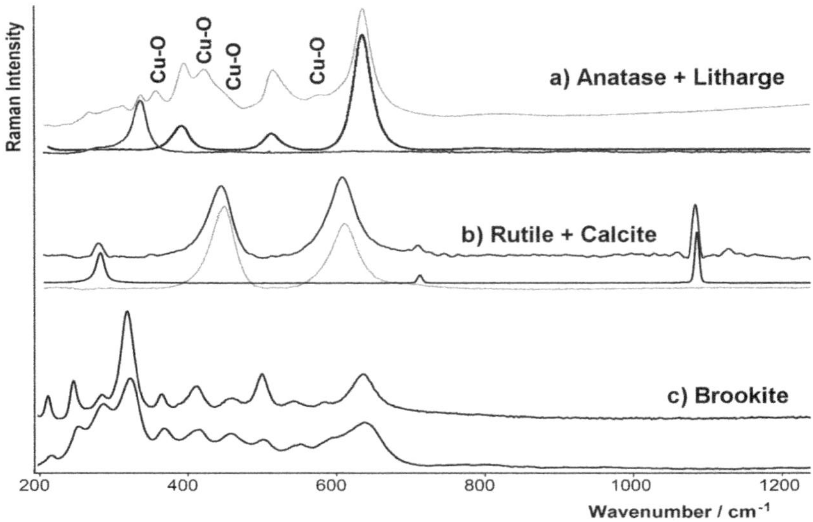

Fig. 3 Raman spectra of some compounds found in roadside samples. Adapted from reference 37 with permission of John Wiley & Sons, Ltd.

silicate of the feldspar family, albite ($NaAlSi_3O_8$), and the iron oxides goethite (FeOOH) and hematite (Fe_2O_3).

Raman analyses were also conducted on vegetation samples found at the roadside.[37] Focusing the laser on different particles coming from the atmospheric deposition found on the leaves, the presence of traffic emitted particles can be detected. However, only quartz (SiO_2) and calcium carbonate ($CaCO_3$) were identified in this study. In this case, the source is the deposition of the dust soil turned over by the wind and passage of vehicles. Broad band at 1300 and 1600 cm^{-1} usually appears when analysing vegetal samples. These bands correspond to amorphous carbon originated by photo-thermo decomposition of the leaves by the laser and special care must be taken, reducing the laser power and/or the exposure time.

2.2 Detecting traffic impact on buildings

Raman spectroscopy has also been utilized to analyse building facades in order to assess their deterioration state for conservation and restoration purposes. That is the case of the study carried out by Maguregui *et al.*[39] The atmospheric pollution is considered the crucial factor in building degradation having caused in some cases a significant loss in the cultural heritage, affecting even to metallic sculptures exposed outdoors.[40] Building materials suffer diverse mechanisms of deterioration due to chemical, physical and biological phenomena that are caused by atmospheric factors. One of the principal results of those kinds of deteriorations is the formation of soluble salts. That effect takes place as a consequence of dissolution-crystallisation and/or hydration-dehydration cycles that contribute to the crystallisation of different salts, which may have different numbers of water molecules, within porous building materials contributing to the porous material decay.[41] The decay pathway starts with the transformation of the atmospheric pollutants (CO_2, SO_x, NO_x) into their respective most oxidised acids (H_2CO_3, H_2SO_4, HNO_3) by interaction with ozone in the presence of humidity or rainwater. The formation of those acids increases the acidity of rainwater in urban areas, which results in reaction between the acid and building materials and contributes to the formation of salts (nitrates, sulphates, carbonates) with different composition depending on the building materials.

Nitrobarite ($Ba(NO_3)_2$) was identified in the external part of a brick sample of a building located very close to a road with high traffic density (more than 20,000 vehicles per day).[39] The Raman study was complemented with μ-XRF analyses and higher concentration of Ba was obtained on the external part of the bricks than in the inner bulk. The barium appearing in the internal part of the brick could have been due to the addition of Ba in the manufacturing process, yielding the formation of BaO. This BaO would be converted to $BaCO_3$ following the same pathway described for other oxides as CaO, which is hydrated to $Ca(OH)_2$ and turns into $CaCO_3$ as a results of the attack from atmospheric or traffic emitted CO_2. However, in this case, the detected salt was $Ba(NO_3)_2$ and only in the external part of the brick. The fact that nitrates of other cations (Ca, Na, Mg) were not found implies that neither NO_x massive attack nor nitrate salt infiltration has happened. Consequently, the presence of $Ba(NO_3)_2$ could be explained by traffic emissions. Airborne particles of barium oxide emitted by traffic as a consequence of tire wear[5] react with atmospheric NO_x aerosol (emitted by traffic exhaust fumes) in its most oxidised acidic form, HNO_3. The barium nitrate formed could then be deposited on the surface of the brick forming the black crust with other airborne contaminants.

With the same aim, Prieto-Taboada *et al.*[42] studied the accumulation of atmospheric pollutants on buildings, which can act as repositories of the past pollution. That work shown a substantial concentration of pollutants even after removing the emission source. Generally, pollutants accumulate on the building surface, in zones frequently soaked by rainwater but not washed out. Grey-to-black crust is the most common formation and it is commonly composed of gypsum crystals and atmospheric depositions, including carbonaceous particles (soot) and heavy metals. By means of Raman spectroscopy, they were able to detect $BaSO_4$ in a mortar sample from an old building in the surroundings of Bilbao (Spain).[43]

2.3 Analysis of diesel exhaust particles

The sensitivity of Raman spectroscopy (which enables one to investigate the composition, phase, crystallinity and crystal orientation), makes it an ideal tool to characterise individual heterogeneous particles in the fine and even ultra-fine particle size ranges. To date, Raman spectroscopy has been used extensively in many applications in both routine and exploratory, but only a limited number of publications on its application to environmental particles in the micrometre and sub-micrometre range could be found, some of them pertaining to the analysis of particles emitted by diesel engines.

In recent decades, air quality has become a very important concern as more and more studies have shown the great impact of atmospheric pollution on the environment, global climate change, and human health.[44] Whereas diesel engines are often resulting in smaller amounts of carbon monoxide and hydrocarbons emissions in comparison with gasoline engines, diesel engines emit a significantly higher amount of particulate matter (PM). Hence, in urban environments, traffic pollution is considered the most important source of particle emission to the atmosphere,[4,45–48] emitting fine (particles with diameters smaller than 25, 10 and 2.5 μm and denoted as PM_{25}, PM_{10} and $PM_{2.5}$) and ultrafine (particles having diameter of less than 100 nm) particles that can easily penetrate deep into the respiratory system. PM may include a broad variety of chemical species, ranging from metals to organic and inorganic compounds. Sulphate, nitrate, and ammonium salts, along with, organic chemicals, biological materials, carbon and metals generally adsorbed on carbonaceous cores are known to be the major chemical components of atmospheric PM.[49] There might be a causal relationship between exposure to diesel emissions from mobile sources and the incidence of cancer, respiratory symptoms and respiratory diseases. Diesel PM have therefore potential environmental impacts, including health effects, climate change, ecological effects and visibility. The health effects of inhaled particulate matter are associated with the size, shape, and chemical toxicity.

The sampling and analysis of airborne particulate matter is complicated by the complexity of the particle size, particle interactions, chemical partitioning between gaseous, liquid, and solid phases, and interaction with sampling media.[44] Besides, analytical methods for determination of the chemical composition of airborne matter require both sophisticated equipment and often very strict and time-consuming sample preparation techniques. One of the most widely studied groups of analytes found in ambient aerosols are polycyclic aromatic hydrocarbons (PAHs), which originate from incomplete combustion, and therefore are often accompanied by the presence of carbonaceous solid microparticles in gasoline and diesel motor exhaust.

In addition to the determination of organic molecules by GC-MS, Koziel *et al.*[50] conducted preliminary experiments to determine the feasibility of single particle matrix investigations with SPME. These experiments were completed using 7-μm PDMS fibre and Raman microspectroscopy measurements. Characteristic bands of Si–O–Si and methyl groups originating from the PDMS could be detected. These bands did not interfere with bands for carbon and hydrocarbon groups that originated from analysed particles. According to the authors, Raman spectra of uncoated (bare) fibres and fibres coated with PDMS were obtained using 50 × magnification with the short focus objective lenses, since a 100 × magnification was necessary only for the very small particles. The laser power had to be adjusted in order to avoid local burning of the sample when attempted to collect a Raman spectrum with the maximum (7.2 mW) laser power. With a laser power lowered to 0.071 mW, (the lowest possible power with the 50 magnification lens) they obtained some Raman spectra, which were characterized by two Raman bands at 1336 and 1600 cm^{-1}, indicating the presence of carbonaceous material as a result of incomplete diesel combustion.

Using this methodology, Odziemkowski *et al.*[51] developed, for the first time, the use of solid-phase microextraction (SPME) fibres for airborne particulate sampling followed by both Raman microspectroscopy and GC-MS analysis. Commercial polydimethylsiloxane (PDMS) 7-μm SPME fibres were used for sampling of airborne particulate matter followed by Raman microspectroscopic analysis of single particulates. SPME fibre coating performed very well in sampling of various exhausts from organic fuel samples. Moreover, sampling with SPME was very simple and fast. Raman analysis indicated that in all investigated cases the PDMS coating was stable and inert toward analytes. No interaction with the sampling media was observed. Raman bands of the PDMS coating were observed exactly at the same positions as for blanks and air samples. The sampling is easy and very effective in the field application as demonstrated by diesel exhaust sampling and Raman analysis.[50] The main advantage of using Raman microscopy for identification of the airborne particulate is that it is suitable for single-particle analysis of less than 1 μm in diameter. According to the work of Odziemkowski *et al.*, combining SPME and Raman microspectroscopy, known to be a powerful method of chemical fingerprinting of molecules, is also a potentially new procedure for the identification, study, and possible numeration of airborne particulate matter.

Air sampling of tailpipe diesel exhaust, from a hot and a cold engine with SPME fibres results in the deposition of the graphitic carbon on the surface of the SPME fibre. In the case of a "cold engine", the entire surface of the SPME fibre is covered with a thin carbon film after 1-min. exposure. However, for "hot engine", the 1-min exposure of SPME fibre to the tailpipe diesel exhaust did not result in a measurable Raman signal.[51] The extension of the exposure time to 5 min was necessary to achieve meaningful Raman measurements. In general, the SPME fibre coating was covered with much fewer black spots compared to the cold engine sample. As it can be seen in Fig. 4, the SPME fibre is covered by randomly distributed black spots with shapes that might be described as regular and irregular ovals with diameters varying from less than 1 to 5 μm (Fig. 4B). The appearance of two characteristic Raman bands indicates the presence of carbonaceous material as a result of incomplete diesel combustion (Fig. 4C).

Based on the work of Odziemkowsky *et al.*,[51] Pacenti *et al.* presented a new methodological procedure based on a sequential mixed application of Raman microspectroscopy and automated SPME Multi Fibre System (MFS)-Fast GC-MS.[52] In their work, the exhaust emissions of 8 vehicles of different categories were analyzed to attempt to differentiate them (four diesel automobiles, a gasoline automobile catalyst-equipped and one heavy-duty diesel truck). The exhaust emissions were sampled keeping the SPME fibres longitudinally near the tailpipe for 150 s, as carried out for the first time by Odziemkowsky *et al.*[51] Whereas Odziemkowski *et al.* stated in their work they performed Raman analyses in the confocal

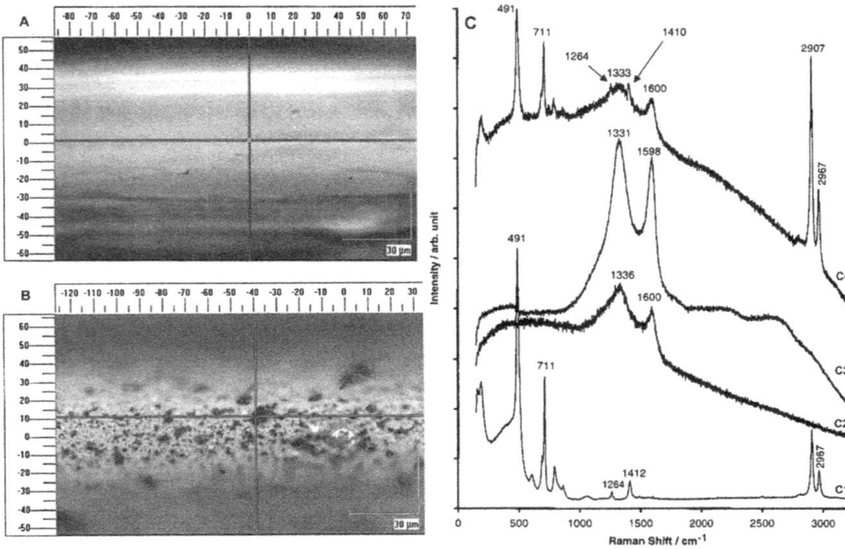

Fig. 4 White light images of SPME fibre coated with the thin layer of PDMS before (A) and after (B) 5-min exposure to hot engine diesel exhaust smoke with a magnification of 50 ×; and Raman spectra of the SPME fibre blank (C1), and those collected at three different locations on the SPME fibre coating that was exposed for 1 min to cold diesel exhaust (C2-C4). Adapted from reference 51 with permission of Analytical Chemistry Society.

apparatus configuration with the aim of diminishing fluorescence, such a configuration was not necessary during this new study because the laser source used for Raman measurements emits light at 785 nm, so that the range of interesting Raman signal is fairly far away from that related to organic matter's fluorescence. After exposure to the exhaust emissions, fluky disseminated black particles of different (regular or irregular) morphology and size appeared on the fibre surface. Every black particle/grain visible to the microscope with 50 × magnification was examined through Raman spectroscopy. Almost all the Raman spectra registered in correspondence of the black spots presented the vibrational bands of the PDMS coating and of the glass fibre underneath, together with two bands located near 1300 cm^{-1} and 1600 cm^{-1} (Fig. 5A). These latter signals are characteristic of carbonaceous material which results from incomplete diesel combustion. They could also observe the presence of calcite, hematite and quartz, which are typical species in dust, in some spectra collected in sporadic locations of the fibre. On the other hand, they obtained a very interesting result once on a black particle: the Raman signal of PAHs. Figure 5B shows some peaks which may be attributed to PAHs. However they do not go into further details.

2.3.1 Analysis of diesel soot. Carbonaceous aerosols have been identified as major contributors to climate change with radiative forcing, (*i.e.* the difference of radiant energy received by the earth and energy radiated back to space) of the same magnitude as greenhouse gases. These aerosols originate in biomass burning or incomplete combustion

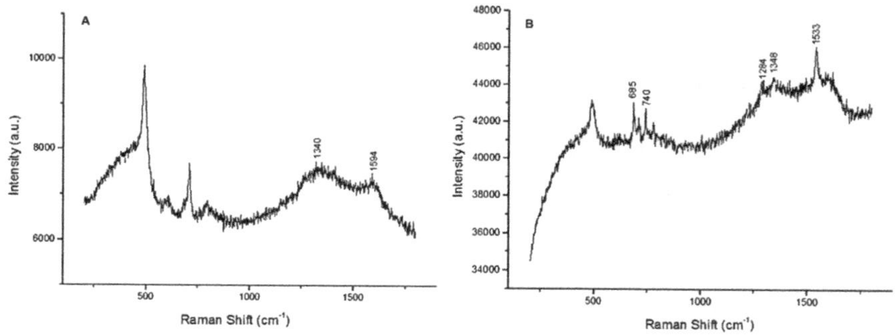

Fig. 5 Raman spectra of A) carbonaceous particle on the SPME fibre after exposure to diesel engine exhausts and B) a black particle; the signals in the high frequency region may be indicative of some PAHs. Taken from reference 52 with permission of Biolife.

of fossil fuel and absorb a part of the solar radiation causing a warming effect on climate. However, the level of scientific understanding on the climatic effect of aerosols is very low. This is due to the lack of knowledge on both the organic/black carbon ratios present in the aerosols, as well as the scattering and absorption cross section of the particulates and, hence, results in a high degree of uncertainty associated with the estimates of the radiative forcing of aerosols. The radiative forcing of carbonaceous aerosols depends on their scattering and absorption properties, which in turn are strongly influenced by morphology and nanostructure (*i.e.*, amorphous *vs.* graphitic carbon content of the soot). Hence, considerable efforts are being spent to characterize the morphology of carbonaceous aerosols and their nanostructure.

On the other hand, atmospheric aerosol particles are of central importance for atmospheric chemistry and physics, climate, and public health. Especially in urban areas, soot particles emitted by diesel engines account for a major fraction of air pollutants. Present and future emission limits require that soot particles must be efficiently removed from diesel engine exhaust. A wide range of particle trapping systems and exhaust after treatment technologies have been proposed and are currently under development. Continuously regenerating traps (CRT) or diesel particulate filters (DPF) which have been applied for this purpose need to be regenerated periodically by oxidation and gasification of the deposited soot. The behaviour of this regeneration step is strongly influenced by the structure and reactivity of the deposited soot particles. Especially the formation of highly reactive soot would make it possible to perform this regeneration step at relatively low temperatures. For this reason, knowledge about the structure of soot can be used also for investigation and optimization of diesel exhaust after treatment (CRT and/or DPF) systems. Moreover, this information can help to understand the influence of soot particles on environmental chemistry, climate, and public health.

Actually, Raman scattering is a highly sensitive spectroscopic technique useful to study atomic microstructures in carbonaceous materials and is sensitive to the structural change or rearrangement of carbon crystallites. With the aim of characterizing the internal microstructure of carbonaceous aerosols and diesel soot, Raman spectroscopy has been applied to investigate and analyse PM samples emitted from different motor vehicles and get detailed information about the reactivity of soot. Raman spectroscopy is sensitive to the graphene layer distribution of the soot. The first-order Raman spectra of soot and related carbonaceous materials typically shows two broad and strongly overlapping peaks with intensity maxima near 1580 cm^{-1} (G or "graphite" peak) and 1350 cm^{-1} (D or "disorder" peak), that are composed of at least five distinct bands. The D peak is known to be characteristic for disordered graphite and its intensity increases relative to the G peak with increasing degree of disorder in the graphitic structure.[53] In contrast to the G peak, the D peak is dispersive, its position shows a shift (from *ca.* 1350 to 1330 cm^{-1}) with the increase of excitation wavelength. The shape of the spectrum, therefore, has the potential to provide several of the "dimensions" needed to characterize a nanostructure.

In this sense, Soewono *et al.*[54] characterised the morphology and Raman spectra of different engine-emitted particles (a light-duty diesel turbo charged direct-injection (LDDTDI) engine and a heavy-duty four-stroke, direct-injection natural gas engine). They tried different powers of excitation laser (1%–100% relative intensity), spot diameter (0%–100% defocusing) and exposure time in order to find the optimum measurement conditions. The spectrometer had to be mostly operated in non-continuous scanning mode to avoid burning the soot samples. The highest quality and reproducibility of the soot spectra were achieved with a fully focused laser beam (diameter of laser spot 1–2 μm), a laser power of 10% ($\sim$2 mW), and an exposure time of 4×30 s. The spectral parameter of the soot was determined by the curve fitting after linear baseline correction. To ensure the reproducibility of the curve fit, the fitting procedure was repeated at least five times for each Raman spectrum. They applied the Raman spectral analysis with both two-band ("G" and "D" at $\sim$1578 and $\sim$1340 cm^{-1}) and five-band (G, D1, D2, D3, and D4 at about 1580, 1350, 1500, 1620, and 1200 cm^{-1}, respectively) combinations to quantify the degree of structural disorder present in soot. They observed that both methods indicated that biodiesel results in less-ordered soot. For soots from both engines, higher loads resulted in apparently more ordered soot. The work concluded that comparing the numerical values of spectral parameters remains very difficult and the spectral fitting procedures can be critical because the two-band model often fits the spectra poorly and the five-band method can be sensitive to the treatment of contamination, noise, and drift. Moreover, the authors suggested that some of the difficulty in interpreting Raman spectra from carbonaceous materials may arise because the complexity of the nanostructure is not adequately represented by a single index corresponding to the degree of graphitization.

Knauer *et al.*[55] studied changes in structure and reactivity of soot during oxidation by oxygen at increasing temperatures. They analysed the correlation between structure and reactivity of GfG soot, EURO VI and

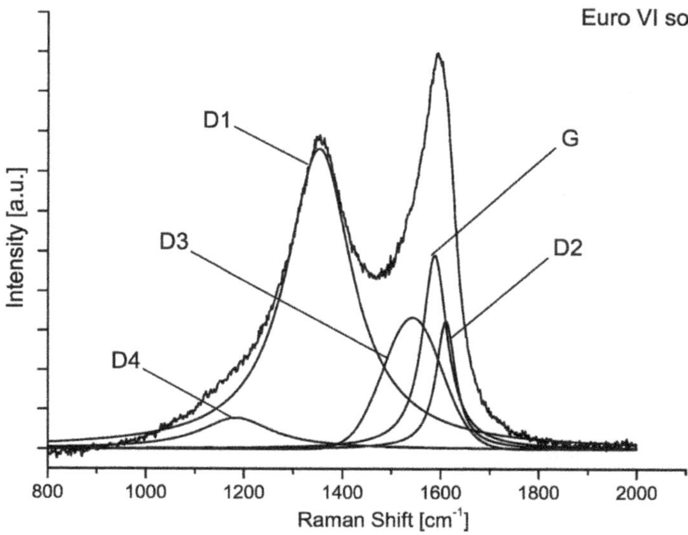

Fig. 6 Typical spectrum ($\lambda_0 = 514\,nm$) of untreated EURO VI soot with five band fits. Taken from reference 54 with permission of Analytical Chemistry Society.

EURO IV diesel soot, and graphite powder with different analytical methods to get an overview on the relationship between structure and reactivity of soot particles in diesel exhaust. GfG soot and graphite powder represent the higher and lower reactivity limits. Spectra of the soot samples were taken before and after the oxidation process with a $50\times$ magnification objective and 10 s integration time. A defocused laser beam (diameter $\sim 40\,\mu m$) and 25% of the source power were applied to avoid laser-induced decomposition of the soot samples. After a multi-point baseline correction, determination of spectral parameters using a five band fitting procedure (G, D1–D4) as well as by evaluation of the dispersive character of the D mode. The Fig. 6 shows an exemplary spectrum of untreated EURO VI soot with five band model fit.

They found that the observed Raman spectra of untreated EURO VI and EURO IV soot samples are quite different in comparison to spectra of GfG soot (Fig. 7). The two Raman peaks are more separated, which implies a more homogeneous structure with a lower content of molecular carbon for EURO VI and EURO IV soots than for GfG soot. Also, the analysis of Raman spectroscopic parameters show a higher degree of disorder and a higher amount of molecular carbon for untreated GfG soot samples than for untreated EURO VI and EURO IV soots. The structural analysis based on the dispersive character of the D mode shows grown differences in spectra of graphite powder and EURO IV, VI, and GfG soots measured at 514 and 633 nm, which can be explained by an increasing degree of disorder and/or molecular content from graphite powder to GfG soot. The obvious changes in dispersion for spectra of GfG soot before and after oxidation suggests changes in the structure of those samples, whereas no significant changes can be found in difference spectra of graphite powder and EURO IV and VI soots samples before and

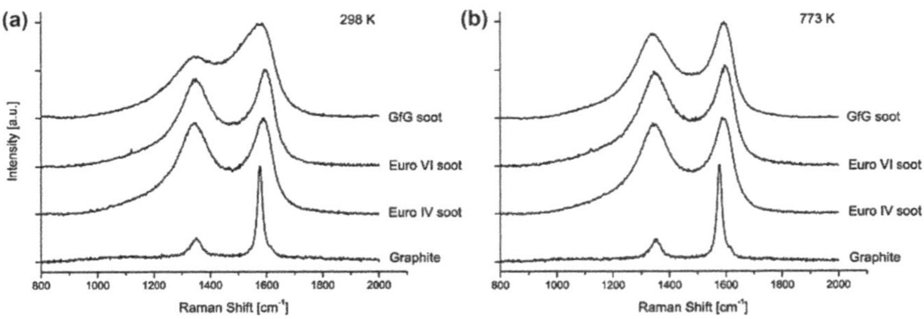

Fig. 7 Raman spectra ($\lambda_0 = 514$ nm) of untreated (a) and oxidized (b) GfG soot, EURO VI soot, EURO IV soot, and graphite powder. Taken from reference 55 with permission of Analytical Chemistry Society.

after oxidation (Fig. 7). Finally, authors concluded that Raman spectroscopy could be established as a rapid analytical tool to predict reactivity of soot by analysing the structure. In particular, the dispersive character of the D mode can be applied for structural analysis of soot and related carbonaceous materials.

3 Applications of SEM-EDX on road dust and environmental particles

Chemical and structural compositions of environmental particles are of prime importance with reference to their impact on the environment and humans, especially if these particles belong to the fraction that is airborne. Particulate matter produced by motor vehicles mainly originates from the combustion of hydrocarbons and from the wear of tires, brakes, and the road surface, producing the road dust. Resuspended road dust can contribute to the atmospheric particulate matter of big cities in a profound manner. Moreover, road dust poses a health concern due to carcinogenic and toxic components potentially present in the micron-sized fractions. Hence, the characterisation of the molecular composition of the fine fraction is evidently of importance for public health. Open literature reports on the elemental concentrations of trace, toxic metals and metalloids present in road dust, but molecular composition is seldom investigated. The application of scanning electron microscopy coupled with energy-dispersive X-ray spectroscopy detection (SEM-EDS) for individual environmental particles appeared to be a useful method for studying individual particles, especially fine particles of atmospheric origin. It allows the characterisation of individual particles. By following this strategy of characterisation in conjunction with bulk elemental analysis, information about the chemical composition and properties of aerosol particles is provided and it significantly improves the knowledge about the atmospheric reactions that these airborne particles can undergo.

Bucko *et al.*[56] analysed the microstructural characteristics of road dust from roadsides in Finland with the aim to study the magnetic pollutants derived from road traffic, since in urban areas, circulation of motor vehicles is considered among the most important sources of magnetic particulate emission into the environment. They took soil samples from two areas with different traffic volumes and performed direct observations of the magnetic extracts by means of SEM-EDS to characterize properties such as grain size, shape, and mineralogical composition. They identified two groups of magnetic materials: angular/aggregate particles (diameter $\sim$2–100 μm) derived from road traffic pollution and magnetic spherules (d $\sim$ 3–15 μm) possibly originated from industrial and domestic heating systems. Traffic derived particles where the most common observed magnetic materials throughout the study and can be grouped into two types according to their compositions: (1) magnetic particles including Ni, Cu, and Bi and (2) iron oxide particles including irregular distribution of different chemical elements, such as Ti, Mn, Cr, Mg, Ca, Cu, Fe, S, V, Ni, Zn, and Si. In addition to magnetite, pyrrhotite (FeS) was a common component of magnetic extracts. The authors attribute the presence of such metals to traffic emissions according to different publications found in the bibliography. These state that vehicles may generally produce non-spherical magnetite particles *via* exhaust emissions, braking systems, and the abrasion or corrosion of vehicle engine and body work.

The same authors published another paper where they performed SEM-EDS analyses on roadside snow samples with the aim to study the applicability of snow surveying in collection, identification and detailed characterization of vehicle-derived magnetic particles.[57] The importance of studying snow samples lies on the fact that snow acts as a natural filter for various chemical elements and particles originated from anthropogenic activities (*e.g.*, industry, road traffic). It appears to be a good collector of organic and inorganic pollutants from the atmosphere and can be used to monitor local airborne pollution from road traffic. Deposition of polyaromatic hydrocarbons, heavy metals, platinum-group elements (Rh, Pt, Pd) and concentrations of particles in different size fractions on snow have been reported. In Nordic countries, snow remains on the ground for several months. Therefore, snow surveying is an ideal method of environmental monitoring, especially when natural filters such as vegetation and soil are not available. In their new study they confirmed that roadside snow samples were enriched in particles with different shape and grain size, and complex mineralogical composition.[57] Angular particles composed of Fe, Cr and Ni were the most abundant. Furthermore, particles containing Zn, $BaSO_4$ and W were also identified (Fig. 8). Enrichment in these elements may be attributed to several emission sources, including non-exhaust emissions. In northern European countries where winter is the longest season, road sanding and the use of studded tires are considered as major sources of the non-exhaust fraction of PM_{10}. Moreover, non-exhaust emissions such as material originating from brakes, tire wear and abrasion of the road surface, are an important source of trace metals in the urban areas. Materials containing iron, steel, chromium, nickel, zinc and copper belong to the major components used in the automotive industry. Processes such as braking and abrasion of vehicle components made from various metal materials

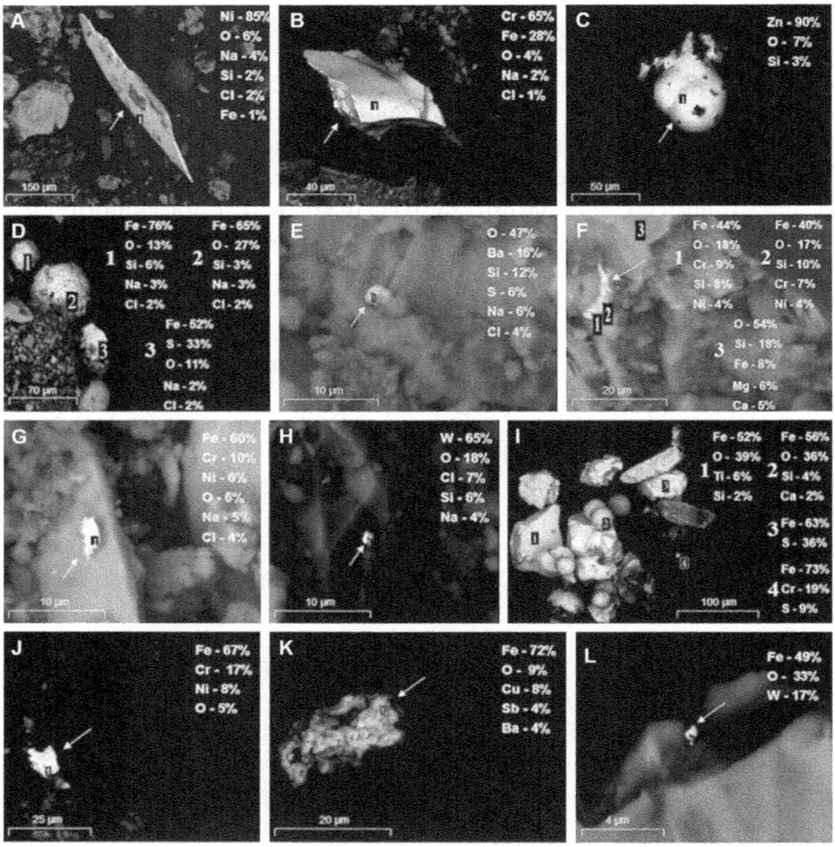

Fig. 8 SEM photographs and chemical composition (based on EDS spectra) of magnetic extracts from road dust accumulated on the roadside snow. Taken from reference 57 with permission of Elsevier.

generate significant amount of particles which are deposited in the near surroundings of roads. SEM and EDS analyses of brake linings and brake dust material generated during the application of brakes performed by Ingo *et al.*,[58] identified the presence of $BaSO_4$-containing particles in both brake lining material and break wear dust samples. $BaSO_4$, which is used as filler in brake linings (to reduce manufacturing costs and improve manufacturability), is considered as a significant source of high Ba concentrations in roadside soil samples.[9] Several tire companies produce tire studs, containing friction-increasing hard metal tips made out of wolfram carbide (WC), thus the W-rich particles identified in the roadside snow may originate from the abrasion of these components during winter. Minerals such as goethite (FeO(OH)), ilmenite ($FeTiO_3$), titanomagnetite ($Fe_3Ti_2O_4$) and pyrrhotite (FeS), which occur in local bedrock, were also identified in the samples. Crushed stone (aggregate), sand and gravel, obtained mostly from local sources, are commonly used for road construction. Therefore, the abrasion of the road surface might be the main source of these minerals.

Potgieter *et al.*[59] combined the use of Raman spectroscopy and SEM-EDS with the aim to further investigate and characterise road dust sediment of Manchester (UK). They sampled road dust from one of the main arteries to Manchester city centre (Oxford Road). The $\leq 2\,mm$ fraction was dry-sieved into potentially inhalable fractions (<38, 38–63 and 63–125 μm). They observed that the sample was frequented by bright yellow, reddish, orange and brown particles. SEM-EDS results of the three different sieved fractions revealed the presence of Si and Al in all the particles and Cr, Pb, Zn, Cu and Fe were found among 3–5.5%, 39–52%, 10–15.5%, 65–85% and 71–80% of the particles of each fraction, respectively. The presence of iron oxides (Fe_xO_y) and aluminosilicates was also reported. SEM-EDS data showed $>50\%$ association of Cr-rich particles with Pb. Subsequent Raman analysis revealed that the yellow particles match to be $PbCrO_4$, often occurring together with calcite, as it is shown in Fig. 9A. Although Cr–Pb–Fe associations were also observed in SEM-EDS, Raman spectroscopy indicated that iron oxy-hydroxides are present as discrete particles. The only other Pb species confirmed by Raman spectroscopy was PbO, often in combination with $PbCrO_4$. Most

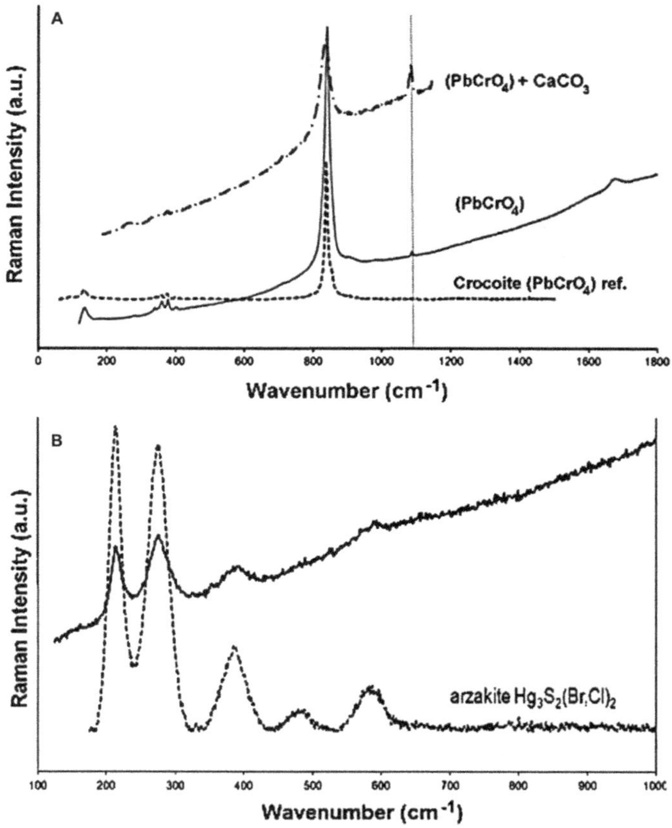

Fig. 9 Raman spectra from lead cromate with and without calcite (A) and from a yellow-coloured particle together with a reference spectrum for arzakite (B). Taken from reference 59 with permission of Elsevier.

of the particles analysed by means of Raman spectroscopy consisted mainly of quartz. Other inorganic salts/minerals such as calcite, dolomite, anatase, feldspar (mainly as amazonite $KAlSi_3O_8$ but also albite $NaAlSi_3O_8$) and various iron oxides, which mostly occurred as mixed haematite, goethite, magnetite and lepidocrocite were also detected. The presence of Mn_xO_y is also suspected due to various spectra portraying a strong band in the area between 580 and 700 cm^{-1} (probably hausmannite or Mn_3O_4). The most remarkable thing they reported was the evidence of a mercury sulphide compound. As it can be seen in Fig. 9B, a very good correlation between the Raman spectrum of the particle analysed and that of arzakite ($Hg_3S_2(Br, Cl)_2$), a yellow mineral, was obtained.

Fujiwara *et al.*[60] performed individual street dust particles characterization by SEM. They collected samples in the city of Buenos Aires during two months at 15 sites grouped in five zones with different urban characteristics and traffic profile. Dust samples were collected from pavement edges using a plastic dustpan and brushes in urban zones with different traffic patterns and urban characteristics but avoiding areas near to site-specific pollution sources. For SEM-EDS examination, a representative portion of sample was sprinkled onto double sided carbon tape mounted on a SEM stub. For major elements, no significant differences between the compositions of particle populations belonging to the five sites were detected. Most of these particles showed a variety of chemical compounds and only occasionally some particles composed of a single pure chemical species were observed. The particles were grouped into three types based on morphology and elemental composition: (i) mineral matter, (ii) spherical combustion products, and (iii) other types. Mineral particles, accounting for ~70–80% of the bulk analysed, were the most abundant, and most likely derived from natural sources, with a potential contribution of anthropogenic activities such as construction. Normally, the mineral grains have irregular shapes but elongated particles were also detected. They performed a size fractionated study of the road dust ($A < 37$ μm, $37 < B < 50$ μm, $50 < C < 75$ μm and $75 < D < 100$ μm).

In Seoul (Korea), more than the 85% of total atmospheric pollutants are emitted from vehicles. Kim *et al.*[61] performed a two-year monitoring of the magnetic particles in the atmosphere. They employed SEM-EDS to characterize the road dust: grain size, shape and composition. In their study included two industrial sites, five high traffic sites, along main roads with heavy traffic loads (>100,000 vehicles per day) and frequent traffic jams, and a park area with no apparent pollution source for comparison. Roadside dust samples were collected monthly during two years by sweeping of the road surface with a brush. For each sampling site, dust samples were obtained from five to seven locations and were intermixed in a plastic bag to yield a composite sample. Thus, the samples collected from each site may represent the net accumulation of roadside dusts in each month. The sampling was carried out at least 6 days after rain event to obtain sufficient quantity of dust samples and to minimize the wash-out effect, removing also coarse particles, such as leaves, glass, trash and small stones. They coated the magnetic extracts with platinum in order to identify carbon since the anthropogenic

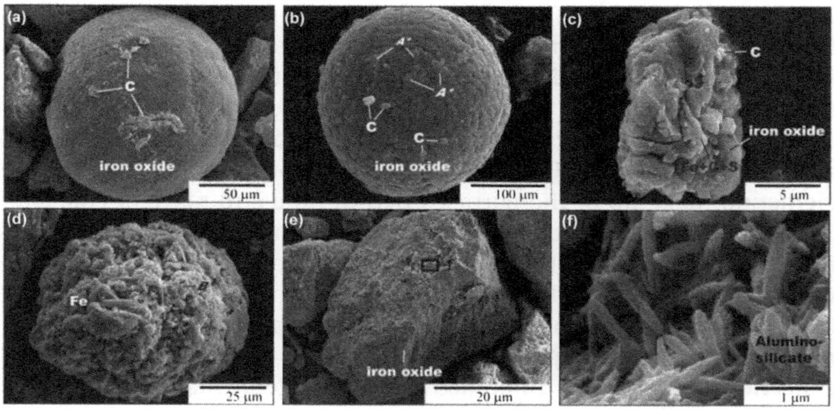

Fig. 10 SEM photographs of magnetic extracts from selected roadside dust samples: (a) an iron-oxide spherule associated with carbon (C), (b) an iron-oxide spherule associated with Al–Ca–Na–Si materials (refer to as *A**) and C, (c) an aggregate of iron-oxides and the particles composed of Fe, C and S, (d) an aggregate of pure iron and Al–Ca–Fe–K–Mg–Si materials (refer to as *B**), (e) an angular iron-oxide particle and (f) the surface of the angular particle (the rectangular frame in (e)) on a magnified scale. Taken from reference 61 with permission of Elsevier.

magnetic materials derived from fossil fuel combustions possibly contain carbon, the major element of the fuel. Based on the morphology, they grouped the observed magnetic materials into three types: spherules, aggregates and angular particles. The most frequently observed magnetic materials were spherules with diameters of 30–250 μm, which were identified as iron-oxides. About 70% of them contained carbon and the other 30% contained also other elements, such as Al, Ca, Na and Si. According to their compositions, two types of aggregates with lengths of 2–15 μm were found: (1) aggregates of iron-oxides and magnetic materials composed of Fe, C and S; and (2) aggregates of pure Fe and materials composed of various chemical elements, such as Al, Ca, Fe, K, Mg and Si. Finally, angular iron-oxides, with silicates adhered to the surface, were only found in a sample from a park area and are likely to be natural in origin. Figure 10 shows different SEM photographs where above mentioned particles were identified. To identify the magnetic materials derived from vehicle emission, they also performed SEM observations on the magnetic extracts from particulates gathered by a dust collector from diesel-powered vehicle emissions. From these observations they could conclude that the observed aggregates of iron-oxides and Fe–C–S materials in roadside dust originate from vehicle emissions, while aggregates of pure Fe and Al–Ca–Fe–K–Mg–Si materials are produced from the abrasion of brake lining surfaces.

In a recent study of Wilkinson *et al.*[62] the chemical composition for particles on a heavily utilized roundabout has been analysed by collection onto filters and employing automated SEM-EDS. Roundabouts are interesting from the viewpoint of being connection points between several roads, *i.e.* heavy traffic, and also because when entering a roundabout vehicles generally need to lose speed in order to manoeuvre safely through it . The loss of speed due to braking is hypothesized to cause brake wearing and so generate PM. In order to assess the health impact of inhalable particles on a local level it is important to monitor the emission source in detail. Taking a roundabout with heavy traffic as an example, it was shown that by collecting particles in and around the roundabout and analysing them with SEM-EDS, differences could be discerned in the elemental make up of the particles tied to time and place of collection. The elemental constituents observed in the particles were mostly those that were to be expected from literature: background (resuspended roadside) dust in the form of aluminosilicates and Ca/K with Na, S, and P. Moreover, the results obtained from this study proved that differences in particle amounts and types occur depending on the placement of particle collection and there is a difference in anthropogenic markers, such as Fe, from these collections.

3.1 Wear particles from brakes

When brakes are applied, friction between pads/linings and rotating counterparts always leads to the release of wear particles and a considerable fraction is released as airborne particulates. The wear rate is determined largely by the properties of the rubbing couple and by the conditions in which the brakes operate (speed, pressure, temperature, and chemicals available in the environment). A characteristic brake lining pad is a multicomponent composite typically formulated of more than 10 constituents and polymer matrix. Several thousand different raw materials have been used in different brands. According to the terminology accepted in automotive industry, "low-metallic" brake pads represent composites with relatively low metal content. The minimal knowledge about the morphology, chemical composition, and toxicity of nano- and micro-sized particles released from automotive brakes is alarming due to the fact that brake pad manufacturers currently do not have to deal with development of eco-friendly formulations. Since the friction process generates heat, and very high pressures are applied on the friction interface, the structure and chemistry of released wear debris often differs considerably from the bulk friction material of the brake. However, it is not known what exactly is being released from brakes, how to properly analyse wear debris, and what is its fate and impact on the environment. The released wear particles can be categorized as airborne particles (released into the air and typically deposited away from the roadside) and non-airborne particles (deposited on vehicle/brake hardware or falling on the road surfaces). Nano-sized airborne particles can easily be inhaled into the respiratory tract, posing hazards related to potential oxidative stress and inflammation due to their increased surface area and higher reactivity with biomolecules and tissues. Hence, the environmental concerns related to brake wear particles have brought more attention in recent years and the number of articles appearing in the bibliography related to the characterization of these particles has increased.

Brake linings contain 1–5% Sb(III) as stibnite (Sb_2S_3), which is employed as a lubricant to reduce vibrations and to improve friction stability.[13] Stibnite is also introduced as an additive in tire vulcanization. During braking, brake pads may indeed reach temperatures, high enough to induce oxidation of Sb. Because of the temperature reached during braking, Sb_2S_3 is first superficially oxidized to Sb_2O_3. Later, at temperatures between 400 °C and 550 °C and in presence of oxygen Sb_2O_3 undergo oxidation to the thermally more stable Sb_2O_5.

Some studies have been conducted to identify the distribution of Sb species in PM. Varrica *et al.*[13] characterised samples of brake linings, brake pad wear residues, road dust, and atmospheric particulate matter PM_{10} and $PM_{2.5}$ by means of SEM-EDS. Their results revealed that brake pads particles contain Sb and S, in relative amount compatible with the presence of stibnite (Sb_2S_3), together with metals such as Fe and Cu. Samples of dust accumulated on wheel rims contain micron-sized particles made up of Si, Al, Cr, Fe, Cu, and Sb. Appreciable amounts of Sb and other metals (Fe, Cu, Cr, and Mn) were also found in road dust and in PM particles of various grain sizes. Figure 11 shows SEM photographs of brake pad (a), brake pad wear residues accumulated on wheel rims (b), road dust (c), and particulate matter (d), together with their EDX spectrum that reveals the presence of above mentioned elements.

Kukutschova *et al.*[63] designed an experiment to characterize the particles emitted by commercial low-metallic brake linings. Wear debris was generated using an automotive brake dynamometer inside an environmental chamber with a controlled filtered air. A Berner Low Pressure Impactor (BLPI) was connected to the dynamometer chamber for the collection of generated wear. The BLPI enabled the sampling of particles with a mean aerodynamic diameter ranging between 37 nm and 9.5 mm.

Fig. 11 SEM images and EDS spectrum of brake pad (a), brake pad wear residues accumulated on wheel rims (b), road dust (c), and particulate matter (d). Taken from reference 13 with permission of Elsevier.

SEM-EDS results of the brake sample confirmed the presence of steel chips, brass (CuZn), Cu-powder, Cu-chips, MgO, metal sulphides, and aluminium silicates. The EDS analysis of the finest fraction of the wear particles (<100 nm) only detected the presence of carbon and iron in this case. It was not possible to detect Cu, Sn and other metals previously detected by elemental analysis or by SEM-EDS of the initial brake pad. However, in contrast to nano-sized particles, most of the elements present in the initial brake pads could be detected in the fine (<2.5 μm) and coarse fraction (>2.5 μm) of the brake dust. The authors attribute this to the fact that the method lacks the sensitivity necessary to detect very low contents of Cu, Zn, Sn, and S. It is well known that EDS requires a critical mass/volume of material present in order to be capable to generate a sufficient amount of secondary electrons, and the sensitivity of this method is limited to micrometre particles only. They also submitted wear particles to Raman analysis and the results revealed the presence of carbon black and graphitic particles.

3.2 Wear particles from tire debris

Detailed knowledge of the size, shape, microstructure, and constituents of the particles from the wear of tires, usually named "tire debris" or "tire dust", eventually serves to control and improve the tire manufacturing process and to estimate the potential impact of particles on human health and on the environment. Camatini *et al.*[64] proved in their work in 2001 that SEM plays a significant role in the characterization and non-destructive identification of tire debris. SEM images showed that tire debris had a typical, warped surface and pores, and that its characteristic elements, detected by EDS, were S and Zn.

Further studies were carried out by Peltola *et al.*[65] and Kupiainen *et al.*[66,67] They analysed street dust samples in Finland in order to characterize wolfram carbide (WC) particles coming from tire studs. Suspended urban street dust is created when vehicle traffic wears the pavement and the applied traction sand. The dust concentrations are further increased by the common use of studded winter tires that effectively wear both the road surface and applied sand, grinding the components into finer particles. Therefore, the street dust consists mainly of minerogenic material from the pavement and traction sand together with particles emitted from tires, brakes, engines (combustion and friction) and corrosion of vehicles. The tire studs, containing a friction increasing hard metal tip made out of WC, also wear and inevitably produce WC particles. In countries or regions where studs are used, elevated wolfram concentrations have been found in several types of sampling media, such as, road runoff, moss or humus samples. The elevated wolfram concentrations were believed to be mainly caused by the wearing of WC from tire studs. Street dust itself has been studied and analysed for many elements, but less information can be found on the concentrations of wolfram and particles derived from studs in street dust.

The size range of the wolfram particles found by Peltola and co-workers[65] was between 0.1 μm (and probably smaller) and 1.4 μm. A more precise chemical analysis of the wolfram (ie tungsten) particles was hindered by their small size, which contaminated the X-ray spectra with information from the surrounding area as well. Thus it was not possible to determine whether the particles consisted specifically of WC rather than metallic wolfram, or of bonding metals used in WC. Although many particle types were found with the SEM, only iron containing particles (some with minor concentrations of chromium) occurred more frequently than the WC particles. Small quantity of Pt, Pb and Ce particles were detected. This result is surprising, considering that there are a number of publications discussing the presence of such metals coming from traffic emissions in the urban environment. However, the high energy used in the SEM analysis made it particularly easy to visually identify dense metallic particles with the backscatter detector. Another reason for not identifying other metallic particles could be that many metals oxidise more easily than WC and Pt. The SEM analysis was not calibrated to visually identify different types of oxides that are probably less visible than the particles identified here.

In Northern latitudes with constant snow coverage during winter months, high particle concentrations have been observed. PM_{10} levels rise drastically especially during the spring in urban areas with high traffic volume. The particles are deposited in snow, and when snow melts, road surfaces dry out, and a proportion of the dust is resuspended by traffic. Mineral dust forms usually a major part of the PM of the springtime episodes. The mineral dust in the particulate mass mainly results from the use of anti-skid methods to enhance traction on snowy or icy road surfaces. Such methods include spreading of traction sand on the road surfaces and equipping tires with metal studs or a special rubber design, and salting the road to prevent sliding. The traction sand is crushed into smaller particles under the tires and the pavement aggregate is worn by interaction with the tires. The study of Kupiainen[66,67] and co-workers take the combined effect of these two methods into account. The SEM-EDS samples were prepared by pressing a tape attached to an aluminium plate onto the filter surface covered with particles. The samples were sputtered with carbon to make the sample surface conductive. The results of this study show that when both traction sand and studded tires were used, the use of traction sand increased the concentrations of PM_{10}, which might support the dominant role of sanding material in the dust. However, a chemical study of the particle showed that a significant part of the particulate matter came from asphalt. This result indicates that the pavement wear is strongly increased by the grinding impact of sand under the tires, which produces dust also from the asphalt aggregate. This phenomenon was named the sandpaper effect. Its understanding is important to reduce harmful effects of springtime road dust in practical winter maintenance of urban roads. Dust emissions and the sandpaper effect were dependent on several factors, such as the mechanical and mineralogical properties of the sanding and pavement aggregates, the quantity of traction sand used, the size distribution of the sand grains as well as the type of tires.

4 Future perspectives

The suitability of stand-alone SEM-EDS and micro-Raman spectroscopy (MRS) analyses for the determination of elemental and molecular profiles of a sample is undisputed. The results and discussions presented in this chapter also suggest that the challenges faced with the analyses of fine heterogeneous particles can be successfully met. SEM provides a way of visualising samples with excellent spatial resolution and a large depth of field. It reveals detailed information about the sample morphology that is not apparent when using optical microscopy. The combination of SEM with X-ray detection yields sub-micrometre scale, high quality elemental analysis. Although most SEMs are routinely fitted with EDS equipment, and whilst this has proven a very valuable technique, it yields only elemental information, making the analysis of complex heterogeneous compounds particularly challenging. MRS, on the other hand, provides chemical, physical and structural information about the materials analysed, but the visual resolution is much poorer than in SEM-EDS. MRS solves a variety of analytical problems across a wide range of applications, one of which is that samples can be analysed "as-received" and *in-situ*, and it seldom requires any sample preparation. The "fingerprinting" of a molecular structure obtained by micro-Raman spectroscopy can be successfully complemented by means of X-ray spot analysis through the application of SEM-EDS. The elemental composition revealed by SEM-EDS is essential for a correct interpretation of the collected Raman spectra.[68] The combination of these two different techniques may result in unambiguous chemical and structural characterisation of a wide range of samples at the micrometre scale and its beneficial application to heterogeneous environmental particles is obvious. Such a combination allows morphological, elemental, chemical and physical analyses without moving the sample between instruments. The "two-in-one unit" undoubtedly offers a number of advantages and provides a solution for several fundamental and analytical problems. First of all, the problem of the object relocation becomes irrelevant, except in the case of sub-micrometre-sized particles. SEM is an established method of examining samples at a variety of magnifications, with two main imaging modes (secondary and backscattered electron imaging: SEI and BEI, respectively) providing the best spatial resolution and deriving contrast either from surface topography or mean atomic number. This means that in the hybrid system the features of interest can be easily located using the SEM. These can then be rapidly identified with MRS. Consequently, the elemental and molecular spectra, the SEI and the white light images, can be acquired from the same sample position. The data which can be acquired includes the (i) morphology and mean atomic number from SEM (SEI and BEI); (ii) elemental composition from EDS analysis; (iii) chemical composition and identification from MRS and (iv) physical structure (crystallographic and mechanical data) from MRS. As an example, Fig. 12 shows the analysis of a heavy mineral sand by means of the combined SEM-EDS/MRS. Both the X-Ray and the Raman spectrum of the same particle grain can be obtain, apart from the SEM image.

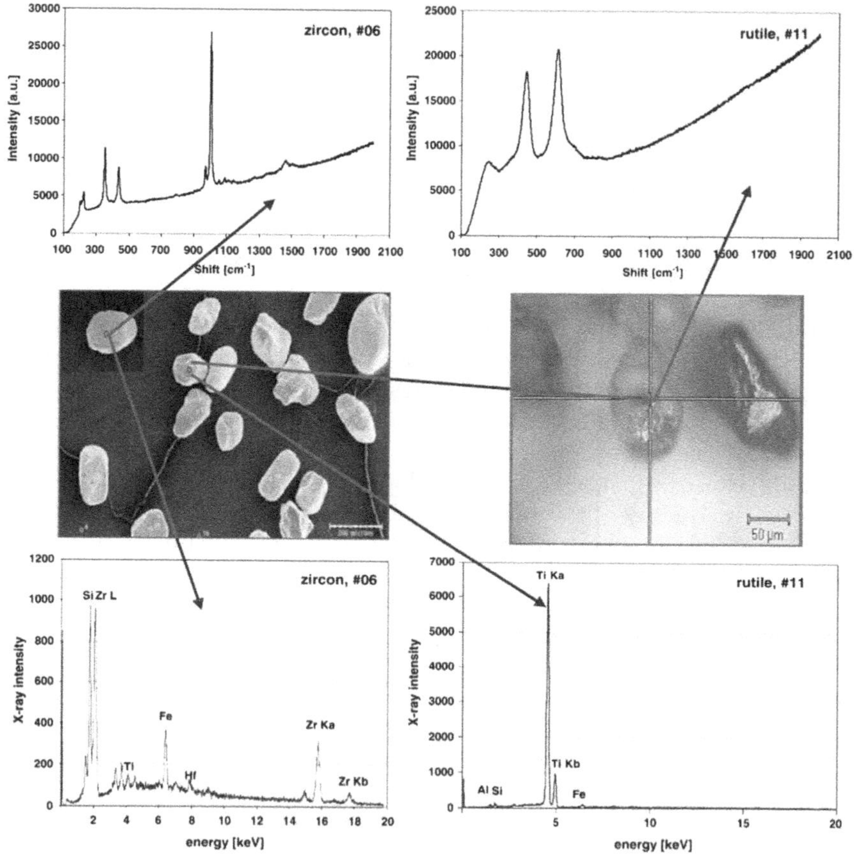

Fig. 12 Heavy mineral sand sample investigated by the two techniques SEM-EDS/MRS. Taken from reference 70 with permission of Elsevier.

Some companies provide an interfaced SEM-EDS/MRS instrument (*e.g.* the Structural and Chemical Analyser, the SCA unit, from Renishaw plc.[69]), but little literature on the analyses of individual environmental particles with such a system was found in peer reviewed journals.[70,71] The first use of a combined SEM and Raman instrument was reported in 2004 by Jarvis *et al.*[72] for bacterial discrimination. Later, in 2007, Black and Brooker published the utilization of the SCA device for the analysis of Clinker cement, whereas Otieno-Alego *et al.*[73] describe some forensic applications using the same instrument in 2007. Recently, a study of characterization of metal-rich particulate matter in an abandoned zinc/lead mine and other characterizing harmful compounds contained in black slags have been published where the SCA was employed.[49,74]

Although such combination of techniques seems very promising, Worobiec *et al.*[70] stated that the measurement technique by an interfaced SEM-EDS/MRS instrument is less obvious than one would expect, especially in the case of analysis of individual fine particles; this analytical approach needs further and more sophisticated optimisation. The most important aspects regard the following: (a) vacuum *versus* ambient sample chamber for MRS; (b) beam damage and molecular changes; (c) deposition of amorphous carbon in SEM-EDS sample chamber; and (d) relocation of fine particles. After taking these aspects into account, the utilization of combined SEM-EDS/MRS will lead to an important improvement for future researches and will contribute to more detailed information about the particles emitted by the traffic.

4.1 Vacuum *versus* ambient sample chamber for MRS

According to Worobiec *et al.*,[70] the Raman intensities of haematite peaks recorded under vacuum conditions are up to 10 times lower than those in ambient air (stand-alone set-up). This can be crucial during spectral identification, especially for particles containing two or more polymorphs of a compound, where subtle differences become important. This is of particular importance in the case of the analysis of the micrometre-sized particles, since if the vacuum has such a major influence on the quality of the spectra, it can lead to some Raman bands not being recognised. The analysis of soil samples rich in humic-type compounds, for example, is a challenge for stand-alone systems due to the small Raman cross section and fluorescence phenomena, and a further reduction in signal could lead to masking of Raman active bands.

4.2 Beam damage

One of the main aspects that needs be taken into account during the analysis of individual particles by combined SEM-EDS/MRS is particle damage and any subsequent molecular changes that might occur under a laser or electron beam.[75] Damage to particles by the electron beam can be minimised, and under ambient conditions carefully choosing the laser power (for a given laser wavelength) will avoid damage in MRS and maintain the ability to record Raman spectra of sufficient quality. The damage to individual particles with the SEM-EDS/MRS interface seems to be more problematic than in stand-alone instruments. It appears that electron beam sensitive particles are not necessarily laser beam sensitive, neither in ambient nor vacuum environments. It is however possible to eliminate the damage caused by the electron beam excitation in a vacuum environment by using a liquid-nitrogen cooled sample stage. Similarly there are several techniques (laser defocusing, dilution of sample, use of thin sections, *etc.*) that can reduce laser damage to a satisfactory degree in MRS. According to Worobiec *et al.*,[70] it became apparent that beam damage in the hybrid system occurs in a totally different manner and the physical and chemical phenomena underlying the damage remain unclear. This aspect is currently under investigation. The obvious factors that will affect the damage processes are the atmosphere (vacuum/air), the laser beam intensity and the presentation (*i.e.* preparation) of the sample.

4.3 Amorphous carbon deposition

The build-up of carbon on sample surfaces in the SEM is frequently observed, but remains relatively poorly understood. In the SEM, where the electron beam impinges upon carbon adsorbed on the sample surface, its effect is to cross-link the CH_x species. This cross-linking effectively polymerises the contamination and stabilises it. Another unfortunate consequence is that, where the electron beam irradiates the sample, the local carbon activity is elevated, and this has the effect of attracting any surface mobile or residual carbonaceous species within the vacuum chamber to the irradiated point in order to reduce the local surface energy. The result is that a carbon layer gradually "grows" on the sample surface. This can frequently be observed as what is often (and erroneously) referred to as a "charge square" on the sample. A dark rectangle is observed where the electron beam was scanning the surface. It is a popular misconception that the amount of carbon build-up on a sample is related to the SEM vacuum cleanliness or the SEM pumping method, but although these are factors, the majority of the carbon available for polymerisation is that which is already adsorbed on the sample. Poor sample preparation or handling will result in higher levels of adsorbed carbon, and this represents the major contributor to the carbon (polymer) growth on the sample in the vicinity of the electron beam.

4.4 Particle relocation

The ultimate solution to the relocation problem is of course to use an interfaced approach, as stated earlier. This is, however, not always the case for the analyses of single particles in the ultra-fine size range, since excitation of precisely the same area is strongly dependent on the precision of the reconfiguration of the laser or electron beam that needs to take place during switching between the two modes of analysis. When the appearances of particles are different due to the difference of depth of field, it becomes nearly impossible to ensure that exactly the same micrometre particle and particle part is being analysed. They pointed out that relocation of the same particle is one of the most difficult tasks due to the completely different properties of images obtained by electron beam scanning and white light reflection. The greater depth of field provided by SEM is very often not achievable by the optical microscope. This is a main problem for large and thick objects, where some features, easily located by means of SEM, appeared invisible in the white light image due to, for example, the surface roughness, which is certainly the case with mineralogical samples. In case of individual particles analysis of geological samples, the problem of relocation is mainly dependent on the sample preparation. Sample preparation should facilitate well-separated and coordinated particles to ensure successful relocation.

References

1 F. Amato, M. Pandolfi, T. Moreno, M. Furger, J. Pey, A. Alastuey, N. Bukowiecki, A. S. H. Prevot, U. Baltensperger and X. Querol, *Atmos. Environ.*, 2011, **45**, 6777–6787.

2 D. Hjortenkrans, B. Bergbäck and A. Häggerud, *Environ. Sci. Technol.*, 2007, **41**, 5224–5230.

3 L. Bozo, H. Eerens, S. Larssen, M. M. Millan, N. Moussiopoulos, S. Papalexiou and Z. Samaras, *Environ. Pollut.*, 1998, **1**, 433–464.
4 G. D'Amato, L. Cecchi, M. D'Amato and G. Liccardi, *J. Investig. Allergol. Clin. Immunol.*, 2010, **20**, 95–102.
5 A. Thorpe and R. M. Harrison, *Sci. Total Environ.*, 2008, **400**, 270–282.
6 K. Adachi and Y. Tainosho, *Environ. Int.*, 2004, **30**, 1009–1017.
7 J. K. Gietl, R. Lawrence, A. J. Thorpe and R. M. Harrison, *Atmos. Environ.*, 2011, **44**, 141–146.
8 M. Keuken, H. Denier van der Gon and K. van der Valk, *Sci. Total Environ.*, 2010, **408**, 4591–4599.
9 J. A. Carrero, I. Arrizabalaga, J. Bustamante, N. Goienaga, G. Arana and J. M. Madariaga, *Sci. Total Environ.*, 2013, **458–460**, 427–434.
10 X. Chen, X. Xia, Y. Zhao and P. Zhang, *J. Hazard. Mater.*, 2010, **181**, 640–646.
11 F.-R. Li, L.-F. Kang, X.-Q. Gao, W. Hua, F.-W. Yang and W.-L. Hei, *Soil Sediment Contam.*, 2007, **16**, 473–484.
12 F. Fujiwara, R. J. Rebagliati, J. Marrero, D. Gómez and P. Smichowski, *Microchem. J.*, 2011, **97**, 62–67.
13 D. Varrica, F. Bardelli, G. Dongarrà and E. Tamburo, *Atmos. Environ.*, 2013, **64**, 18–24.
14 B. D. Garg, S. H. Cadle, P. A. Mulawa, P. J. Groblicki, C. Laroo and G. A. Parr, *Environ. Sci. Technol.*, 2000, **34**, 4463–4469.
15 E. R. McKenzie, J. E. Money, P. G. Green and T. M. Young, *Sci. Total Environ.*, 2009, **407**, 5855–5860.
16 E. Smolders and F. Degryse, *Environ. Sci. Technol.*, 2002, **36**, 3706–3710.
17 T. B. Councell, K. U. Duckenfield, E. R. Landa and E. Callender, *Environ. Sci. Technol.*, 2004, **38**, 4206–4214.
18 C. L. S. Wiseman, F. Zereini and W. Püttmann, *Sci. Total Environ.*, 2013, **442**, 86–95.
19 B. Skrbic, S. Milovac and M. Matavulj, *Ecol. Indic.*, 2011, **13**, 168–177.
20 G. Nabulo, H. Oryem-Origa and M. Diamond, *Environ. Res.*, 2006, **101**, 42–52.
21 J. Li, Y. Lu, H. Shim, X. Deng, J. Lian, Z. Jia and J. Li, *J. Environ. Monit.*, 2010, **12**, 466–471.
22 H. Zhao and X. Li, *J. Hazard. Mater.*, 2013, **246–247**, 267–276.
23 M. Rucandio, M. Petit-Domínguez, C. Fidalgo-Hijano and R. García-Giménez, *Environ. Sci. Pollut. Res.*, 2011, **18**, 51–63.
24 E. Smith and G. Dent, *Modern Raman Spectroscopy – A Practical Approach*, John Wiley & Sons Ltd, England, 2005.
25 Ian R. Lewis and H. G. M. Edwards, *Handbook of Raman Spectroscopy. From the Research Laboratory to the Process Line*, Marcel Dekker Inc, New York, 2001.
26 D. A. Long, *The Raman Effect. A Unified Treatment of the Theory of Raman Scattering by Molecules*, John Wiley & Sons Ltd, New York, USA, 2002.
27 John R. Ferraro, K. Nakamoto and C. W. Brown, *Introductory Raman Spectroscopy*, Academic Press, 2003.
28 N. Goienaga, N. Arrieta, J. A. Carrero, M. Olivares, A. Sarmiento, I. Martinez-Arkarazo, L. A. Fernández and J. M. Madariaga, *Spectrochim. Acta, Part A*, 2011, **80**, 66–74.
29 J. A. Carrero, N. Goienaga, M. Olivares, I. Martinez-Arkarazo, G. Arana and J. M. Madariaga, *J. Raman Spectrosc.*, 2012, **43**, 1498–1503.
30 Yimei Zhu and H. Inada, in *Encyclopedia of Nanotechnology*, ed. B. Bhushan, Springer, New York, USA, 2012.

31 Weilie Zhou, Robert Apkarian, Zhong Lin Wang and D. Joy, in *Scanning Microscopy for Nanotechnology*, eds. Weilie Zhou and Z. L. Wang, Springer, New York, USA, 2006.
32 G. H. Michler, *Electron Microscopy of Polymers*, Springer, Berlin, 2008.
33 Joseph I. Goldstein, Dale E. Newbury, Patrick Echlin, David C. Joy, Charles E. Lyman, Eric Lifshin, Linda Sawyer and J. R. Michael, *Scanning Electron Microscopy and X-ray Microanalysis*, Springer, Berlin, 2003.
34 P. Echlin, *Handbook of Sample Preparation for Scanning Electron Microscopy and X-Ray Microanalysis*, Springer, New York, USA, 2009.
35 U. Villanueva, J. C. Raposo and J. M. Madariaga, *Microchem. J.*, 2013, **106**, 107–120.
36 U. Villanueva, J. C. Raposo, K. Castro, A. de Diego, G. Arana and J. M. Madariaga, *J. Raman Spectrosc.*, 2008, **39**, 1195–1203.
37 J. A. Carrero, G. Arana and J. M. Madariaga, *Spectrosc. Eur.*, 2012, **24**, 11 12, 14–1512, 6–15.
38 G. MacKinnon, A. B. MacKenzie, G. T. Cook, I. D. Pulford, H. J. Duncan and E. M. Scott, *Sci. Total Environ.*, 2011, **409**, 5010–5019.
39 M. Maguregui, A. Sarmiento, I. Martinez-Arkarazo, M. Angulo, K. Castro, G. Arana, N. Etxebarria and J. M. Madariaga, *Anal. Bioanal. Chem.*, 2008, **391**, 1361–1370.
40 J. Aramendia, L. Gomez-Nubla, K. Castro, I. Martinez-Arkarazo, D. Vega, A. Sanz Lopez de Heredia, A. Garcia Ibanez de Opakua and J. M. Madariaga, *J. Raman Spectrosc.*, 2012, **43**, 1111–1117.
41 E. Doehne and C. A. Price, *Stone Conservation: An Overview of Current Research*, The Getty Conservation Institute, Los Angeles, USA, 2010.
42 N. Prieto-Taboada, I. Ibarrondo, O. Gomez-Laserna, I. Martinez-Arkarazo, M. A. Olazabal and J. M. Madariaga, *J. Hazard. Mater.*, 2013, **248–249**, 451–460.
43 N. Prieto-Taboada, M. Maguregui, I. Martinez-Arkarazo, M. A. Olazabal, G. Arana and J. M. Madariaga, *Anal. Bioanal. Chem.*, 2011, **399**, 2949–2959.
44 F. Zereini and C. L. S. Wiseman, *Urban Airborne Particulate Matter*, Springer, Berlin, 2011
45 X. Han and L. P. Naeher, *Environ. Int.*, 2006, **32**, 106–120.
46 R. J. Laumbach and H. M. Kipen, *J. Allergy Clin. Immunol.*, 2012, **129**, 3–11.
47 G. Martini, B. Giechaskiel and P. Dilara, *Biomarkers*, 2009, **14**, 29–33.
48 A. U. Raysoni and W.-W. Li, *J. Environ. Res. Dev.*, 2009, **4**, 421–429.
49 N. Goienaga, A. Sarmiento, M. Olivares, J. A. Carrero, L. A. Fernández and J. M. Madariaga, *Anal. Chem.*, 2013, **85**, 7173–7181.
50 J. A. Koziel, M. Odziemkowski and J. Pawliszyn, *Anal. Chem.*, 2001, **73**, 47–54.
51 M. Odziemkowski, J. A. Koziel, D. E. Irish and J. Pawliszyn, *Anal. Chem.*, 2001, **73**, 3131–3139.
52 M. Pacenti, C. Lofrumento, S. Dugheri, A. Zoppi, I. Borsi, A. Speranza, P. Boccalon, G. Arcangeli, A. Antoniucci, E. M. Castellucci and V. Cupelli, *Eur. J. Inflammation*, 2009, **7**, 25–37.
53 S. K. Sze, N. Siddique, J. J. Sloan and R. Escribano, *Atmos. Environ.*, 2000, **35**, 561–568.
54 A. Soewono and S. Rogak, *Aerosol Sci. Technol.*, 2011, **45**, 1206–1216.
55 M. Knauer, M. E. Schuster, D. Su, R. Schloegl, R. Niessner and N. P. Ivleva, *J. Phys. Chem. A*, 2009, **113**, 13871–13880.
56 M. Bućko, T. Magiera, L. Pesonen and B. Janus, *Water, Air, & Soil Pollut.*, 2010, **209**, 295–306.
57 M. S. Bućko, T. Magiera, B. Johanson, E. Petrovský and L. J. Pesonen, *Environ. Pollut.*, 2011, **159**, 1266–1276.

58 G. M. Ingo, M. D'Uffizi, G. Falso, G. Bultrini and G. Padeletti, *Thermochim. Acta*, 2004, **418**, 61–68.
59 S. Potgieter-Vermaak, G. Rotondo, V. Novakovic, S. Rollins and R. Grieken, *Environ. Geochem. Health*, 2012, **34**, 689–696.
60 F. Fujiwara, R. J. Rebagliati, L. Dawidowski, D. Gómez, G. Polla, V. Pereyra and P. Smichowski, *Atmos. Environ.*, 2011, **45**, 1497–1505.
61 W. Kim, S.-J. Doh, Y.-H. Park and S.-T. Yun, *Atmos. Environ.*, 2007, **41**, 7627–7641.
62 K. E. Wilkinson, J. Lundkvist, J. Netrval, M. Eriksson, G. A. Seisenbaeva and V. G. Kessler, *Environ. Pollut.*, 2013, **182**, 364–370.
63 J. Kukutschova, P. Moravec, V. Tomasek, V. Matejka, J. Smolk, J. Schwarz, J. Seidlerova, K. Safarova and P. Filip, *Environ. Pollut.*, 2011, **159**, 998–1006.
64 M. Camatini, G. F. Crosta, T. Dolukhanyan, C. Sung, G. Giuliani, G. M. Corbetta, S. Cencetti and C. Regazzoni, *Mater. Charact.*, 2001, **46**, 271–283.
65 P. Peltola and E. Wikstroem, *Boreal Environ. Res.*, 2006, **11**, 161–168.
66 K. Kupiainen and H. Tervahattu, *Environ. Monit. Assess.*, 2003, **93**, 287–300.
67 K. Kupiainen, H. Tervahattu and M. Räisänen, *Sci. Total Environ.*, 2003, **308**, 175–184.
68 E. A. Stefaniak, A. Worobiec, S. Potgieter-Vermaak, A. Alsecz, S. Török and R. Van Grieken, *Spectrochim. Acta, Part B*, 2006, **61**, 824–830.
69 http://www.renishaw.com/en/sem-raman-system-6639.
70 A. Worobiec, S. Potgieter-Vermaak, A. Brooker, L. Darchuk, E. Stefaniak and R. Van Grieken, *Microchem. J.*, 2010, **94**, 65–72.
71 S. Potgieter-Vermaak, G. R. Van and J. H. Potgieter, *Spectrosc. Eur.*, 2010, **22**, 12–16.
72 R. M. Jarvis, A. Brooker and R. Goodacre, *Anal. Chem.*, 2004, **76**, 5198–5202.
73 V. Otieno-Alego, *J. Raman Spectrosc.*, 2009, **40**, 948–953.
74 L. Gómez-Nubla, J. Aramendia, S. F.-O. de Vallejuelo, K. Castro and J. M. Madariaga, *J. Raman Spectrosc.*, 2013, **44**, 1163–1171.
75 A. Worobiec, L. Darchuk, A. Brooker, H. Potgieter and R. Van Grieken, *J. Raman Spectrosc.*, 2011, **42**, 808–814.

Vibrational spectroscopy of N-donor ligand metal complexes: probing excited states

Anastasia B. S. Elliott, Holly van der Salm and Keith C. Gordon*

DOI: 10.1039/9781782621485-00211

This report reviews work on vibrational spectroscopy of N-donor ligand metal complexes from 2008–2013. The particular focus is on the examination of excited states using resonance Raman spectroscopy, transient resonance Raman spectroscopy and time-resolved infrared spectroscopy. The report describes some of the basic concepts around each technique and then highlights work from the review period that exemplifies the utility of each method. In addition tables are included that summarise the work using these methods over the review period.

1 Introduction

N-donor ligand or polypyridyl ligand metal complexes play a key role in a variety of important chemical applications. These include:

1. solar energy systems;
2. photocatalysis;
3. analyte sensing.

Metal polypyridyl complexes are at the heart of the dye-sensitised solar cell, developed by Grätzel.[1,2] These complexes have also found utility in photocatalysis[3] for CO_2 remediation or hydrogen evolution.[4] There is also significant research into the use of these complexes in analyte sensing including with DNA[5–7] and in the detection of anions.[8–11] Key to the success of each of these applications are the excited states, for it is the excited states that are the critical reactive reagents in these applications.

In this chapter we review papers from 2008–2013 that cover the use of vibrational spectroscopic methods that provide insight in to the nature of excited states. In this regard it is important to carefully consider the "types" of excited states that are important. Metal complexes may have a variety of excited states and these will be discussed shortly. An alternative and useful way to describe types of excited states is by considering timescale. These differing domains are depicted in Fig. 1. Upon photo-excitation the system is excited to the Franck-Condon state; this can be considered as an ensemble of vibrational levels of the upper state or as a wavepacket deposited on the excited state surface (FC Fig. 1). Subsequent to this initial state relaxation processes occur (I, Fig. 1) – typically on ps timescales that lead to the population of the lowest energy thermally equilibrated state, or THEXI state as termed by Adamson, which typically

Chemistry Department and MacDiarmid Institute, University of Otago, Dunedin, New Zealand. E-mail: keith.gordon@otago.ac.nz

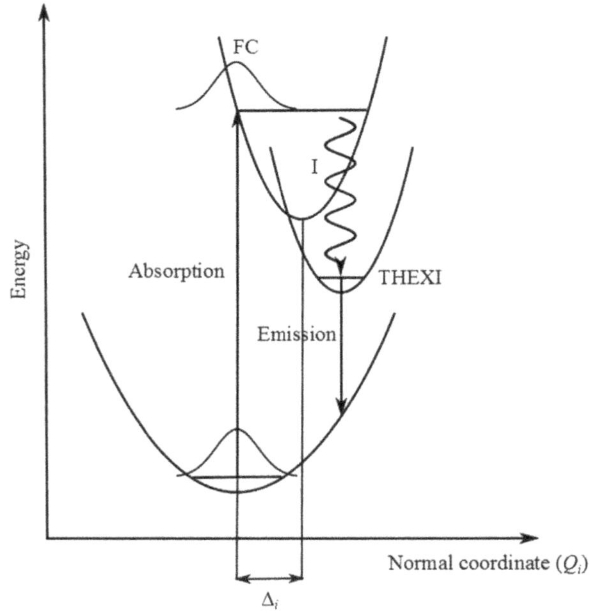

Fig. 1 Potential energy surfaces showing Franck-Condon (FC) and THEXI states and intermediate relaxation (I). The structural distortion between ground and resonant surface along the normal coordinate (Q_i) is given as Δi.

has a lifetime of nanoseconds (THEXI, Fig. 1).[12] In many applications it is the THEXI state that is the active reagent; however this is not always the case and the intermediate states may also play a role particularly if fast electron transfer occurs as in DSSCs.[2]

Excited state classification

The types of excited states that play a role in metal polypyridyl complexes have been summarised by Crosby.[13] If one assumes an octahedral ligand field the metal d-orbitals split into t_{2g} (dπ) and e_g (dσ). The types of excited states are then:

1. Ligand field (LF or d-d) – originating from a dπ → dσ transition. The ε for this is low because the transition is Laporte forbidden. Excited states of this type can lead to photochemistry;
2. Ligand-centred (LC) – originating from a ligand-based $\pi_L \rightarrow \pi^*_L$ transition. These typically have $\varepsilon \sim 10^4$ to 10^5 M^{-1} cm^{-1} and they are generally photochemically stable but may lead to emission;
3. Metal-to-ligand charge-transfer (MLCT) – a $d\pi \rightarrow \pi^*_L$ transition. These have $\varepsilon \sim 10^3$–10^4 M^{-1} cm^{-1} and are generally stable and can result in strong emission;
4. Ligand-to-metal charge-transfer (LMCT). These occur for d^5 configurations and can be observed in excited state spectra of complexes, such as $[Ru(bpy)_3]^{2+}$ or Ru(III) complexes;[14] the $\varepsilon \sim 10^3$ M^{-1} cm^{-1} is lower than MLCT transitions because of the diminution of wavefunction overlap between the donor and acceptor orbital (Fig. 2).

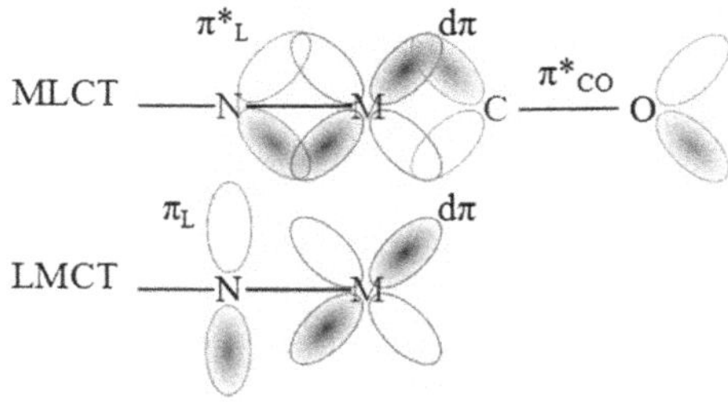

Fig. 2 MO overlap for MLCT and LMCT transitions. The MO for π^*_{CO} is included here to assist latter discussion.

This report deals with three well-used vibrational spectroscopic techniques in understanding photoexcitation and excited states, namely resonance Raman spectroscopy, time-resolved resonance Raman spectroscopy and time-resolved infrared spectroscopy.

One issue that has become ubiquitous in papers dealing with these techniques is the use of quantum calculations, particularly density function theory (DFT). In almost of the examples herein DFT has been used directly or indirectly in the interpretation of the data. An excellent review of the DFT calculations employed is available and we do not give a detailed account here.[15]

2 Resonance Raman spectroscopy

Resonance Raman spectroscopy has been shown to be a powerful tool in probing the electronic structure of polypyridyl compounds[16–18] and has recently been reviewed.[19] The utility can be attributed to the way that resonance Raman can inform on the nature of specific electronic transitions by tuning the exciting laser wavelength. Resonance Raman spectroscopy is unique from the non-resonant case in that the exciting laser wavelength is coincident with an electronic transition. The molecule is consequently excited into an existing electronic state rather than the virtual state of non-resonant Raman spectroscopy. A striking enhancement, of up to six orders of magnitude,[20] of select vibrational modes is observed in a resonant spectrum. What makes the technique so useful is that the nature of this enhancement is not random; instead, it is greatest for modes that mimic the resonant electronic transition or those modes that show a large Δ_i (Fig. 1). This behaviour is exemplified by comparing the resonance Raman spectra to the normal Raman spectrum of a metal polypyridyl complex which has ancillary CO ligands, such as $Re(pqx)(CO)_3Cl$ (Fig. 3).[21] The normal Raman spectrum of this complex shows bands from the pqx ligand and from the ancillary CO ligands. As the CO ligands exhibit a facial arrangement, 3 bands are observed and these lie at 1870, 1919 and 2021 cm^{-1}. This complex has a strong MLCT transition at about 450 nm. If one measures the resonance Raman spectrum at this wavelength then the observed signal differs significantly from that of the non-resonance data; most strikingly only the 2021 cm^{-1} CO band is observed (Fig. 3). As the resonant transition is MLCT in nature the resulting state has a $(d\pi)^5(\pi^*_L)^1$ configuration. The diminution of electron density about the $d\pi$ orbitals means that the backbonding to the

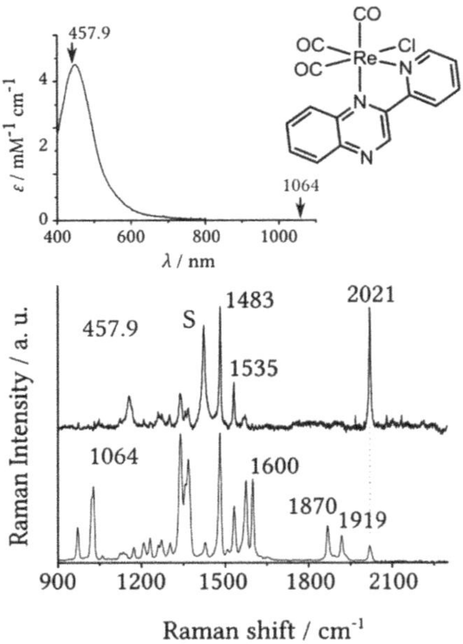

Fig. 3 Electronic absorption spectrum, structure and resonance Raman (upper trace, S denotes solvent band) and normal Raman (lower trace) spectra of Re(pqx)$(CO)_3Cl$.[21]

CO ligands is reduced and these ligand bonds strengthen.[22] The enhancement of only the highest frequency band can be rationalised if one considers the nature of the normal coordinates associated with each of these modes. The mode associated with the band at 1870 cm^{-1} is a stretch of equatorial CO ligands with compression of the axial CO. The MLCT excitation results in the bonds of the CO ligands being strengthened and thus this asymmetric vibration does not mimic the excitation and has a small Δ (Fig. 1). The mode associated with the 1919 cm^{-1} vibration is an asymmetric stretch of the equatorial CO ligands and this also does not mimic the excitation; only the mode associated with the 2021 cm^{-1} vibration does this as it is a symmetric stretch in which the CO bonds in all three ligands are displaced in phase. This mode does mimic the excited state and thus is enhanced.

The CO vibrations are an example in which some empirical knowledge of the normal modes is sufficient to provide insight; however with the advent of computational chemistry the detailed nature of the ligand normal modes are readily available and these can be used in interpreting the structure of the MLCT excited state in terms of the acceptor orbitals on the ligand, in this case pqx. This allows one to map out the change in electron density during each transition independently.

Wavepacket modelling may be used to further interpret resonance Raman spectra. It is particularly useful if the spectra have been taken at a series of excitation wavelengths across the absorption band to give a resonance Raman excitation profile; a plot of resonance Raman cross section for bands versus excitation wavelength. Wavepacket modelling is based on time-dependent theory developed by Heller and coworkers.[23–26] This models a wavepacket propagating on an excited state surface and informs on the magnitude of displacement for chosen vibrational modes (Δ_i). Additionally, mode-dependent reorganisation energies λ_v/cm^{-1} can also be calculated from this method using the relationship: $\lambda_v = (\nu_i/2)\Delta_i^2$, in which ν_i is the wavenumber of the ith mode and Δ_i is the displacement along coordinate i (Fig. 1).

In addition to its use for assigning the MLCT transition, resonance Raman spectroscopy may also be used to identify other transitions, the nature of the chromophores, for identifying reaction products and intermediates and the nature of oxidised or reduced molecules through the use of spectroelectrochemistry. Even when narrowed to the investigation of polypyridyls a number of studies have utilised resonance Raman spectroscopy to achieve the above aims. Table 1 collects such references together for the last 5 years, as well as indicating the exciting wavelengths used and what type of study was performed: Chromophore identification (CI), wavepacket analysis (WP), spectroelectrochemistry (SE), product characterisation (PC) intermediate characterisation (IC) and transient Raman (TR). The TR section is discussed in Section 3.

A full review of all work in Table 1 would be of considerable size and rather than detail each paper a number of highlights of types of systems (solar cell materials and dipyridophenazine complexes) are noted and specific examples of types of experiments (chromophore identification, spectroelectrochemistry and wavepacket analysis) are described.

Solar cell materials

Resonance Raman spectroscopy has been used to try to understand the interactions between substrates and dyes; particularly in dye-sensitised solar cells. Likodimas *et al.*[27] studied the effects of light stress and voltage bias on the solar cell ruthenium complex (N719, Ru(L-R1)(L-R2)$(NCS)_2$, Table 1) on TiO_2. They characterised the bound dye on TiO_2 and found that in cell ageing experiments the TiO_2 showed band broadening. In a related study Zedler *et al.*[63] examined the spectrum of N719 on Au nanoparticles. Through a comparison of the resonance Raman spectra and the surface-enhanced resonance Raman (SERRS) spectra they determined that the dye was only slightly electronically affected by binding to the Au surface. Kavan *et al.* used similar dyes on single-walled nanotubes (SWNT).[32] They found that hole injection into the SWNT rapidly bleached the dye; however electron injection did not adversely affect the dye on the surface. Interactions with dyes and TiO_2 were also examined, using wavepacket analysis, by Giokas[28] who found that the use of phosphate binding groups gave negligible charge-transfer interaction between the dye and TiO_2.

Finally the interaction of carboxylate substituted bpy complexes of the type $RuL_2(CNS)_2$ in solution were examined in regards to deprotonation and reduction.[31] From analysis of the bpy-band shifts in the resonance Raman spectra (supported by DFT calculations) it was shown that the reduction weakens the bpy structure and that the CNS ligands are not spectator units but play a key role with the metal dπ orbitals as the donor MO set in the MLCT transition. Such findings are consistent with time-resolved IR studies of these types of complexes that point to an MLCT that involves the CNS ligands in the donor wavefunction.[64–67]

Table 1 Complexes studied in the 2008–2013 period pertaining to resonance Raman spectroscopy. Techniques include: Chromophore identification (CI); wavepacket analysis (WP); spectroelectrochemistry (SE); product characterisation (PC); intermediate characterisation (IC); transient Raman (TR).

Ligand	Complex	Technique	λ_{exc}/nm	Reference
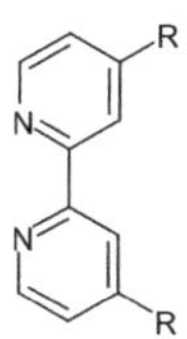	$Ru(L\text{-}R_1)(L\text{-}R_2)(NCS)_2$	IC	514, 785	27
	$R_1 = C_9H_{19}, R_2 = CO_2H$	SE		
	$[Ru(L)_n(bpy)_{3\text{-}n}]^{2+}$	CI	457, 488, 514	28
	n = 1,2,3	IC		
	$R = H_2O_3P$, $H_2O_3PCH_2$	WP		
	$[Ru(bpy)_3]^{2+}$ $[Ru(bpy)_2(L)]^{2+}$	TR	350, 475	29
	$L = CO_2H$			
	$[Re(L)(CO)_3Cl]$			
	$[Re(L)(CO)_3(py)]^+$			
	$[Cu(L)(PPh_3)_2]^+$			
	R= 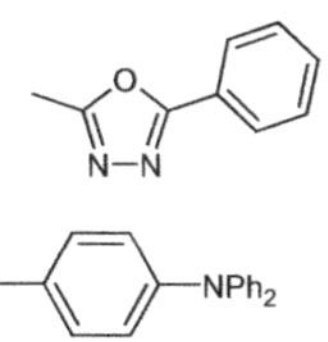	CI	350, 406.7, 413.1, 444.3, 457.9, 488, 514.5	30
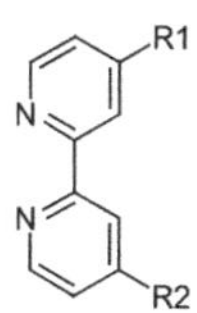	$[Ru(NCS)_2L_2]^{2-}$	SE	458	31
	$R1 = CO_2H$, $R2 = Bu_4N^+CO_2^-$	PC		
	$Ru(L1)(L2)(NCS)_2$	SE	514.5, 568	32
	L1: $R1 = R2 = C_9H_{19}$			
	L2: $R1 = CO_2H$ $R2 = CO_2Na$			

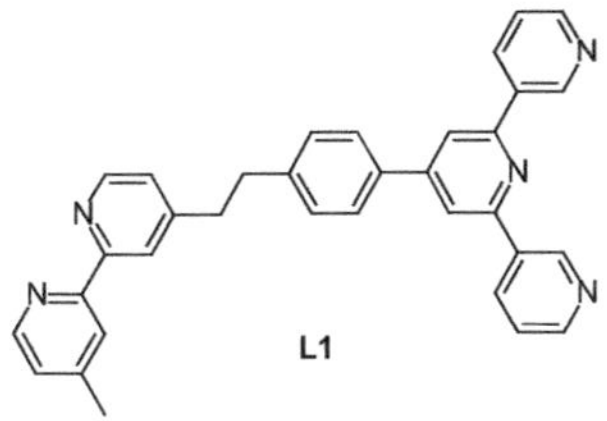 L1 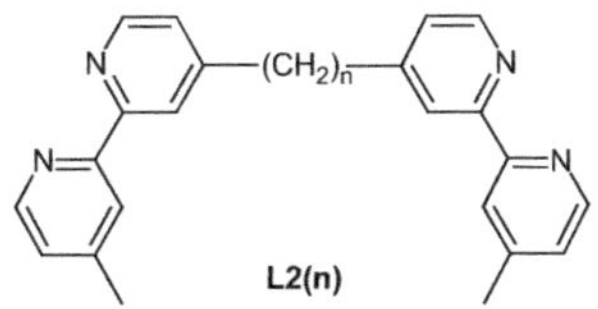 L2(n)	$[Ru(bpy)_2(L1)]^{2+}$, $[\{Ru(bpy)_2(L1)\}_2Fe]^{6+}$, $[Ru(bpy)_2(L1)Fe(bpy)_2]^{4+}$, $[\{Ru(bpy)_2(L2(n))\}_3Fe]^{8+}$ where n = 2, 4, 6	CI TR	355, 532	33
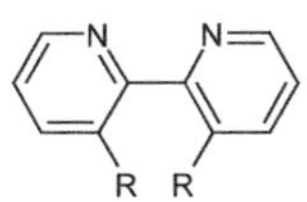 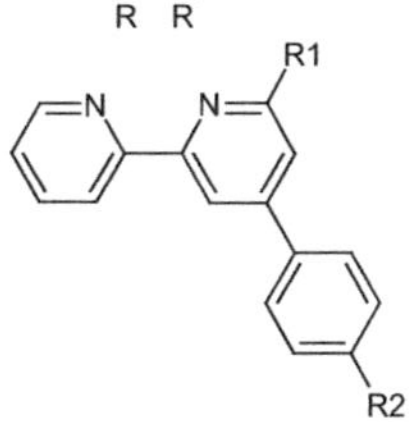	$[Ru(bpy)_2(L)]_2^+$ $Re(CO)_3(L)$ 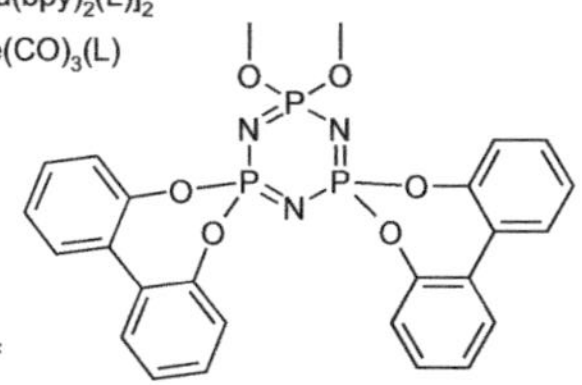 R= R1 = H, Ph 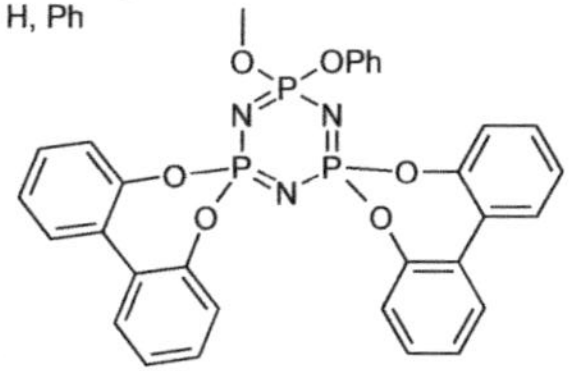 R2 = 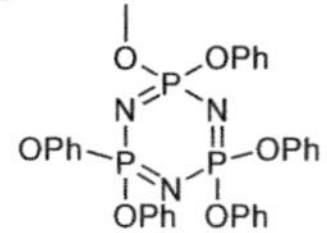	CI TR	355	34

Table 1 (*Continued*)

Ligand	Complex	Technique	λ_{exc}/nm	Reference
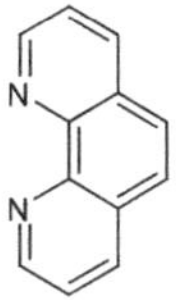	$[Cu(PPh_3)_2(L)]^+$	TR	355	35
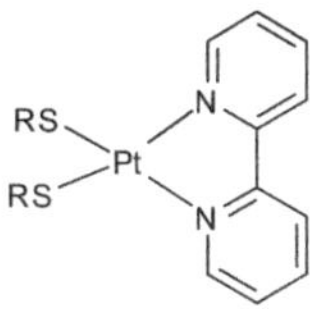	R = 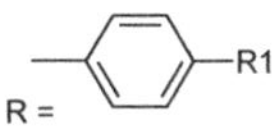 R1 = H, OCH_3, NO_2	TR	493	36
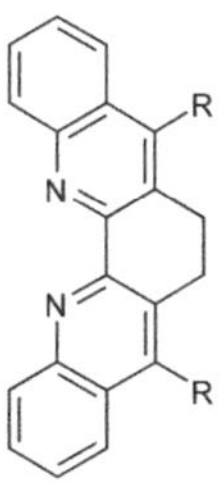	$[Cu(bpy(Mes)_2)(L)]^+$ L = bpy, phen, biq R = CH_3, Ph	CI	406, 448, 458, 488, 515, 532, 568, 594	37
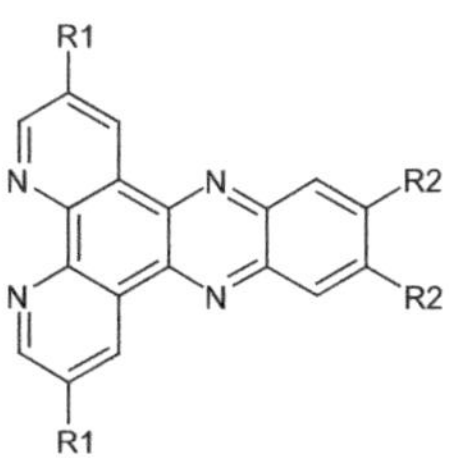	$[Ru(4,4'\text{-}{}^tBu\text{-}bpy)_2(L)]^{2+}$ R1 = Br, R2 = H R1 = H, R2 = Br	CI	458, 476, 488	38

Ligand	Complex	Technique	λ / nm	Ref.
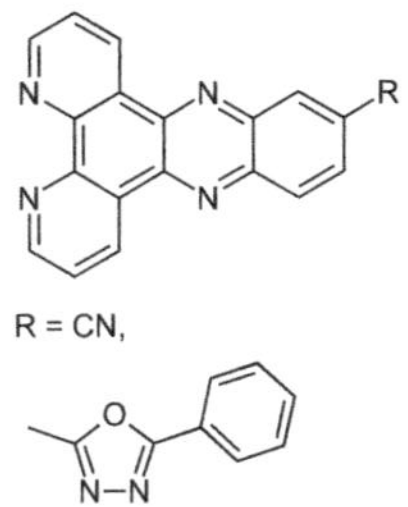	$[Re(L)(CO)_3Cl]$ $[Cu(L)(PPh_3)_2]^+$ $[Ru(bipy)_2(L)]^{2+}$ $[Ir(ppy)_2(L)]^{2+}$	CI TR	356, 413	39
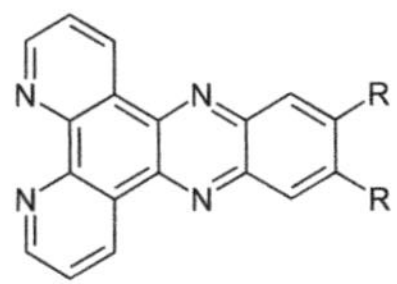	$[Cubpy(Mes)_2(L)]^+$ $[Re(L)(CO)_3Cl]$ $R = SCN, SC_{10}H_{21}$ 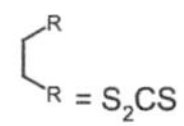	CI	351, 356, 406, 413, 448, 457, 488, 515	40 41
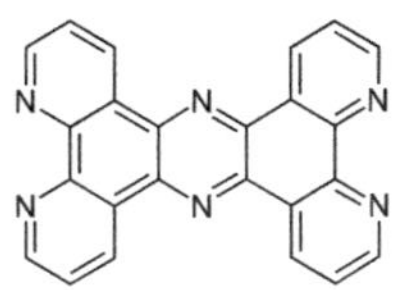	$[Ru(4,4'\text{-}^tBu\text{-}bpy)_2(\mu\text{-}L)PdCl_2]^{2+}$	CI	458	4, 42
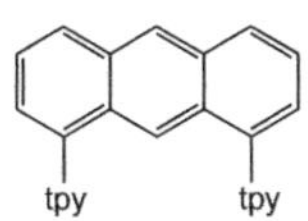	$[Ru_2Cl(bpy)_2(L)]^{3+}$	SE IC	633	43

Table 1 (Continued)

Ligand	Complex	Technique	λ_{exc}/nm	Reference
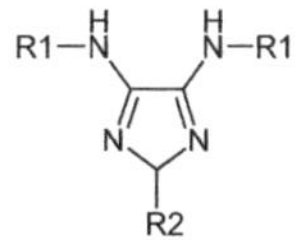	$[RuCl(tpy\text{-}(^tBu)_3)L]$ R1 = tolyl, Ph-NMe_2, Ph-CO_2Et; R2 = Ph	IC WP CI	458, 476, 488	44–46
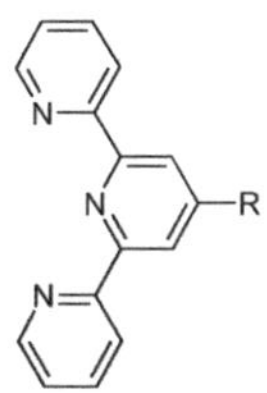	$[RuL_2]^{2+}$ R = H, 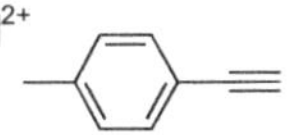 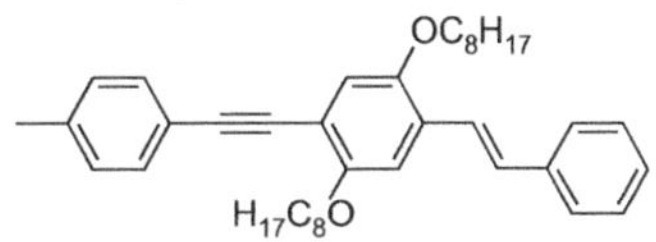	CC	476	47
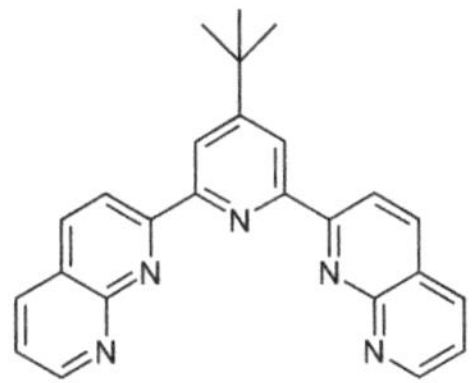	$[Ru(4\text{-methyl pyridine})_2(H_2O)L_2]^{2+}$	IC SE	532	48
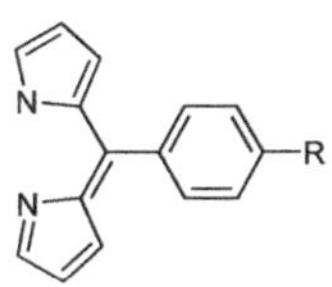	$[Ru(bpy)_2L]^{+/0}$	WP CI	413, 444, 458, 488, 514.5, 532	49

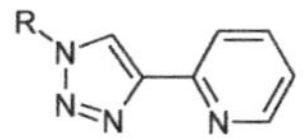	$ReCl(CO)_3(L)$ $R = C_8H_{17}$, FC, 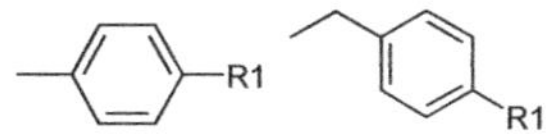 $R1 = H, OCH_3, NO_2$	CI	351	50
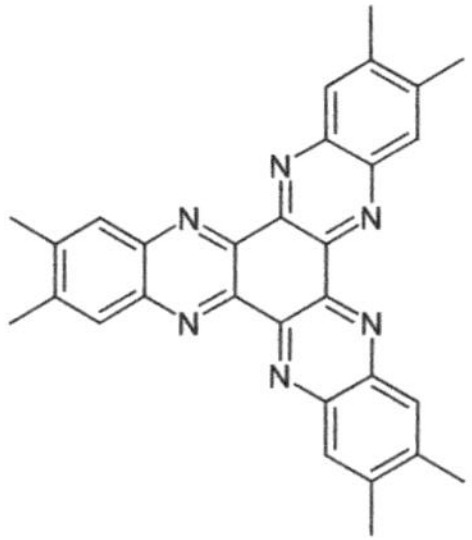	$ReCl(CO)_3(L)$ *anti/syn*-$[(ReCl(CO)_3)_3(L)]$ *anti/syn*-$[(ReCl(CO)_3)_2(L)]$ $[(Cu(PPh_3)_2)_3L]^{3+}$	CI CI WP	444, 568, 647 406, 413, 444, 458, 488, 515, 532, 568, 647	51 52
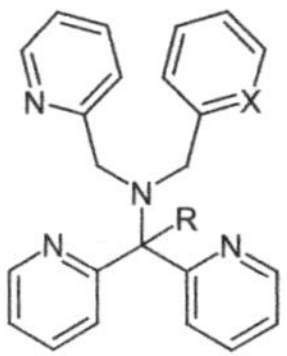	FeLZ $X = N, R = H, Z = CH_3CN$ $X = N, R = CH_3, Z = CH_3CN$ $X = N, R = H, Z = Cl$	CI IC	355, 401, 449, 473	53
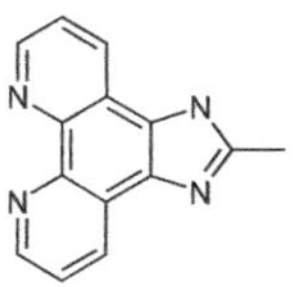	$[Ru(dppz)_2(L)]^{2+}$ R = 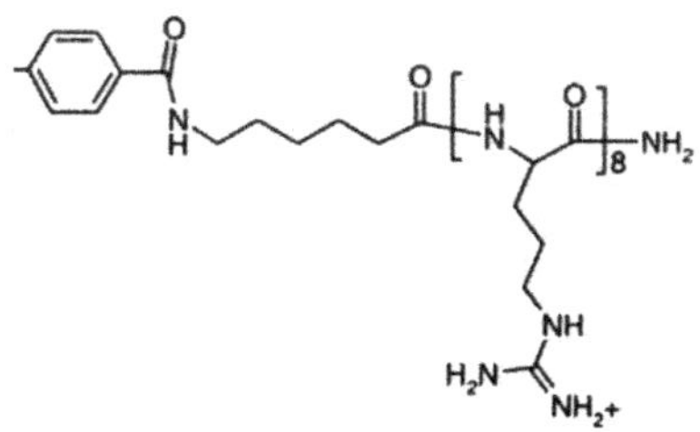	CI	458	54

Table 1 *(Continued)*

Ligand	Complex	Technique	λ_{exc}/nm	Reference
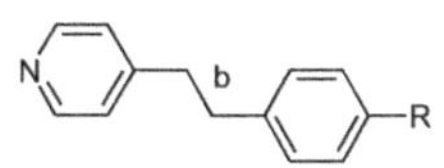 Where the bond order of b = 1,2,3.	$[Re(CO)_3(phen)(L)]^+$ R = NMe_2, NPh_2	CI	351, 413, 444, 458	55
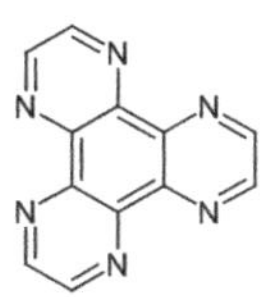 	$[Ru(bpy)_2(L)]^{2+}$	WP	458, 480, 515, 530	56
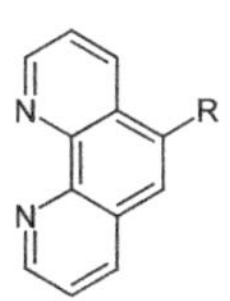 	$[Ru(bpy)_2(L)]^{2+}$ R = β-cyclodextrin	CI	458, 488, 515	57

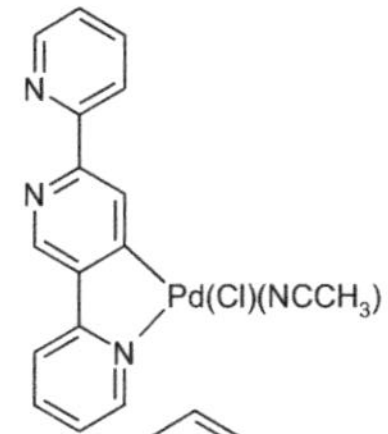	$[Ru(bpy)_2(L)]^{2+}$	CI	355, 450, 473	58
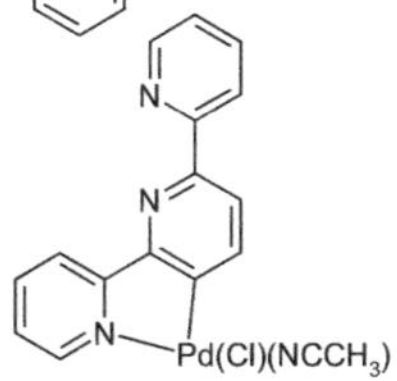				
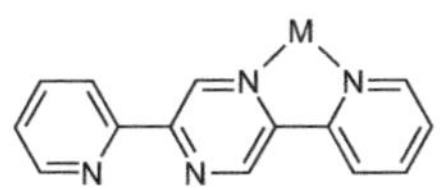	$[Ru(bpy)_2(L)]^{2+}$ M = nothing, $\{Ru(bpy)_2\}$, $PdCl_2$, $PtCl_2$	CI TR	355, 450, 532, 561	59
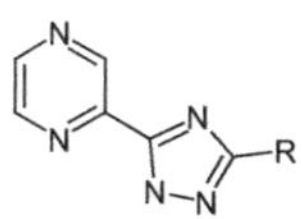	$[Ru(bpy)_2(L)]^{+}$ R = Ph	CI TR	355, 401, 449, 473, 532, 561	29, 60
	$[(bpy)_2(H_2O)Ru^{III}ORu^{III}(OH_2)(bpy)_2]^{4+}$	CI SE IC	458, 466, 477, 488, 497, 502, 515, 568, 647, 676	61, 62

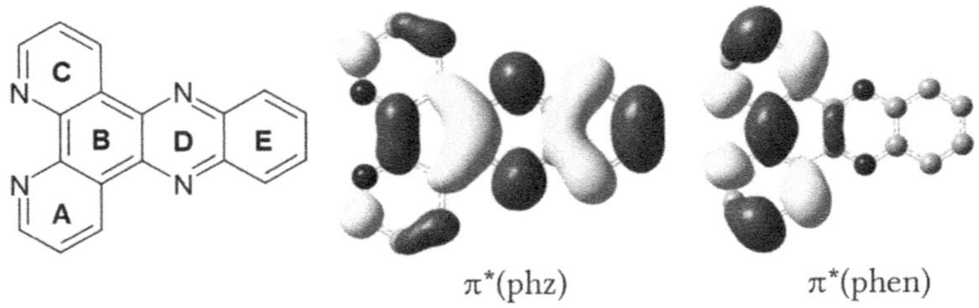

Fig. 4 Structure of dppz with ring labels, wavefuctions for frontier unoccupied MOs.

Dipyridophenazine systems

Dipyridophenazine (dppz) systems have been studied using resonance Raman, transient Raman and time-resolved infrared spectroscopies. They are of interest because they possess a number of low-lying excited states with differing electronic character. These states have significantly different spectral and photophysical properties. The electronic structure of dppz is such that there are two MOs that lie close in energy but have quite different wavefunction disposition. There is a phen-based MO π*(phen) in which wavefuction amplitude is across the rings A, B and C of the dppz framework (Fig. 4); there is significant amplitude at the chelating nitrogen atoms but none of the nitrogen atoms at ring D (the phenazine N atoms). Conversely the π*(phz) MO has nodes at the chelating N atoms but amplitude at the phenazine N atoms.

It is possible to selectively tune the energies of these two MOs independently using appropriate substituents. Kuhnt *et al.*[38] studied $\{Ru(bpy)_2\}^{2+}$ with substituted dppz ligands. They found that with Br substituents on the phen portion (2,7-position) the optical spectrum was broadened and the red edge of the absorption showed enhancement of the phen modes of dppz[68] which are shifted because of the Br substituents; this is similar to an earlier study in which $[Ru(bpy)_{3-n}(Br_2\text{-}bpy)_n]^{2+}$ (where n = 0, 1, 2, and 3) was investigated and found to have lower lying Ru → Br_2bpy transitons than Ru → bpy.[69]

Lundin *et al.*[39] studied dppz with strongly electron withdrawing groups on the phz portion (oxadiazole and CN). By examining the resonance Raman spectrum of the ligand and comparing that to the complex it is possible to gauge the perturbation on the system caused by complexation. For the dppz-oxadiazole the spectrum with 356 nm excitation is dominated by a band at 1540 cm^{-1} associated with the oxadiazole; this is consistent with a π → π* transition in which the π MO is delocalised and the π* MO localised on the phz portion of dppz. This enhancement pattern is also observed in the rhenium(I) complex with 356 nm excitation but on tuning to lower energy (λ_{exc} = 413 nm) phen modes become enhanced and the metal CO band also gains in intensity as the MLCT (dπ → π*(phen)) comes into resonance.

Chromophore identification

An excellent example of chromophore identification was reported by Siebert *et al.*[47] In this study they used resonance Raman spectroscopy to ascertain the effect on the MLCT transition in a series of ruthenium complexes with substituted terpyridyl ligands with extended ethyne phenyl substituents (Fig. 5).

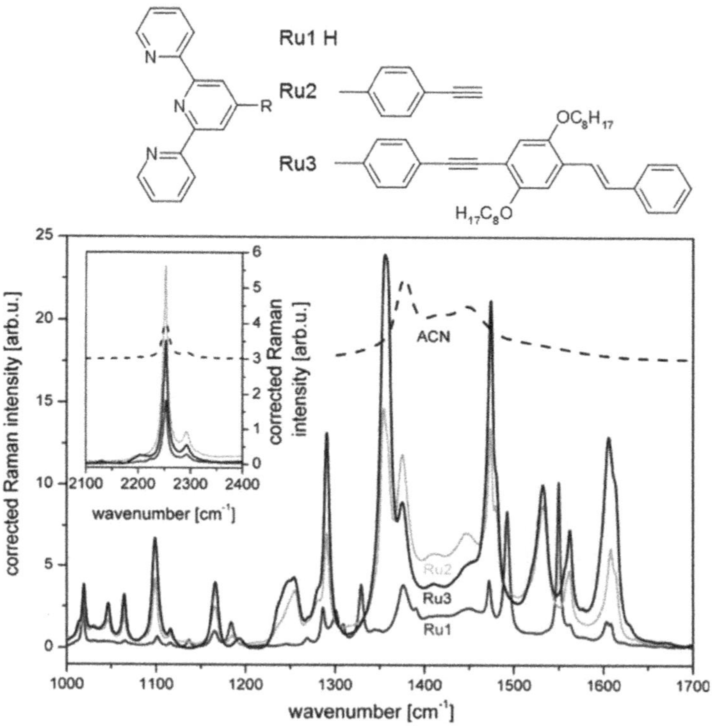

Fig. 5 Ligand structures and resonance Raman spectra of complexes in MeCN (λ_{exc} = 476 nm) from Siebert *et al.*[47]

Electronic absorption spectra of the complexes showed that as the substituent in the 4′-position increases in length there is a bathochromic shift of the MLCT band from 475 to 492 to 497 nm. This is indicative of extended π-conjugation and a greater delocalization of the MLCT state. In order to further characterise this MLCT state Raman spectra were collected with λ_{exc} = 476 nm, in resonance with the transition (Fig. 5).

The spectra of Ru2 and Ru3 are remarkably similar to each other but significantly differ from the spectrum of Ru1. Vibrational modes enhanced in the resonance Raman spectrum of Ru1 are assigned to various terpyridine stretches consistent with a LUMO on the ligand. Ru2 and Ru3 display many extra enhanced bands assigned to terpyridine stretches as well as a vibration involving the directly linked phenyl ring. The similarity between Ru2 and Ru3 in addition to the absence of a C≡C vibration (at around 2200 cm^{-1}) suggests that the excited state extends only to the first phenyl ring on the substituent and no more. Additionally the authors provide a detailed investigation of the band at 1355 cm^{-1}, previously shown in closely related zinc(II) complexes to be an indicator for the conjugation between the terpyridine sphere and the adjacent phenyl ring.[70,71] The authors deconvoluted the band into two separate peaks and plotted the peak areas which in both cases were greatest (by around 1/2) for Ru3. It follows that for Ru3 the transition must be more localised on the molecule where the 1355 cm^{-1} vibration occurs than for Ru2.

This indicates that upon increasing the length of the substituent at the 4′-position the conjugation between the moieties is increased.

Spectroelectrochemistry

Resonance Raman may also be used in conjunction with electrochemistry (spectroelectrochemistry) in order to identify the nature of reduced or oxidised species. This technique was used by Wada *et al.*[43] in order to understand the mechanism of water splitting with the ruthenium catalyst shown in Fig. 6. The method by which O–O bond formation occurs (an essential step to evolution of O_2 and H_2) has been a point of contention among researchers with several possible reaction mechanisms proposed.[61,62,72] This paper describes data that are interpreted as showing the first identification of O–O bond formation prior to the evolution of O_2 through the use of resonance Raman spectroelectrochemistry.

Oxidation of the complex in water leads to a decrease in the observed absorption band at 468 nm and the growing in of a band at 688 nm (Fig. 7(B)). When the resultant solution was left without applying voltage, oxygen was evolved while a new band at 475 nm grew in and bleaching of the 688 nm band occurred (Fig. 7(C)). The authors therefore assigned the

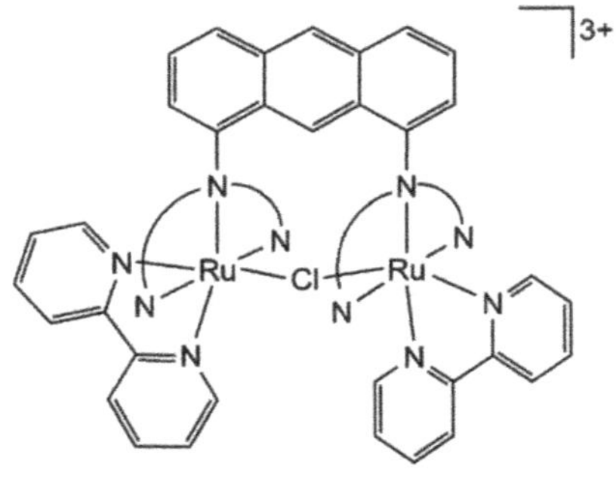

$[1]^{3+}$

Fig. 6 Structure of the ruthenium containing water-splitting complex investigated by Wada *et al.*[43]

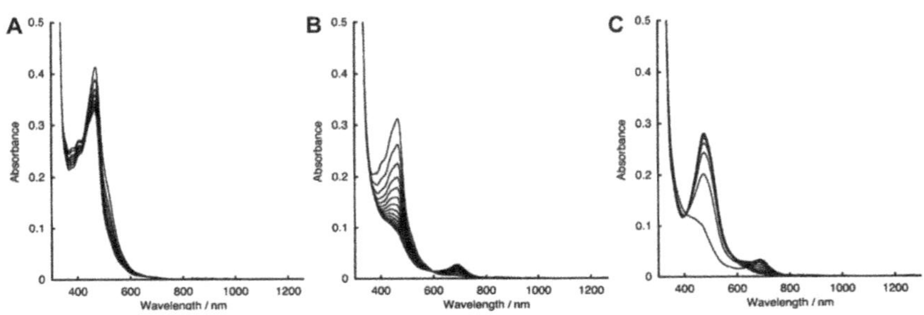

Fig. 7 UV-Vis-NIR spectrum of $[1]^{3+}$ ions in water during electrolysis at A) E = +0.85 V and B) +1.4 V (*vs* Ag/AgCl) and C) after electrolysis had stopped. Used with permission from reference 43.

absorption at 688 nm as being due to the precursor to O_2 evolution, which contains the O–O bond of interest. Additionally the observation of the band at 475 nm after O_2 evolution suggests that the original {Ru^{II}-(bpy)(trpy)} framework of the catalyst is maintained.

Resonance Raman spectra were obtained at 633 nm upon oxidation of $[1]^{3+}$ in H_2O, in order to be in resonance with the 688 nm absorption. Two Raman bands were observed at 442 and 824 cm^{-1} when the solvent used was $H_2{}^{16}O$, which shifted to 426 and 780 cm^{-1} for $H_2{}^{18}O$. These isotope shifts (16 and 44 cm^{-1}) are consistent with the bands being due to the Ru–O and O–O stretching modes respectively. From previous studies[73] the authors assigned the O–O band as being due to an end on μ-O_2. The $(=O)_2$ binding motif was also considered, through the oxidation of a $[Ru(bpy)(tpy)(OH_2)]^{2+}$ complex. The resultant resonance Raman spectrum showed a new band at 819 cm^{-1} in $H_2{}^{16}O$ which shifted to 785 cm^{-1} in $H_2{}^{18}O$, confirming the R=O stretch mode through the calculated isotope ratio (34 cm^{-1}). This species however was not seen in the Raman spectrum of oxidised $[1]^{3+}$. The authors reasoned that after the $[Ru^{IV}(bpy)_2$-(btpyan)$(=O)_2]^{4+}$ ions are formed upon oxidation the two {Ru^{IV}=O} moieties would be so close together that the repulsion of the two oxo groups would be overcome by the coupling reaction of the two to give the O–O bond. The 688 nm band is therefore assigned to the $[Ru^{III}{}_2(bpy)_2$(btpyan)-(μ-O_2)$]^{4+}$ ion.

Wavepacket modelling

Wavepacket modelling is a powerful method that uses resonance Raman intensity analysis to calculate specific mode displacements upon electronic excitation. McLean *et al.*[49] reported a wavepacket analysis of the complexes displayed in Fig. 8. The electronic absorption spectra of these complexes involve a strong ligand-centred (LC) π-π* band at around 464 nm and a weaker broader MLCT band at about 524 nm. Resonance Raman spectra were therefore obtained at 457, 488 and 514.5 nm in order to gain a better knowledge of these transitions. A quantitative analysis of the enhancements of bands seen for the different excitation wavelengths allowed the authors to assign the π-π* band as being localised on the

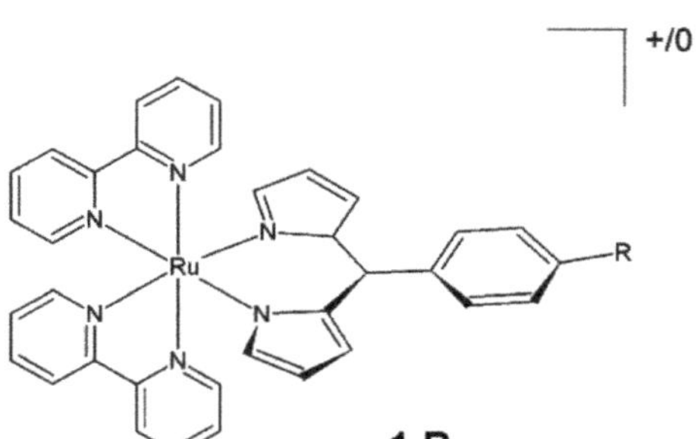

1-R

R = CO_2^-; CO_2H; CO_2CH_3

Fig. 8 Structure of the complexes (1-R) reported by McLean.[49]

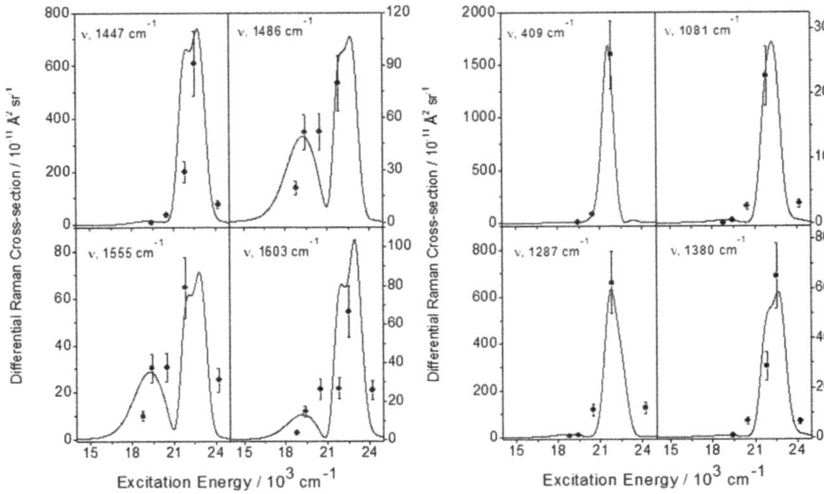

Fig. 9 Raman excitation profiles of [1-CO_2]. Solid lines are simulated profiles with experimental value points at specific excitation wavelengths overlayed. Used with permission from reference 49.

dipyrrin ligand while the MLCT band involves a LUMO on the bipyridyl moiety.

Wavepacket modelling was performed over the two transitions with the largest oscillator strengths (the LC and MLCT transitions) as these would be expected to affect the resonance Raman intensities the most. This assumption is confirmed by the reasonable match seen in Fig. 9 of the simulated Raman excitation profiles with the experimental Raman intensity data points. One mode (q_{41}) was observed to dominate with a dimensionless displacement (Δ) almost double that of any other mode at 0.47 for the LC transition. This indicates that the mode is extremely important in the Franck-Condon region for that transition. The mode, at 409 cm^{-1}, is a bending mode based on the dipyrrin phenyl group. Modes are generally found to have Δs of between 0.05 and 0.25 which, for stretching vibrations, equates to a cartesian bond length increase of 0.15 Å following excitation. The total inner-sphere reorganisation energies, calculated by summing all vibrational energies together for each transition, were found to be very similar for both the transitions. The solvent reorganisation however was found to dominate for the MLCT transition compared to the LC one.

The excitation profiles show two distinct patterns, either a mode is enhanced solely with excitation of the LC band (only one peak seen in Fig. 9) or it is enhanced with both bands (two peaks observed). The type of bands showing only enhancement at the LC transition are all based on the dipyrrin, as noted previously. This pattern of enhancement is not consistent with results obtained from the TD-DFT calculations of [1-CO_2], where the relevant transition is based largely on the bipyridyl ligands. On the other hand TD-DFT calculation of the reduced species [1-CO_2H]$^+$ predicts an intense band here that is dominated by amplitude changes on the dipyrrin ligand alone. Additionally the transition predicted for the

MLCT band involves electron density changes on the ruthenium and bipyridine orbitals (as expected) while that for [1-CO_2] involves dipyrrin. Consequently, the electronic structure of the Ru(bpy)$_2$dipyrrin species seems to be modelled better by [1-CO_2H]$^+$ than by [1-CO_2].

The nature of structural changes that occur in the complex upon π-π* excitation were also proposed by the authors based on the enhanced vibrational modes, specifically q_{41}. The mode is a transannular torsional mode on the dipyrrin phenyl ring. In the ground state this ring is non-planar, the strong enhancement of the mode suggests a large change in this angle upon excitation. Based on previous work[74] the authors suggest that the dihedral angle in the excited state could be 180° which would lead to a significant resonance Raman enhancement in the associated torsional mode, q_{41}, which is consistent with observation.

The effect of changing the R group was considered. A considerable decrease in the resonance Raman enhancement of the q_{41} mode was observed as R was protonated and then esterified. This indicates that the excited-state mode displacement decreases. The authors suggest that this could be due to the increased stacking effect when the phenyl ring substituent is uncharged, as in the protonated and esterified cases, inhibiting the torsional motion of the phenyl ring in the excited state.

It is worth noting that newly developed techniques now allow for a calculation of resonance Raman excitation profiles and these techniques will clearly have a bearing on future research in this area.[56,75]

3 Time-resolved resonance Raman

Time-resolved or excited resonance Raman spectroscopy may be used to obtain spectral information on various moieties present in the transient states of metal polypyridyl complexes. There are two protocols that may be employed in deriving the transient resonance Raman spectra: firstly, one can use the single-colour, or single pulse, pump-probe method. The sample is irradiated with a train of pulses (of ns or ps duration) and the resulting resonance Raman scattering is collected. This method was used in the early experiments on $[Ru(bpy)_3]^{2+}$ pioneered by Woodruff and Dallinger.[76,77] The experiment gives Raman scattering from the excited state because the leading edge of the pulse rapidly establishes excited state in the irradiated volume as the photon to molecule ratio is high. As the laser pulse is often 10 ns in duration diffusion is not an issue and the relaxation of the transient species within the irradiated volume is also negligible. Indeed a pulsed laser may not be required if the excited state is sufficiently long-lived as demonstrated by Hester *et al.* who used a chopped CW argon ion laser beam to obtain the excited state vibrational spectrum of $[Ru(bpy)_3]^{2+}$.[78] This is often referred to as TR^2, transient resonance Raman.

One limitation of the single colour method is that it provides information on the transient species that exists in the irradiated volume during the laser pulse; it is not strictly time-resolved as the experiment gives no ability to change when the Raman scattering is to be measured relative to initial excitation. To do this one can use the two-colour pump-probe method in which the sample is irradiated with two trains of laser pulses; one provides the pump pulse and the second, suitably delayed, provides the probe pulse. Using this type of experiment the resonance Raman spectra from the irradiated volume may be measured as a function of time-delay between excitation and probe. This is referred to as TR^3, standing for time-resolved resonance Raman.

Transient resonance Raman has been used effectively to determine the nature of various species present in metal polypyridyl complex excited states; however the resonance effect dominates the spectra obtained. Thus the vibrational data collected is a strong function of the absorption properties of the excited state.

Two-colour time-resolved studies

An example of TR^3 spectroscopy in which the kinetics of an excited state relaxation are measured is work by Henry *et al.*[29] in which they study the spectra of $[Ru(bpy)_3]^{2+}$ and two substituted analogues using 1 ps pulse widths and a Kerr gate detection system to reject fluorescence.[79,80] From these studies they determined that the population of the ^{3}MLCT (THEXI) state occurs in ~20 ps on the basis of the growth of the $bpy^{\bullet-}$ bands; interestingly in 1987 Brus *et al.* reported that the single-colour transient Raman spectrum with 355 nm (6 ps) pulsed excitation showed $bpy^{\bullet-}$ features.[81] The results are in agreement with earlier transient electronic absorption studies.[82,83]

In a study of multi-nuclear complexes, containing Fe and Ru, it was found that the excited state electronic spectra showed biphasic behaviour (Fig. 10).[33] The assemblies contained both terminal and bridging bpy units and the electronic spectroscopy was unable to differentiate between radical anion species of each. TR^3 spectroscopy was used to show that the MLCT excited state formed on the terminal bpy ($Ru^{III}bpy^{\bullet-}$) was longer-lived than that formed on the bridging bpy, even though the bridge was unconjugated.

Single-colour studies

Single-colour studies have been used to determine the nature of the excited state in cases where transient electronic absorption signatures cannot give unequivocal interpretation. Brennan used TR^2 to show that the excited state of a $[Ru(bpy)_2(\text{pyrazin-2'-yl-1,2,4-triazolato})]^+$ was ($Ru^{III}bpy^{\bullet-}$) despite perturbations (protonation and methylation) to the pyrazin-2′-yl-1,2,4-triazolato ligand.[60] Horvath *et al.*[34] showed that incorporation of $\{Ru(bpy)_2\}^{2+}$ and $\{Re(CO)_3Cl\}$ moieties on to a cyclotriphosphazene unit did not affect the nature of the excited state which showed characteristic $bpy^{\bullet-}$ features.

The characterisation of new types of excited states has also been accomplished using TR^2. In a study of substituted dppz systems[39] the ^{3}LC states were characterised by measuring the TR^2 of the ligand. This was then compared to the complexes to see if the same type of spectrum was observed. In the case of $\{Cu(PPh_3)_2\}^+$ complexes a characteristic ^{3}LC signature was seen – however for the corresponding $\{Re(CO)_3Cl\}$ complex

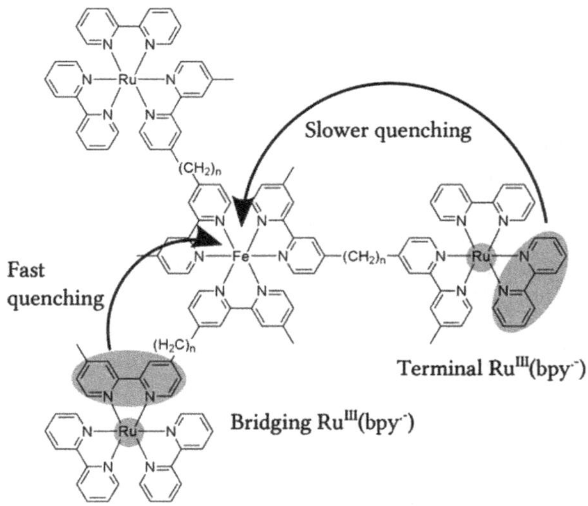

Fig. 10 Structure of multinuclear complexes (n = 2, 4, 6) reported in reference 33, depiction for the two types of Ru(bpy)-based MLCT excited states.

the spectrum appeared to contain phenazine radical anion features suggesting a mixed LC/MLCT excited state.

Finally in a study of $Pt(bpy)(4\text{-}X\text{-}PhS)_2$ ($X{=}NO_2$, OCH_3 and H) Mel'nikov[36] showed using single-colour TR^2 methods that with 493 nm excitation a bpy$^{\bullet-}$ feature is formed – however when the higher energy 400 nm excitation is used the transient spectrum is altered and a low frequency feature corresponding to a sulphur radical anion is observed.

4 Time-resolved infrared spectroscopy (TRIR)

Time-resolved infrared techniques have been widely utilised to probe excited state structural changes in metal polypyridyl complexes on the ps and ns timescale.

Polypyridyl ligands tend to show vibrations in the fingerprint region (1000–1700 cm^{-1}), but these can be difficult to interpret. Often CO, CN or C≡C groups are used as TRIR markers, as these tend to be easily interpreted in terms of band shifts upon excitation, and they are observed in isolated spectral regions. Common to all of these groups, which makes their behaviour easy to interpret, is a π system. Formation of an excited state, depending on its nature, can populate the π^*_{CO} MO (Fig. 3), which leads to a weakening of the bond and thus a decrease in vibrational frequency, or reduce electron density, strengthening the bond and raising the frequency. A common TRIR motif, $Re(CO)_3(L)(diimine)$, may form an MLCT excited state in which the Re centre is formally oxidised from Re(I) to Re(II) and the diimine is reduced; this withdraws electron density from the CO (although population of the diimine π* does produce a backbonding effect, hence the shift in ν(CO) is more significant if the electron on diimine$^{\bullet-}$ is localised far from the metal centre) and an increase in ν(CO) relative to the ground state is observed, while a ligand-centred (LC) π,π^* state causes a decrease in ν(CO) because the increased electron density in the diimine π^*_L system donates electron density *via* backbonding into the π^*_{CO}.

Platinum complexes

Adams *et al.*[84] examined square planar Pt complexes of MesBIAN (see Fig. 11) with two acetylide co-ligands, and examined the C≡C IR region. IR of the oxidised species suggests the HOMO is 32% based on the actylide π-system, and that of the reduced species indicates MesBIAN-based reduction. TRIR showed a decrease in ν(C≡C) for the transient species relative to the ground state (1820, 2028 cm^{-1} and 2124, 2115 cm^{-1} respectively), consistent with a decrease in electron density of the C≡C π system (as opposed to population of π*). Since no bands in the fingerprint region are consistent with the formation of $MesBIAN^{\bullet-}$, then the transient bands are proposed to be due to 'oxidised' $Pt(C{\equiv}C)_2$ as similar bands have been observed for compounds like this where the structure becomes more C=C-like. The excited state is assigned as charge-transfer in nature in which electron density is delocalised from the formally triple-bonded acetylide to MesBIAN, altering the bond order and forming a pseudo-cumulenic structure (R–C=C=C–R′).

Best *et al.*[85] studied Pt complexes; these had one or two metal centres with catechol ligands (Fig. 11, where R=$CONEt_2$ is named bpyam and R = ^{t}Bu named bpy-^{t}Bu). In all cases, catechol vibrations (assigned by DFT and comparison to control Pt(cat)(PPh_2-Et-PPh_2)) dominated the IR spectra. IR spectroelectrochemistry showed for all cations the 1600 cm^{-1} ν(CO)/ν(CC) band shifted to 1575 cm^{-1} – this is a marker for the catechol to semiquinone conversion. For $(Pt(bpyam))_2$(cat) and Pt(bpyam)(cat) the 1639 cm^{-1} amide C=O bleach and 1625 cm^{-1} transient show that bpyam π* is populated. For $(Pt(bpy\text{-}tBu)_2)$(cat) and Pt(bpy-tBu)(cat) bleach of the 1289 cm^{-1} C-O band and the 1624 cm^{-1} CN band, and the formation of 1600 cm^{-1} CN transient, the 1564–1575 cm^{-1} semiquinone marker and $bpy^{\bullet-}$ bands show that the catechol is oxidised and bpy reduced and thus are indicative of an LL'CT state populating bpy π*.

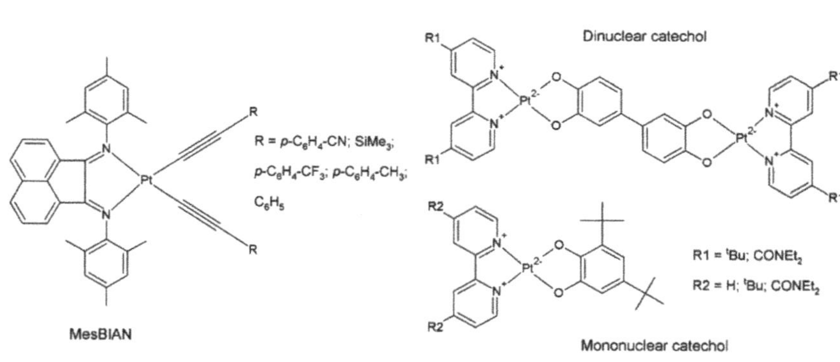

Fig. 11 Complexes studied by Adams[84] (left hand side) and Best[85] (right hand side).

Isomerisation and permanent photoproducts

Busby *et al.*[86] study cis-trans isomerisation about a C=C bond in the excited state for the ligands Medpe$^+$ and stpy (Fig. 12). Re(CO)$_3$-Cl(Medpe)$_2$ shows a large increase in ν(CO) upon excitation (A'(1) 2028 cm^{-1} in the ground state and 2080 cm^{-1} transient), indicating oxidation of Re and suggesting that the electron resides on the pyridyl ring furthest away from the metal centre, because backbonding is reduced.

The bpy analogue (Re(CO)$_3$(bpy)(Medpe)) shows more complicated behaviour, upon excitation initially ground state CO bands at 1929 and 2035 cm^{-1} bleach and transient CO A'(1) bands appear at 2055, 2085 and 2027 cm^{-1}. This indicates the formation of three different excited states: the 2085 cm^{-1} band is assigned to MLCT(Medpe) as it has a similar shift to the complex with two Medpe ligands, while the 2055 cm^{-1} band is assigned as MLCT(bpy) because it also shows an increase in ν(CO) consistent with an MLCT state, but the shift is smaller indicating there is more backbonding from the now-populated polypyridyl π* to CO π*, and finally the 2027 cm^{-1} band, which is decreased in frequency relative to the ground state, is assigned as a ligand-centred ^{3}π,π* state (*trans*-configuration). With τ = 0.6 ps the MLCT(Medpe) band decays and the π,π* band increases in intensity, with τ = 5–10 ps the MLCT(bpy) band decays, and the 2027 cm^{-1} π,π* band decays with τ = 21.3 ps into a 2032 cm^{-1} band assigned to the *cis*-conformer in a direct process (since an isosbestic point is observed).

Mixed-valent systems

Complexes with multiple metal centres may show different types of redox and excited state behaviour: metal-to-ligand charge transfer or one-electron oxidation can produce a mixed-valent system which may be localised or delocalised. These forms of mixed valence are labelled as class I, II and III according to Robin and Day.[87,88] In class III both metal centres are 'half'-oxidised. TRIR is a useful tool to elucidate such behaviour, since the number of bands is sensitive to the symmetry of the metal centre, which is exemplified by Wragg *et al.*[89] Here, either one or two {Ru(CN)$_4$}$^{2-}$ centres are coordinated to bppz (Fig. 13) and probed using IR spectroelectrochemistry and TRIR of CN bands. For [Ru(CN)$_4$(bppz)]$^{2-}$, one-electron oxidation gives an increase in ν(CN) from 2090, 2063, 2053 cm^{-1} in the ground state to 2106, 2096, 2083 cm^{-1}, as expected when Ru(II) is oxidised to Ru(III). 355 nm excitation in D$_2$O or DCM leads to a transient spectrum in which 2056, 2061 cm^{-1} bands are bleached and transient bands are observed at higher frequency: 2102 cm^{-1} in D$_2$O and 2093 cm^{-1} in DCM, which is consistent with MLCT(bppz) depopulating CN π*.

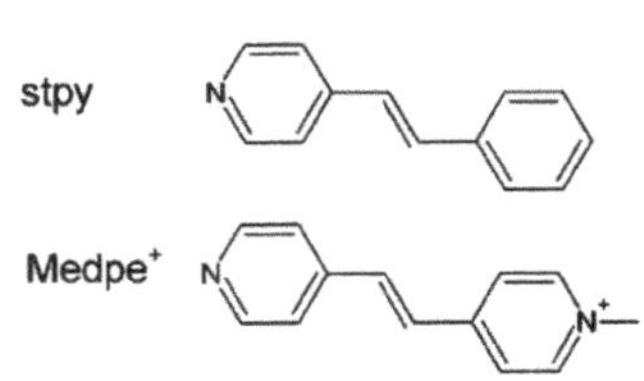

Fig. 12 Structures of isomerisable ligands studied by Busby *et al.*[86]

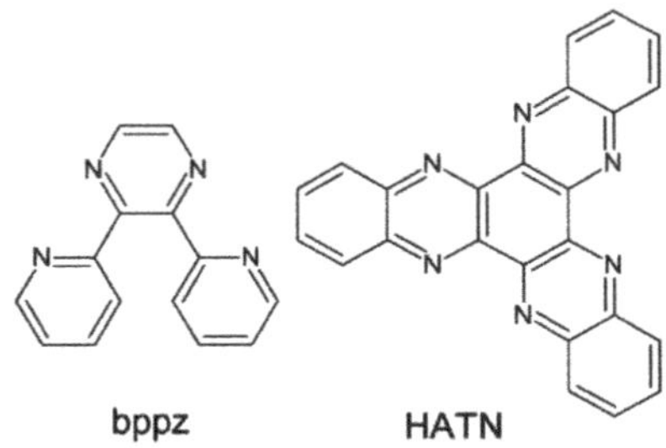

Fig. 13 Structures of bppz[89] and HATN ligands.[51]

$[(Ru(CN)_4)_2(\mu\text{-bppz})]^{4-}$ CN bands are observed at 2088, 2064, 2053 cm^{-1}; one-electron oxidation causes these bands to shift to higher frequency (2094, 2079, 2068 cm^{-1}). The shifts for this complex are approximately half of those observed for the mononuclear complex, and there is only one set of CN bands, therefore the metal centres must both be in the same environment, and on this timescale the complex is in a $(Ru^{2.5+})_2$ state. The two-electron oxidation product CN bands are observed at 2110 and 2095 cm^{-1} which is consistent with both centres being Ru(III). One-electron reduction (which must be bppz-based) predictably causes a decrease in ν(CN). In D_2O, 355 nm excitation bleaches the ground state CN bands, and produces transients of the 2088 cm^{-1} band both higher (2090 cm^{-1}) and lower (2028 cm^{-1}) in frequency. Consequently, on the TRIR timescale the system is mixed-valent since there are now two types of CN environment, the higher frequency band must be formed by MLCT (Ru to bppz) oxidising one Ru(II) to Ru(III) because the electron density of CN π* decreases, while the lower frequency band arises from bppz$^{\bullet-}$ populating the π* of CN groups on the remaining Ru(II) *via* π-back-bonding. Therefore, two different techniques show different behaviours for the same system.

Fraser *et al.*[51] studied a bridging ligand with three metal-binding sites and investigated the behaviour of hexaazatrinapthalene (HATN, Fig. 13) coordinated to one, two or three $Re(CO)_3Cl$ centres; *syn* and *anti* isomers behave identically. $[Re(CO)_3Cl(HATN)]$ shows a bleach of ground state CO bands and the appearance of higher frequency transient CO bands with 400 nm ps excitation. This is consistent with formation of an MLCT state: Re is oxidised and HATN reduced, strengthening CO bands by depopulation of π* CO. $[(Re(CO)_3Cl)_2(HATN)]$ also shows ground state bleaching upon excitation and the appearance of transient bands both higher (2064 cm^{-1}) and lower (2010, 1892 cm^{-1}) in frequency. This is consistent with a mixed-valent MLCT state: CO groups bound to the oxidised Re centre give the higher ν(CO) while those on the non-oxidised Re centre shift to lower frequency as the CO π* accepts electron density from HATN$^{\bullet-}$. Similarly, $[(Re(CO)_3Cl)_3(HATN)]$ shows transient bands at 2073 cm^{-1} (oxidised Re) and 2014 cm^{-1} (two non-oxidised Re centres), therefore in the HATN system MLCT excited states are localised to one metal centre. Two interesting points to note are that the intensity patterns of the bands are consistent with the behaviour described – for $[(Re(CO)_3Cl)_2(HATN)]$ the two transient bands are equal in intensity while for [$(Re(CO)_3Cl)_3$(HATN)] the higher:lower ratio is 1:2 – and that the highest frequency transient shifts to higher frequency from $Re(CO)_3Cl$(HATN) (2058 cm^{-1}) to [$(Re(CO)_3Cl)_2$(HATN)] (2064 cm^{-1}) to [$(Re(CO)_3Cl)_3$(HATN)] (2073 cm^{-1}). This phenomenon arises since in all cases a single electron is transferred from one Re centre to HATN, and this single electron is then shared equally with all of the metal centres *via* π-backbonding, so for two metals then each gets half, and for three each one-third, leading to a decrease overall in electron density populating CO π* in this manner. The ^{3}MLCT state lifetime decreases as the number of metal centres is increased from 1.3 ns to 510 ps to 220 ps.

Solvent dynamics and effects

As well as studying the excited state behaviour of novel complexes, TRIR has also been used to study solvent dynamics by examining changes in excited state behaviour of well-studied compounds such as [$Re(CO)_3$-X(diimine)] and its environmentally sensitive ν(CO) shifts. Blanco-Rodriguez *et al.*[90] compared $[Re(CO)_3(4\text{-Etpy})(bpy)]^+$ in three ionic liquids and MeCN, and found that TRIR showed all ionic liquids to initially have ^{3}LC and ^{3}MLCT features (populated simultaneously from a ^{1}MLCT state), with ^{3}LC also decaying to ^{3}MLCT, while in MeCN only the ^{3}MLCT state was observed thus decay was shorter than the timescale of the measurement. Consequently, ionic liquids slowed the relaxation step; and these showed triexponential dynamic shifting in which the medium and slow components showed proportionality to solvent properties.

Blake *et al.*[91] used TRIR to probe a solid-state structure, where [$Re(CO)_3Cl$(bpy)] was immobilised in a Mn-carboxylate MOF scaffold. No emission was observed although this compound is emissive in solution, which is consistent with the LC π,π* state being lower in energy than the MLCT state (emission from MLCT). TRIR shows initial formation of the MLCT which decays to a ^{3}LC state, contrary to behaviour in solution in which the MLCT state is lowest energy and is observed to form in TRIR.

Kuimova *et al.*[92] observed different excited states in TRIR for [$Re(CO)_3Cl(dppzX_2)$] depending on substituent X (Me, H, F, Cl, CF_3) and on the solvent. Dppz is a unique ligand, which has low-lying UMOs localised to phen or phz, thus different MLCT states are possible: MLCT(phen) and MLCT(phz) as well as a ligand-centred π,π* dppz state (Fig. 4). As the phen part coordinates to the metal while the phz part is further away, a large shift in ν(CO) (approx. 55 cm^{-1}) is consistent with phz population, as the effect of backbonding from dppz to CO is reduced, while phen population causes a smaller shift (40 cm^{-1} or less). Here electron donating (CH_3) or neutral (H) groups show CO bands both higher and lower than in the ground state, indicative of a mixed MLCT/IL state; the 40 cm^{-1} shift higher is consistent with the MLCT being phen localised. When X is a strong electron-withdrawing group (Cl, CF_3) transient CO bands are only observed higher than the ground state, by 60 cm^{-1}, thus MLCT(phz) is the excited state in DCM, while in BuCN (and MeCN) the highest transient ν(CO) is observed at 2072 cm^{-1} (+48 cm^{-1} corresponds to phen) and later also at 2087 cm^{-1} (+63 cm^{-1} corresponds to phz), showing initially populated MLCT(phen) decays to an equilibrium.

Finally when X=F, ν(CO) of the transient appears at 2064 and 2078 cm^{-1} (2024 cm^{-1} in the ground state), which is similar to above in that MLCT(phen/phz) states are in equilibrium in DCM, while in BuCN, MeCN, BuOH and MeOH only MLCT(phz) (νCO 2090 cm^{-1}) is observed.

Biological substituents

Dppz is also widely utilised for its ability to intercalate into the base pairs of DNA, and $[Re(CO)_3(dppz)(py\text{-}X)]^+$ was used by Olmon *et al.*[93] to observe DNA-mediated oxidation of guanine, where X contained an amide bond to polyAT or polyGC. Both the C≡O region (Re centre) and C=O region (DNA) were studied on the ps and ns scale. Both showed ground state bleaching and vibrationally hot ^{3}IL π,π* (phen) that relaxed to ^{3}IL π,π* (phz) (2031, 1915–1935 cm^{-1}). For polyGC, the 2031 cm^{-1} transient signal was much weaker than for polyAT, and combined with a decrease in emission intensity on the same timescale was assigned as a quenching of the ^{3}IL state by a charge transfer ($dppz^{\bullet-}$, $G^{\bullet+}$) state.

Blanco-Rodriguez *et al.*[94] studied $Re(CO)_3$(diimine)L complexes, where diimine was bpy or phen, and L either pyridine or imidazole linked to the amino acids Phe, Tyr or Trp. They found that for Phe and Tyr-containing compounds, transient CO bands are higher frequency (2073, 2013, 1971 cm^{-1}) than the ground state (2040, 1930 cm^{-1}) consistent with breaking of the pseudo-C_{3v} symmetry upon MLCT (Re → diimine); the amino acid does not alter the properties. However, for $Re(CO)_3$(bpy)(py-Trp) while initially the same thing is observed (ground state bleach at 2035, 1930 cm^{-1}; CO transients 2073, 2013, 1971 cm^{-1}, ^{3}MLCT) these transients decay to different ones at 2010, 1895 cm^{-1}. These new transients are down-shifted compared to the ground state, which indicates a reduced ligand as backbonding to CO π* is increased, and this could be either a reduced state ($Re(CO)_3(NN^-)(L)$) or a charge separated state ($Re(CO)_3(NN^-)(L^+)$). To elucidate this, the fingerprint region was also investigated: at short times the bpy bands are bleached as is the amide I band (1678 cm^{-1}) which is shifted to higher frequency (1685 cm^{-1}) consistent with ^{3}MLCT (Re → bpy); at longer times (50 ns) these bands have decayed and the 1521 cm^{-1} amide II/indole band is bleached and transient bands assigned to Trp^+ at 1278 and 1497 cm^{-1} are observed, indicating the formation of a ^{3}CS state.

Molecular dyads

TRIR has been used to probe photoinduced energy transfer in which structural information is desired. For example, Easun *et al.*[95] synthesised a molecular dyad containing an $Ru(bpyam)_2$ centre and an $Re(CO)_3Cl$ centre linked by Mebpy-Et-Mebpy (denoted L (L2 in Fig. 14)) which coordinates to both centres; Re(C≡O) and amide C=O bands are strong IR reporters and Re and Ru both show ns triplet MLCT states, with Re the higher energy of the two. On the ps scale, increase in ν(C≡O) relative to the ground state is indicative of MLCT at the Re centre (since this is what the carbonyls report on, and must be to L), and this state decays in 20 ps. The amide ν(C=O) bleach is observed at 1636 cm^{-1} and transient band at 1619 cm^{-1} which is assigned to population of bpyam π*, and thus indicative of ^{3}MLCT (Ru → bpyam). The Re MLCT decays with $\tau = 22$, 830 ps,

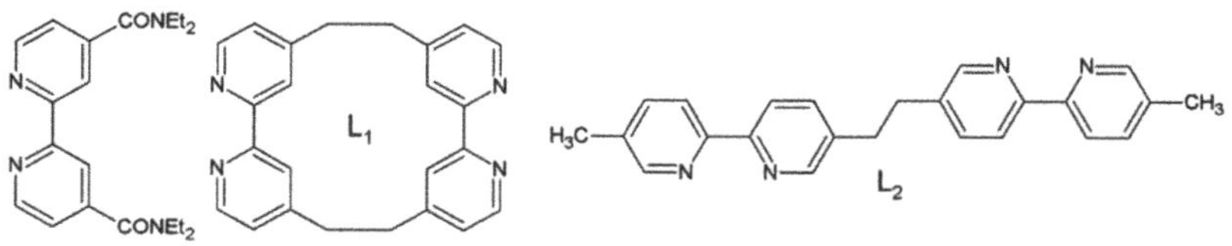

Fig. 14 Structures of ligands used in molecular dyads reported by Easun *et al.*[95]

and the Ru MLCT grows in with $\tau = 750$ ps; the authors suggest the Re MLCT decays to Ru MLCT, as would be expected based on relative energy levels. In addition, a 1649 cm^{-1} amide C=O band appears within 13 ps and persists to 2 ns; this is assigned as Ru to L MLCT, the presence of this second state formed also from the Re MLCT on a faster timescale, is thought to arise from different conformers showing different transfer rates. This conformational flexibility issue is addressed in a subsequent paper which also explores the use of $\{Ru(CN)_4\}^{2-}$ which is very solvatochromic which enables the switching in energy of the two MLCT states, and thus the energy transfer direction, with solvent.[96] The CN groups report on the Ru centre they are bound to, while the bpyam ligand is probed using ν(C=O). In D_2O, $\{Ru(CN)_4\}^{2-}$ is higher energy than $Ru(bpyam)_2$, while in MeCN this is reversed. When the linker is L1 in D_2O, transient CN bands are observed at higher and lower in frequency than the ground state bleach. The higher energy band is attributed to a MLCT(Ru(CN)→L1) state, since backbonding to CN is reduced and thus bonds strengthened, but as evidenced by the presence of a lower energy band and complicated kinetics, there is more than one state. The amide (C=O) band also shows transients higher and lower than the ground state, indicative of multiple states; the higher energy is assigned to MLCT (Ru(bpyam)→L1) since Ru is transiently oxidised but bpyam is not reduced (less electron density in C=O π* since some is shifted onto the linker), while the lower energy is obviously attributed to the population of bpyam π* in a MLCT (Ru(bpyam)→bpyam) state. Decay kinetics suggest that MLCT (Ru(CN)→L1) decays to MLCT (Ru(bpyam)→L1) with $\tau = 20$ ps, 500 ps (there are two conformers), which then itself decays to the lowest energy MLCT (Ru(bpyam)→bpyam) state. In MeCN, ν(CN) higher in frequency than the ground state is attributed to MLCT (Ru(CN)→L1) and the lower frequency transient is assigned as MLCT (Ru(bpyam)→L1); the amide C=O of bpyam shows ground state bleaching and a lower frequency transient at 1616 cm^{-1} on the ps scale, which must be attributed to MLCT (Ru(bpyam)→bpyam). MLCT (Ru(CN)→L1) forms rapidly from the two other MLCT states, as shown by the relative kinetics. When a more conformationally flexible linker ligand L2 is used, in D_2O the MLCT (Ru(bpyam)→L2) state is not observed, but the initially formed MLCT (Ru(CN)→L2) state decays to MLCT (Ru(bpyam)→bpyam) with two τ values due to two conformers. In MeCN, again the MLCT (Ru(bpyam)→L2) state is not observed and MLCT (Ru(bpyam)→bpyam) decays to MLCT (Ru(CN)→L2).

Other TRIR studies from 2008 to 2013 are summarised in Table 2. Abbreviations used are MLCT (metal-to-ligand charge transfer), LC

Table 2 Complexes studies in the 2008–2013 period using time-resolved infrared not already discussed in the text, excitation wavelengths are 355 nm for ns studies and 400 nm for ps studies unless otherwise noted.

Compound	Substituents or reactants, solvent	Excited state	τ	References
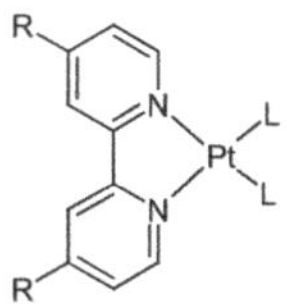	L = Cl, R = $CO_2{}^tBu$, DCM	^3MLCT	16 ps	97
	L = 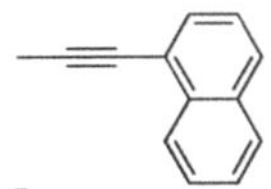 R = $CO_2{}^tBu$,DCM or THF	^3MLCT, $^3LL'CT$	23 ns	97
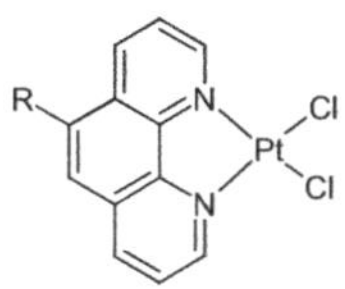	R = 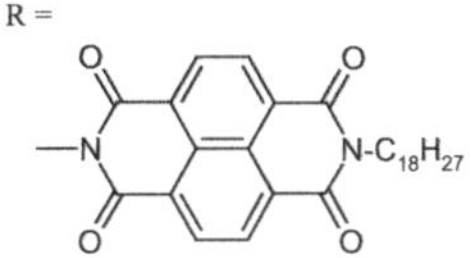 , DCM	$^3MLCT(phen)$, $^3LC(R)^a$	3 ps (MLCT), 520 μs (LC)	98
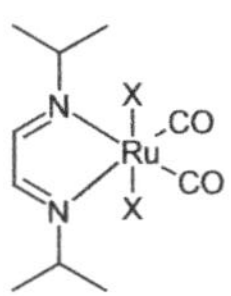	X = Cl, MeCN	$^3MLCT/^3XLCT$, ^3LC	7.2 ns	99
	X = I, MeCN	$^3MLCT/^3XLCT$, ^3LC	>1 ns	99
	X = CH_3, I, MeCN	$^3MLCT/^3XLCT$, ^3LC	>1 ns	99
	X = Cl, $SnPh_3$, MeCN	$^3MLCT/^3XLCT$, ^3LC	>1 ns	99
Re(CO)$_3$Br(bpy)	DMF		51 ns	100
	Co(dmgH)$_2$ + TEOA, DMF		15 ns (MLCT), >10 μs (reduced complex – τ dependent on TEOA conc.)	100

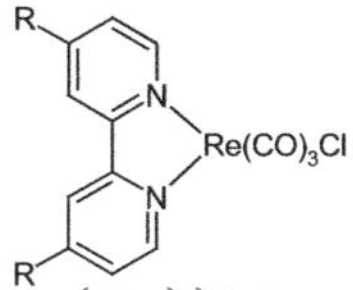 $R = (CH_2)_3)C_6F_{13}$	MeCN Supercritical CO_2	^{3}MLCT[b] ^{3}MLCT[b]		101 101
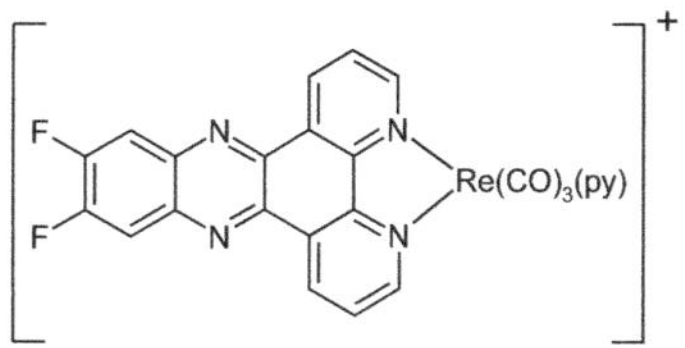	PolyAT, D_2O PolyGC, D_2O	^{3}LC $Re(CO)_3(py)(dppzF_2^{\bullet -})$, $G^{\bullet +}$	8, 210 ns 30, 317 ns	102 102
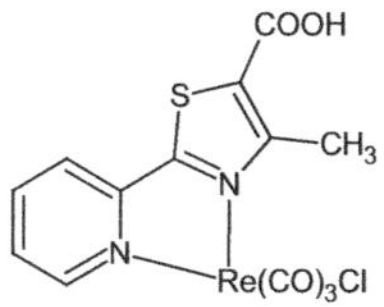 [Re(Pmt-H)(CO_3)Cl]	THF	^{3}MLCT (CO, PMT-H)[c]	10 ns	103
trans-$Mo_2(T^iPB)_2[Re(PMT)(CO)_3Cl]_2$	T^iPB = 2,4,6-Triisopropylbenzene, THF	^{1}MLCT, $^3\delta\delta^*$[c]		103
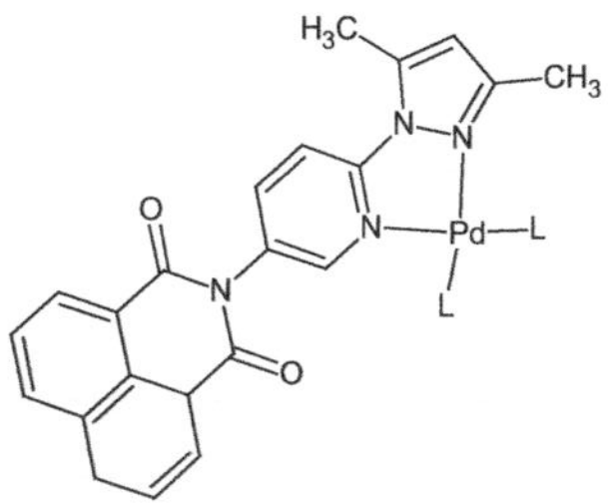	L = 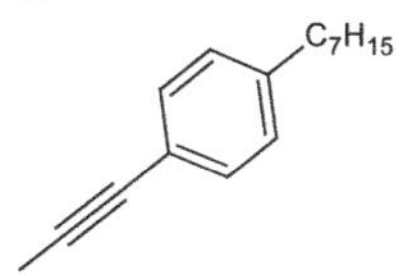 DCM	^{3}MLL'CT (Pt(CC) to Nap), ^{3}LC(Nap)	150 ps (MLL'CT)	104

Table 2 *(Continued)*

Compound	Substituents or reactants, solvent	Excited state	τ	References
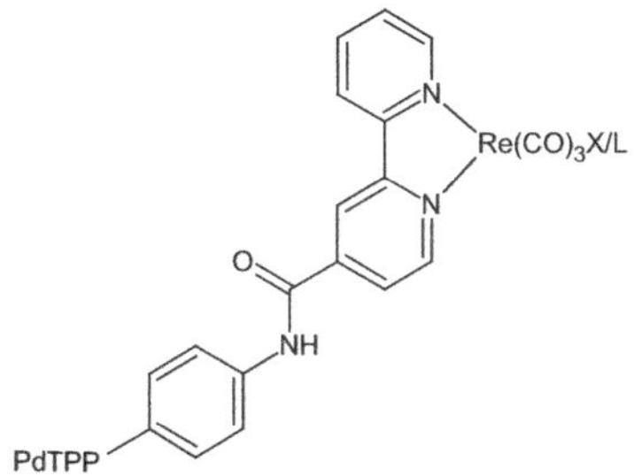	DCM	CS[d]	30–45 ps, 10 μs	105
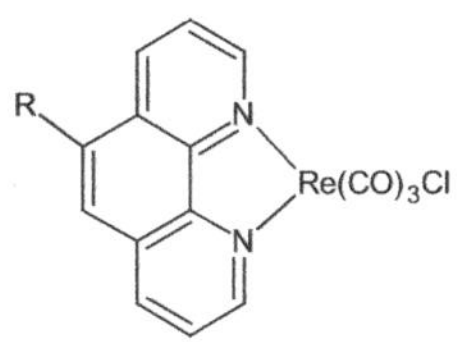	R = 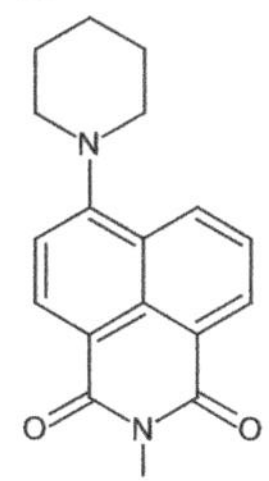 THF	^{3}MLCT(phen)[b]	15 ns (MLCT), 651 μs (LC)	106
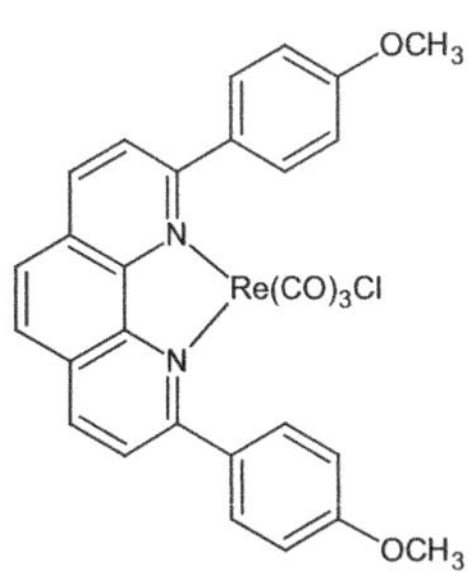	DCM	^{3}MLCT	10 ps	107

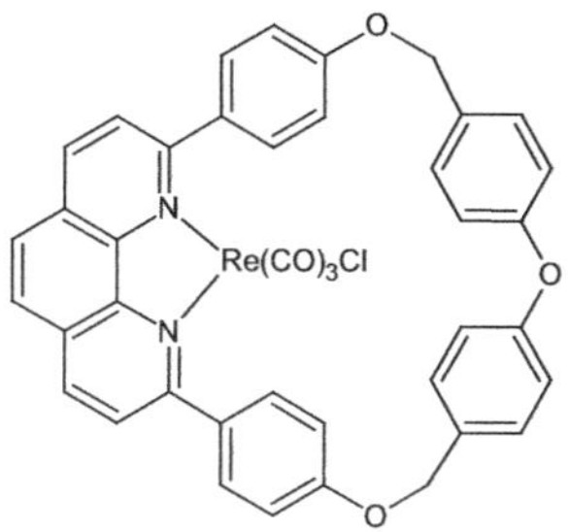	DCM	^{3}MLCT	13 ps	107
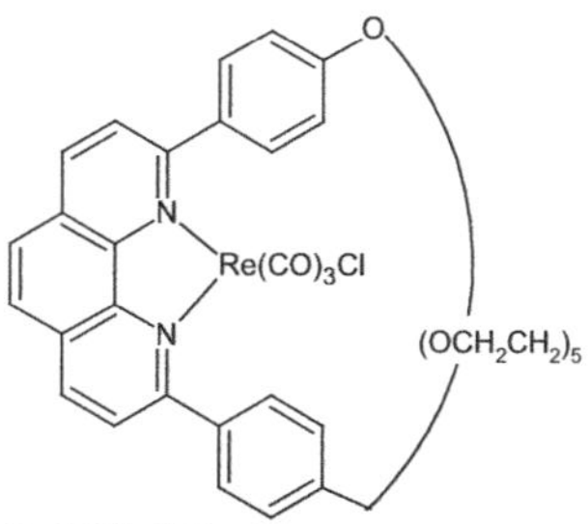	DCM	^{3}MLCT	22 ps	107
$Re(CO)_3Cl(bpy)$	DMF	^{3}MLCT, ^{3}LC		108
$Re(CO)_3Cl(phen)$	DMF	^{3}MLCT, ^{3}LC		108
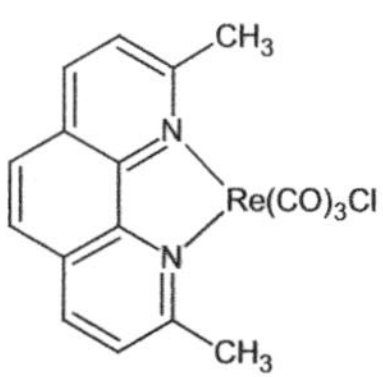	DMF	^{3}MLCT, ^{3}LC		108

Table 2 *(Continued)*

Compound	Substituents or reactants, solvent	Excited state	τ	References
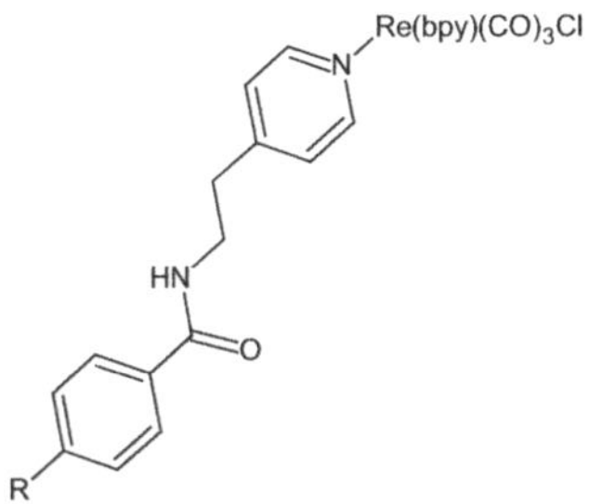	R = 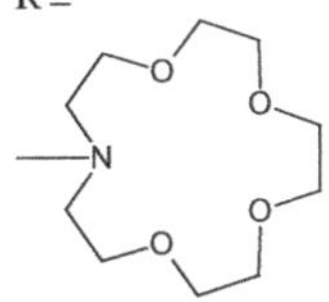	^{3}MLCT(bpy), ^{3}LL'CT	1 ns	109
	THF R = NMe_2, THF	^{3}MLCT(bpy)	150 ns	109
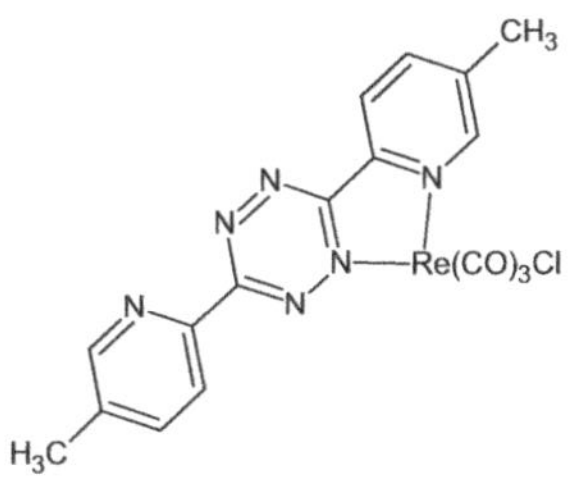	$CHCl_3$	^{3}MLCT(tetrazine)^e	225 ps	110
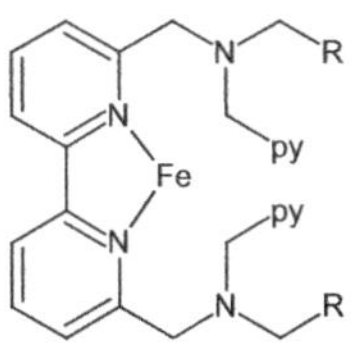	R = py, acetone	HS^f	9.4 ps	111
	R = Ph, acetone	HS^f	12.7 ps	111

^{a}395 nm; ^{b}410 nm; ^{c}various; ^{d}532 nm; ^{e}500 nm; ^{f}387 nm.

(ligand-centered, ππ*), LL'CT (ligand-to-ligand charge transfer), MLL'CT (metal-ligand-to-ligand charge transfer), XLCT (X to ligand charge transfer (where X is usually a halogen)), CS (charge separated) and HS (high spin).

5 Conclusions

The report surveys studies on metal polypyridyl complexes conducted in the last five years that have studied excited state properties using either resonance Raman spectroscopy, to probe the Franck-Condon state, or time-resolved resonance Raman and infrared spectroscopies to examine intermediate and THEXI states.

References

1 B. O'Regan and M. Graetzel, *Nature*, 1991, **353**, 737–740.
2 M. Graetzel, *Acc. Chem. Res.*, 2009, **42**, 1788–1798.
3 A. J. Morris, G. J. Meyer and E. Fujita, *Acc. Chem. Res.*, 2009, **42**, 1983–1994.
4 S. Tschierlei, M. Presselt, C. Kuhnt, A. Yartsev, T. Pascher, V. Sundstrom, M. Karnahl, M. Schwalbe, B. Schafer, S. Rau, M. Schmitt, B. Dietzek and J. Popp, *Chem. Eur. J.*, 2009, **15**, 7678–7688.
5 J. K. Barton, *Science*, 1986, **233**, 727–734.
6 Y. Jenkins, A. E. Friedman, N. J. Turro and J. K. Barton, *Biochemistry*, 1992, **31**, 10809–10816.
7 M. H. Lim, H. Song, E. D. Olmon, E. E. Dervan and J. K. Barton, *Inorg. Chem.*, 2009, **48**, 5392–5397.
8 R. B. P. Elmes and T. Gunnlaugsson, *Tetrahedron Lett.*, 2010, **51**, 4082–4087.
9 H. J. Mo, Y. L. Niu, M. Zhang, Z. P. Qiao and B. H. Ye, *Dalton Trans.*, 2011, **40**, 8218–8225.
10 C. Bazzicalupi, A. Bencini, S. Puccioni, B. Valtancoli, P. Gratteri, A. Garau and V. Lippolis, *Chem. Commun.*, 2012, **48**, 139–141.
11 J. A. Kitchen, E. M. Boyle and T. Gunnlaugsson, *Inorg. Chim. Acta*, 2012, **381**, 236–242.
12 A. W. Adamson, *J. Chem. Educ.*, 1983, **60**, 797–802.
13 G. A. Crosby, *J. Chem. Educ.*, 1983, **60**, 791–796.
14 S. Záliš, C. Consani, A. E. Nahhas, A. Cannizzo, M. Chergui, F. Hartl and A. V. Jr., *Inorg. Chim. Acta*, 2011, **374**, 578–585.
15 T. J. Dines, in *Spectroscopic Properties of Inorganic and Organometallic Compounds: Techniques, Materials and Applications, Volume 43*, The Royal Society of Chemistry, 2012, vol. 43, pp. 229–256.
16 G. F. Strouse, J. R. Schoonover, R. Duesing, S. Boyde, W. E. Jones and T. J. Meyer, *Inorg. Chem.*, 1995, **34**, 473–487.
17 J. A. Treadway, B. Loeb, R. Lopez, P. A. Anderson, F. R. Keene and T. J. Meyer, *Inorg. Chem.*, 1996, **35**, 2242–2246.
18 L. A. Worl, R. Duesing, P. Y. Chen, L. Dellaciana and T. J. Meyer, *J. Chem. Soc.-Dalton Trans.*, 1991, 849–858.
19 M. Wächtler, J. Guthmuller, L. González and B. Dietzek, *Coord. Chem. Rev.*, 2012, **256**, 1479–1508.
20 R. J. H. Clark and T. J. Dines, *Angew. Chem. Int. Edn*, 1986, **25**, 131–158.
21 M. R. Waterland, S. L. Howell and K. C. Gordon, *J. Phys. Chem. A*, 2007, **111**, 4604–4611.
22 D. C. Grills, J. J. Turner and M. W. George, *Comprehensive Coordination Chemistry II*, 2004, **2**, 91–101.

23 E. J. Heller, R. Sundberg and D. Tannor, *J. Phys. Chem.*, 1982, **86**, 1822–1833.
24 S. Y. Lee and E. J. Heller, *J. Chem. Phys.*, 1979, **71**, 4777–4788.
25 A. B. Myers, R. A. Mathies, D. J. Tannor and E. J. Heller, *J. Phys. Chem.*, 1982, **77**, 3857–3866.
26 D. J. Tannor and E. J. Heller, *J. Phys. Chem.*, 1982, **77**, 202–218.
27 V. Likodimos, T. Stergiopoulos, P. Falaras, R. Harikisun, J. Desilvestro and G. Tulloch, *J. Phys. Chem. C*, 2009, **113**, 9412–9422.
28 P. G. Giokas, S. A. Miller, K. Hanson, M. R. Norris, C. R. K. Glasson, J. J. Concepcion, S. E. Bettis, T. J. Meyer and A. M. Moran, *J. Phys. Chem. C*, 2013, **117**, 812–824.
29 W. Henry, C. G. Coates, C. Brady, K. L. Ronayne, P. Matousek, M. Towrie, S. W. Botchway, A. W. Parker, J. G. Vos, W. R. Browne and J. J. McGarvey, *J. Phys. Chem. A*, 2008, **112**, 4537–4544.
30 R. Horvath, M. G. Fraser, S. A. Cameron, A. G. Blackman, P. Wagner, D. L. Officer and K. C. Gordon, *Inorg. Chem.*, 2013, **52**, 1304–1317.
31 L. Zedler, J. Guthmuller, I. R. de Moraes, S. Krieck, M. Schmitt, J. Popp and B. Dietzek, *J. Phys. Chem. C*, 2013, **117**, 6669–6677.
32 L. Kavan, O. Frank, M. Kalbac and L. Dunsch, *J. Phys. Chem. C*, 2009, **113**, 2611–2617.
33 R. Horvath, J. Lombard, J. C. Lepretre, M. N. Collomb, A. Deronzier, J. Chauvin and K. C. Gordon, *Dalton Trans.*, 2013, **42**, 16527–16537.
34 R. Horvath, C. A. Otter, K. C. Gordon, A. M. Brodie and E. W. Ainscough, *Inorg. Chem.*, 2010, **49**, 4073–4083.
35 S. L. Howell and K. C. Gordon, *J. Raman Spectrosc.*, 2008, **39**, 813–826.
36 M. Y. Mel'nikov and J. A. Weinstein, *High Energy Chemistry*, 2008, **42**, 287–289.
37 M. G. Fraser, H. van der Salm, S. A. Cameron, A. G. Blackman and K. C. Gordon, *Inorg. Chem.*, 2013, **52**, 2980–2992.
38 C. Kuhnt, M. Karnahl, S. Tschierlei, K. Griebenow, M. Schmitt, B. Schafer, S. Krieck, H. Gorls, S. Rau, B. Dietzek and J. Popp, *Phys. Chem. Chem. Phys.*, 2010, **12**, 1357–1368.
39 N. J. Lundin, P. J. Walsh, S. L. Howell, A. G. Blackman and K. C. Gordon, *Chem. -Eur.J.*, 2008, **14**, 11573–11583.
40 M. G. Fraser, H. van der Salm, S. A. Cameron, J. E. Barnsley and K. C. Gordon, *Polyhedron*, 2013, **52**, 623–633.
41 M. G. Fraser, A. G. Blackman, G. I. S. Irwin, C. P. Easton and K. C. Gordon, *Inorg. Chem.*, 2010, **49**, 5180–5189.
42 S. Tschierlei, M. Karnahl, M. Presselt, B. Dietzek, J. Guthmuller, L. Gonzalez, M. Schmitt, S. Rau and J. Popp, *Angew. Chem. Int. Edn*, 2010, **49**, 3981–3984.
43 T. Wada, H. Ohtsu and K. Tanaka, *Chem.-Eur. J.*, 2012, **18**, 2374–2381.
44 M. Wachtler, S. Kupfer, J. Guthmuller, S. Rau, L. Gonzalez and B. Dietzek, *J. Phys. Chem. C*, 2012, **116**, 25664–25676.
45 S. Kupfer, J. Guthmuller, M. Wachtler, S. Losse, S. Rau, B. Dietzek, J. Popp and L. Gonzalez, *Phys. Chem. Chem. Phys.*, 2011, **13**, 15580–15588.
46 S. Kupfer, M. Wachtler, J. Guthmuller, J. Popp, B. Dietzek and L. Gonzalez, *J. Phys. Chem. C*, 2012, **116**, 19968–19977.
47 R. Siebert, F. Schlutter, A. Winter, M. Presselt, H. Gorls, U. S. Schubert, B. Dietzek and J. Popp, *Cent. Eur. J. Chem.*, 2011, **9**, 990–999.
48 D. E. Polyansky, J. T. Muckerman, J. Rochford, R. F. Zong, R. P. Thummel and E. Fujita, *J. Am. Chem.Soc.*, 2011, **133**, 14649–14665.
49 T. M. McLean, D. M. Cleland, S. J. Lind, K. C. Gordon, S. G. Telfer and M. R. Waterland, *Chem.-Asian J.*, 2010, **5**, 2036–2046.

50 T. Y. Kim, A. B. S. Elliott, K. J. Shaffer, C. J. McAdam, K. C. Gordon and J. D. Crowley, *Polyhedron*, 2013, **52**, 1391–1398.
51 M. G. Fraser, C. A. Clark, R. Horvath, S. J. Lind, A. G. Blackman, X. Z. Sun, M. W. George and K. C. Gordon, *Inorg. Chem.*, 2011, **50**, 6093–6106.
52 S. J. Lind, T. J. Walsh, A. G. Blackman, M. I. J. Polson, G. I. S. Irwin and K. C. Gordon, *J. Phys. Chem. A*, 2009, **113**, 3566–3575.
53 A. Draksharapu, Q. Li, H. Logtenberg, T. A. van den Berg, A. Meetsma, J. S. Killeen, B. L. Feringa, R. Hage, G. Roelfes and W. R. Browne, *Inorg. Chem.*, 2012, **51**, 900–913.
54 L. Cosgrave, M. Devocelle, R. J. Forster and T. E. Keyes, *Chem. Commun.*, 2010, **46**, 103–105.
55 D. M. Cleland, G. Irwin, P. Wagner, D. L. Officer and K. C. Gordon, *Chem. - Eur. J.*, 2009, **15**, 3682–3690.
56 J. Guthmuller, B. Champagne, C. Moucheron and A. Kirsch-De Mesmaeker, *J. Phys. Chem. B*, 2010, **114**, 511–520.
57 M. Atmeh, N. R. Russell and T. E. Keyes, *Polyhedron*, 2008, **27**, 1690–1698.
58 G. S. Bindra, M. Schulz, A. Paul, R. Groarke, S. Soman, J. L. Inglis, W. R. Browne, M. G. Pfeffer, S. Rau, B. J. MacLean, M. T. Pryce and J. G. Vos, *Dalton Trans.*, 2012, **41**, 13050–13059.
59 M. Schulz, J. Hirschmann, A. Draksharapu, G. S. Bindra, S. Soman, A. Paul, R. Groarke, M. T. Pryce, S. Rau, W. R. Browne and J. G. Vos, *Dalton Trans.*, 2011, **40**, 10545–10552.
60 C. Brennan, A. Draksharapu, W. R. Browne, J. J. McGarvey, J. G. Vos and M. T. Pryce, *Dalton Trans.*, 2013, **42**, 2546–2555.
61 J. W. Jurss, J. J. Concepcion, J. M. Butler, K. M. Omberg, L. M. Baraldo, D. G. Thompson, E. L. Lebeau, B. Hornstein, J. R. Schoonover, H. Jude, J. D. Thompson, D. M. Dattelbaum, R. C. Rocha, J. L. Templeton and T. J. Meyer, *Inorg. Chem.*, 2012, **51**, 1345–1358.
62 D. Moonshiram, J. W. Jurss, J. J. Concepcion, T. Zakharova, I. Alperovich, T. J. Meyer and Y. Pushkar, *J. Am. Chem. Soc.*, 2012, **134**, 4625–4636.
63 L. Zedler, F. Theil, A. Csaki, W. Fritzsche, S. Rau, M. Schmitt, J. Popp and B. Dietzek, *RSC Adv.*, 2012, **2**, 4463–4471.
64 J. B. Asbury, R. J. Ellingson, H. N. Ghosh, S. Ferrere, A. J. Nozik and T. Lian, *J. Phys. Chem. B*, 1999, **103**, 3110–3119.
65 J. B. Asbury, Y. Wang and T. Lian, *J. Phys. Chem. B*, 1999, **103**, 6643–6647.
66 R. J. Ellingson, J. B. Asbury, S. Ferrere, H. N. Ghosh, J. R. Sprague, T. Lian and A. J. Nozik, *J. Phys. Chem. B*, 1998, **102**, 6455–6458.
67 K. S. Finnie, J. R. Bartlett and J. L. Woolfrey, *Langmuir*, 1998, **14**, 2744–2749.
68 C. Kuhnt, S. Tschierlei, M. Karnahl, S. Rau, B. Dietzek, M. Schmitt and J. Popp, *J. Raman Spectrosc.*, 2010, **41**, 922–932.
69 P. A. Mabrouk and M. S. Wrighton, *Inorg. Chem.*, 1986, **25**, 526–531.
70 M. Presselt, B. Dietzek, M. Schmitt, J. r. Popp, A. Winter, M. Chiper, C. Friebe and U. S. Schubert, *J. Phys. Chem. C*, 2008, **112**, 18651–18660.
71 A. Winter, C. Friebe, M. Chiper, U. S. Schubert, M. Presselt, B. Dietzek, M. Schmitt and J. Popp, *ChemPhysChem*, 2009, **10**, 787–798.
72 J. J. Concepcion, J. W. Jurss, M. K. Brennaman, P. G. Hoertz, A. O. T. Patrocinio, N. Y. M. Iha, J. L. Templeton and T. J. Meyer, *Acc. Chem. Res.*, 2009, **42**, 1954–1965.
73 M. T. Khan and G. Ramachandraiah, *Inorg. Chem.*, 1982, **21**, 2109–2111.
74 H. L. Kee, C. Kirmaier, L. Yu, P. Thamyongkit, W. J. Youngblood, M. E. Calder, L. Ramos, B. C. Noll, D. F. Bocian, W. R. Scheidt, R. R. Birge, J. S. Lindsey and D. Holten, *J. Phys. Chem. B*, 2005, **109**, 20433–20443.

75 M. Wachtler, J. Guthmuller, L. Gonzalez and B. Dietzek, *Coord. Chem. Rev.*, 2012, **256**, 1479–1508.
76 R. F. Dallinger and W. H. Woodruff, *J. Am. Chem. Soc.*, 1979, **101**, 4391–4393.
77 P. G. Bradley, N. Kress, B. A. Hornberger, R. F. Dallinger and W. H. Woodruff, *J. Am. Chem. Soc.*, 1981, **103**, 7441–7446.
78 M. Forster and R. E. Hester, *Chem. Phys. Lett.*, 1981, **81**, 42–47.
79 P. Matousek, M. Towrie, C. Ma, W. M. Kwok, D. Phillips, W. T. Toner and A. W. Parker, *J. Raman Spectrosc.*, 2001, **32**, 983–988.
80 P. Matousek, M. Towrie, A. Stanley and A. W. Parker, *Appl. Spectrosc.*, 1999, **53**, 1485–1489.
81 P. J. Carroll and L. E. Brus, *J. Am. Chem. Soc.*, 1987, **109**, 7613–7616.
82 W. Henry, C. G. Coates, C. Brady, K. L. Ronayne, P. Matousek, M. Towrie, S. W. Botchway, A. W. Parker, J. G. Vos, W. R. Browne and J. J. McGarvey, *J. Phys. Chem. A*, 2008, **112**, 10703–10704.
83 N. H. Damrauer and J. K. McCusker, *J. Phys. Chem. A*, 1999, **103**, 8440–8446.
84 C. J. Adams, N. Fey, Z. A. Harrison, I. V. Sazanovich, M. Towrie and J. A. Weinstein, *Inorg. Chem.*, 2008, **47**, 8242–8257.
85 J. Best, I. V. Sazanovich, H. Adams, R. D. Bennett, E. S. Davies, A. J. H. M. Meijer, M. Towrie, S. A. Tikhomirov, O. V. Bouganov, M. D. Ward and J. A. Weinstein, *Inorg. Chem.*, 2010, **49**, 10041–10056.
86 M. Busby, F. Hartl, P. Matousek, M. Towrie and A. Vlcek, *Chem. Eur. J.*, 2008, **14**, 6912–6923.
87 M. B. Robin and P. Day, *Adv. Inorg. Chem. Radiochem.*, 1968, **10**, 247–422.
88 C. Creutz, *Prog. Inorgan. Chem.*, 1983, **30**, 1–73.
89 A. B. Wragg, S. Derossi, T. L. Easun, M. W. George, X.-Z. Sun, F. Hartl, A. H. Shelton, A. J. H. M. Meijer and M. D. Ward, *Dalton Trans.*, 2012, **41**, 10354–10371.
90 A. M. Blanco-Rodriguez, K. L. Ronayne, S. Zalis, J. Sykora, M. Hof and A. Vlcek, *J. Phys. Chem. A*, 2008, **112**, 3506–3514.
91 A. J. Blake, N. R. Champness, T. L. Easun, D. R. Allan, H. Nowell, M. W. George, J. Jia and X.-Z. Sun, *Nat Chem*, 2010, **2**, 688–694.
92 M. K. Kuimova, W. Z. Alsindi, A. J. Blake, E. S. Davies, D. J. Lampus, P. Matousek, J. McMaster, A. W. Parker, M. Towrie, X. Z. Sun, C. Wilson and M. W. George, *Inorg. Chem.*, 2008, **47**, 9857–9869.
93 E. D. Olmon, P. A. Sontz, A. M. a. Blanco-Rodriguez, M. Towrie, I. P. Clark, A. n. Vlcek and J. K. Barton, *J. Am. Chem. Soc.*, 2011, **133**, 13718–13730.
94 A. M. Blanco-Rodriguez, M. Towrie, J. Sykora, S. Zalis and A. Vlcek, *Inorg. Chem.*, 2011, **50**, 6122–6134.
95 T. L. Easun, W. Z. Alsindi, M. Towrie, K. L. Ronayne, X.-Z. Sun, M. D. Ward and M. W. George, *Inorg. Chem.*, 2008, **47**, 5071–5078.
96 T. L. Easun, W. Z. Alsindi, N. Deppermann, M. Towrie, K. L. Ronayne, X.-Z. Sun, M. D. Ward and M. W. George, *Inorg. Chem.*, 2009, **48**, 8759–8770.
97 E. A. Glik, S. Kinayyigit, K. L. Ronayne, M. Towrie, I. V. Sazanovich, J. A. Weinstein and F. N. Castellano, *Inorg. Chem.*, 2008, **47**, 6974–6983.
98 I. V. Sazanovich, M. A. H. Alamiry, J. Best, R. D. Bennett, O. V. Bouganov, E. S. Davies, V. P. Grivin, A. J. H. M. Meijer, V. F. Plyusnin, K. L. Ronayne, A. H. Shelton, S. A. Tikhomirov, M. Towrie and J. A. Weinstein, *Inorg. Chem.*, 2008, **47**, 10432–10445.
99 A. Gabrielsson, M. Towrie, S. Zalis and A. Vlcek, *Inorg. Chem.*, 2008, **47**, 4236–4242.
100 B. Probst, C. Kolano, P. Hamm and R. Alberto, *Inorg. Chem.*, 2009, **48**, 1836–1843.
101 M. D. Doherty, D. C. Grills and E. Fujita, *Inorg. Chem.*, 2009, **48**, 1796–1798.

102 Q. Cao, C. M. Creely, E. S. Davies, J. Dyer, T. L. Easun, D. C. Grills, D. A. McGovern, J. McMaster, J. Pitchford, J. A. Smith, X.-Z. Sun, J. M. Kelly and M. W. George, *Photochem. Photobiol. Sci.*, 2011, **10**, 1355–1364.
103 S. E. Brown-Xu, M. H. Chisholm, C. B. Durr, T. L. Gustafson and T. F. Spilker, *J. Phys. Chem. A*, 2013, **117**, 5997–6006.
104 I. V. Sazanovich, M. A. H. Alamiry, A. J. H. M. Meijer, M. Towrie, E. S. Davies, R. D. Bennett and J. A. Weinstein, *Pure & Appl. Chem.*, 2013, **85**, 1331–1348.
105 J. Schneider, K. Q. Vuong, J. A. Calladine, X.-Z. Sun, A. C. Whitwood, M. W. George and R. N. Perutz, *Inorg. Chem.*, 2011, **50**, 11877–11889.
106 J. E. Yarnell, J. C. Deaton, C. E. McCusker and F. N. Castellano, *Inorg. Chem.*, 2011, **50**, 7820–7830.
107 A. M. Blanco-Rodriguez, M. Towrie, J.-P. Collin, S. Zalis and A. Vlcek Jr., *Dalton Trans.*, 2009, 3941–3949.
108 A. El Nahhas, C. Consani, A. M. a. Blanco-Rodriguez, K. M. Lancaster, O. Braem, A. Cannizzo, M. Towrie, I. P. Clark, S. Zalic, M. Chergui and A. Vlcek, *Inorg. Chem.*, 2011, **50**, 2932–2943.
109 J. D. Lewis, M. Towrie and J. N. Moore, *J. Phys. Chem. A*, 2008, **112**, 3852–3864.
110 G. Li, K. Parimal, S. Vyas, C. M. Hadad, A. H. Flood and K. D. Glusac, *J. Am. Chem. Soc.*, 2009, **131**, 11656–11657.
111 M. M. N. Wolf, R. Groβ, C. Schumann, J. A. Wolny, V. Schunemann, A. Dossing, H. Paulsen, J. J. McGarvey and R. Diller, *Phys. Chem. Chem. Phys.*, 2008, **10**, 4264–4273.

Nuclear quadrupole resonance spectroscopy

Keith B. Dillon

DOI: 10.1039/9781782621485-00248

This chapter reviews new publications on pure nuclear quadrupole resonance (NQR) spectroscopy of inorganic and organometallic compounds, up to the end of 2012.

1 Introduction

This chapter reports and comments on publications concerning pure nuclear quadrupole resonance (NQR) spectroscopy of inorganic and organometallic compounds which have appeared since volume 44 in this series,[1] up to the end of 2012. There has been a slight decrease in the number of new articles compared with the previous two years, with again some changes of emphasis. Indium-115 seems to be one of the most popular nuclei for study this year, reflecting ongoing interest in indium-based superconductors. There is also a welcome reappearance of some lanthanide results, and the first occasion since this series of reviews started of data for ^{97}Mo, although there are earlier results in the literature for this nucleus. There are no new results for ^{209}Bi nuclei this year, however.

One trend which should perhaps be discouraged is the publication of purely theoretical papers in which NQR parameters for various nuclei are calculated, usually by density functional theory (DFT) methods, but with no experimental data for comparison. Hence the validity of any conclusions reached cannot be verified. Such papers are mentioned briefly here, but are not described in detail because of these limitations. This is a completely different approach to that used in other series of papers, where comparison is always made between the results of DFT calculations and experimental data, either new or from the literature.

Correlation analysis has been applied to the relationship between polarisability effects and NQR parameters in various types of donor-acceptor complexes, including H-bonded species, charge-transfer complexes and transition metal derivatives.[2] The contribution of polarisability effects was deduced to range from 0% (for classic series) up to 75%. NQR and NMR studies on iron-based pnictide superconductors have been reviewed, particularly those using ^{75}As nuclei for NQR.[3] NQR and NMR investigations of stripe materials, including cuprates, have been surveyed.[4] Iron pnictides were also briefly discussed. Progress in heavy-fermion superconductivity for $CeMIn_5$ (M = Co, Rh or Ir) and related materials, as revealed by various physical techniques including ^{115}In NQR spectroscopy,

Chemistry Department, University of Durham, South Road, Durham DH1 3LE, UK. E-mail: k.b.dillon@durham.ac.uk

has been reviewed.[5] New methods for the detection of ^{14}N NQR frequencies have been discussed, as applied to various inorganic and organic compounds, including NH_4NO_3.[6]

The normal format is followed in the more detailed sections comprising the remainder of this review, with results for main group elements preceding those for transition metals and lanthanides.

2 Main group elements

2.1 Group 1 (Hydrogen-2, Deuterium)

NQR parameters (the quadrupolar coupling constant e^2Qq/h and the asymmetry parameter η of the electric field gradient (efg)) for ^2H and ^{17}O nuclei have been calculated by DFT methods at the B3LYP level with the 6-311^{++}G (d,p) standard basis set for several bisphosphonate derivatives.[7] (These compounds are registered as drugs for inhibition of various types of cancer.) Some substituents were found to have a strong effect on the ^{17}O NQR parameters, in contrast with those for ^2H. The presence of a bromine atom in the ring was thought to increase activity. There was no comparison with experimental results, however. ^2H, ^{14}N and ^{17}O quadrupolar coupling constants were similarly calculated for some newly-synthesised phosphoric triamides.[8] Although these compounds were characterised by various physical methods, including multinuclear NMR spectroscopy, IR, mass spectrometry and elemental analysis, no experimental NQR spectra were recorded, again making comparison with the calculated values impossible.

2.2 Group 2 (Barium-137)

Barium, copper and indium NQR spectra have been recorded for the paper-chain compound $Ba_3Cu_3In_4O_{12}$.[9] Two ^{137}Ba signals from non-equivalent positions were seen at 4.2 K, which were assigned by comparison with the structure. The ^{135}Ba resonances were expected to occur in a region where strong, broad ^{63}Cu and ^{65}Cu signals were found, however, and these were not detected. Unexpectedly, no magnetic splitting of any of the NQR signals was observed at 4.2 K. The temperature (T)-dependence of the two ^{115}In resonances up to 50 K showed a sharp kink in the vicinity of the Néel temperature T_N (around 12.5 K), but below T_N the two signals shifted rapidly in opposite directions. This behaviour was ascribed to fluctuations of a dynamic efg tensor, coexisting with long-range magnetic order. The $^{63,65}Cu$ pure NQR spectrum was observed in the range 24–30 MHz. Extensive NMR data were also presented. The results showed unambiguously that one of the two types of copper ion was in a nonmagnetic spin state below T_N. The pure NQR spectrum was attributed to Cu^{II} ions forming non-magnetic spin-singlet dimers. NQR and NMR results pointed to the existence of at least two nonequivalent sets of the copper trimers, probably due to displacement of the Cu^I ions along the cavity in the c direction. Hence the paper chains in this compound and in $Ba_3Cu_3Sc_4O_{12}$ were suggested to be a system of Cu^I-$2Cu^{II}$ trimers.

2.3 Group 13 (Boron-11, Aluminium-27, Gallium-69 and 71, and Indium-115)

The three papers from the same research group on boron-11 nuclei describe the computation by DFT methods of ^{11}B NQR parameters (among others) for Ge-doped,[10] Si-doped[11] and SiC-doped[12] (4,4) armchair single-walled boron phosphide nanotubes. While some interesting conclusions have been drawn from the results, there was no comparison with any experimental data. This limits their usefulness and applicability, as mentioned in the Introduction.

^{27}Al NQR measurements have been described for $CeOs_2Al_{10}$, which exhibits a novel phase transition at $T_0 = 29$ K.[13] The NQR parameters for all the Al sites at ambient pressure (*P*) were compared with those in $CeRu_2Al_{10}$ ($T_0 = 27$K) and $CeFe_2Al_{10}$, with no phase transition. Under $P = 0.66$ GPa, which altered T_0 to 32.5 K, a distinct NQR splitting was observed just below T_0, showing a homogeneous transition. Despite the increase of T_0 under *P*, the spin-lattice relaxation rate (SLR) was suppressed over the whole *T* range by application of *P*. The compound was thought to be in an intermediate state between $CeRu_2Al_{10}$ and $CeFe_2Al_{10}$, suggesting an approach to a Kondo semiconductor in a valence fluctuation regime. Various physical methods including ^{27}Al NQR have been used to study YFe_2Al_{10} over a wide *T*-range.[14] The *T*-dependence of the SLR, together with those of the magnetic susceptibility and specific heat, followed a weak power law, a signature of critical fluctuations of Fe moments. The results suggested the existence of ferromagnetic correlation in this system. No magnetic ordering was found down to 50 mK. A crossover from quantum critical to Fermi-liquid behaviour was observed with increasing magnetic field, *i.e.* on going from NQR to NMR. An erratum to this paper in 2013 did not affect the scientific content.[15] Similar calculations to those for ^{11}B nuclei have been carried out for ^{27}Al nuclei in aluminium nitride and aluminium phosphide nanotubes,[16] and for ^{27}Al and ^{14}N nuclei in Si-doped (6,0) zigzag single-walled aluminium nitride nanotubes.[17] In both cases there were no experimental results for comparison.

^{69,71}Ga NQR measurements at 15 K including the SLR have been recorded for a series of intermetallic solid solutions $Fe_{1-x}Co_xGa_3$.[18] Single crystal X-ray diffraction data were also obtained for $x = 0.5$, and DFT calculations were carried out. The results showed that Fe-Fe and Co-Co dumbbells were preferred to Fe-Co dumbbells in the crystal structure, although the latter were present in significant amounts. The band structure showed a gap of about 0.4 eV, irrespective of the Co content. The solid solution was metallic for $x > 0.025$. The SLR was very sensitive to the Co concentration, and correlated well with the square of the density of states at the Fermi level. It was highest for $x = 0.25$. The effect of low-*T* annealing on the NQR of ^{69}Ga and ^{115}In nuclei respectively in layered GaSe and InSe single-crystal semiconductors has been investigated.[19] As the annealing *T* was lowered to room temperature (R*T*), the quality of the samples (and hence the spectra) improved, due to a decrease in the concentration of defects and to ordering in a system of polytypes. The results were extended to a *p*-GaSe-*n*-InSe heterogeneous structure for various annealing temperatures. Optimal improvement was obtained by annealing for 4h between 150 and 200 °C.

The structure of the heavy-fermion compound $Ce_{1-x}La_xCoIn_5$ has been probed *via* ^{115}In NQR, using a number of single crystal samples varying in the amount of La present.[20] The NQR spectrum of the In(1) site was composed of five peaks, described by a binomial distribution of the nearest-neighbour La atoms. The results implied that analysis of the NQR spectra would yield a more accurate estimate of the La doping level than energy dispersive X-ray spectroscopy. New ^{115}In NQR data have been obtained for the 9/2 ↔ 7/2 transition of the In(2) site in the heavy fermion compound $CePt_2In_7$ between 6 K and 1.5 K.[21] The sub-lattice magnetisation in the antiferromagnetic (AF) state was extracted from the results. Approximately half of the In(2) sites appeared to experience no static hyperfine field. The ^{115}In NQR spectrum of a single crystal sample of $CePt_2In_7$ was recorded at 1.6 K at both ambient P and at 2.4 GPa.[22] The data revealed the coexistence of commensurate and incommensurate AF orders at ambient P, and showed that the commensurate ordering was stabilised by increasing P. The SLR for In(3) sites indicated the localised nature of f electrons far above T_N (5.2 K at ambient P). In the paramagnetic state, the SLR values decreased on applying P, but the residual values well below T_N were increased by application of P. NQR measurements as a function of T have been recorded for Ce_2PdIn_8, another heavy-fermion superconductor.[23] Above the Kondo coherence temperature ~30 K, the SLR was independent of T, whereas at lower T down to the onset of superconductivity at $T_c = 0.64$ K it was nearly proportional to $T^{\frac{1}{2}}$ Below T_c, there was no coherence peak, and the SLR decreased as T^3 at least down to 75 mK. These findings indicated that the compound was close to the AF quantum critical point, and that the superconducting state had an unconventional character with line nodes in the superconducting gap. Theoretical calculations have been applied to the heavy fermion compound $CeRhIn_5$, where ^{115}In NQR measurements had shown evidence of a coexistent phase with commensurate antiferromagnetism and *d*-wave superconductivity.[24] Around the transition line between the AF and coexistent states, low-energy incommensurate spin-excitations were found to develop, due to Fermi surface nesting. The SLR was calculated, and gave a reasonable explanation of the T-dependence for the NQR SLR in a system with a coexistent ground state. ^{115}In NQR data including the SLR for $CeIr(In_{1-x}Cd_x)_5$ have revealed that the antiferromagnetism induced by Cd doping emerged locally around the Cd dopants, but that superconductivity was suddenly induced at $T_c = 0.7$ and 0.9 K for 2.34 and 2.75 GPa respectively.[25] The superconducting characteristics, with a large fraction of residual density of states at the Fermi level which increased with T_c, differed from those for anisotropic superconductivity mediated by AF correlations. Superconductivity in this system could be mediated by valence fluctuations.

2.4 Group 15 (Nitrogen-14, Arsenic-75, and Antimony-121 and -123)

The new papers on ^{14}N NQR involve the use of either $NaNO_2$ or NH_4NO_3 as test compounds, similar to results described in previous years. NQR signals from both single crystal and powder samples containing spin-1 nuclei under the Carr-Purcell-Meiboom-Gill pulse sequence have been modelled numerically.[26] The single crystal clearly showed the effects of dipolar coupling for on-resonant pulses, while the powder did not. For certain off-resonant conditions, the powder sample exhibited the same response as the single crystal. This corresponded to the observation of rapid decay under these experimental conditions. Experiments were carried out for powdered $NaNO_2$ samples at R*T* near the 4.64 MHz transition. The functional form of the echo train, when dipolar coupling was not refocussed, was clearly different from the decay for an on-resonance sequence, and could be used to characterise dipolar coupling. Two novel double-resonance conditions between spin-1 and spin-$\frac{1}{2}$ nuclei in a crystalline solid have been demonstrated, using NH_4NO_3 as the test sample.[27] By using a magnetic field oscillating at the spin-$\frac{1}{2}$ Larmor frequency, the NQR frequency of the spin-1 nucleus was matched to the Rabi (RDR) or Rabi plus Larmor (RLDR) frequency, as opposed to the conventional use of the Larmor frequency only. The oscillating field was used to increase the signal-to-noise ratio per time by a factor of 3.5 for NQR detection of the ^{14}N resonances in this compound. The advantages of the new methods were discussed. Attempts have been made to improve the signal detection range for the ^{14}N NQR signals of NH_4NO_3 by using interference suppression algorithms.[28] Such techniques were considered to be critically important for practical applications such as the detection of explosives, and the experiments were to be extended from the laboratory to outdoor environments. Aspects of the effective detection of ammonium nitrate-based explosives (such as a mixture of ammonium nitrate and fuel oil) by pulsed NQR methods have been examined by measuring the SLR, resonance frequencies and signal intensities of the ^{14}N signal as a function of *T*.[29] Best results were achieved using a multipulse sequence with a composite preparation pulse. The NQR signal level was significantly enhanced by use of the cross-polarisation effect between the quadrupolar ^{14}N nuclei and protons.

^{75}As NQR (together with ^{75}As and ^{7}Li NMR) data have been reported for self flux grown LiFeAs and 5% Co-doped LiFeAs single crystals, and for a polcrystalline LiFeAs sample.[30] The different samples could be distinguished *via* their NQR frequencies. All the materials were superconducting, but differences were found in the Knight shift and SLR for samples differing in NQR frequency. Possible causes of this behaviour were discussed. ^{75}As NQR (and NMR) spectra have been recorded for single-crystal $Ba_{1-x}K_xFe_2As_2$, with x between 0.27 and 1.00.[31] The NQR frequency increased linearly with increasing x. In contrast, the SLR in the normal state had a strong *T*-dependence, indicating the existence of large AF spin fluctuations for all values of x. The *T*-dependence of the SLR showed gaplike behaviour below *ca.* 100 K for $0.6<x<0.9$. In the superconducting state, the *T*-dependence of the SLR suggested a multiple-superconducting-gap feature. There was no coherence peak just below T_c. The results were analysed using a simple two-gap model. Systematic ^{75}As NQR measurements on high quality samples of $LaFeAsO_{1-x}F_x$ ($x=0.03$, 0.04, 0.06, 0.08, 0.10 or 0.15) have been carried out.[32] AF spin fluctuations seen above the magnetic ordering temperature $T_N = 58$ K for $x = 0.03$ persisted in the regime $0.04 \leq x \leq 0.08$, where superconductivity set in. A dome-shaped x dependence of T_c was found, with the highest $T_c = 27$ K at $x = 0.06$, and significant AF spin fluctuations. This behaviour was similar to that observed in $La_{2-x}Sr_xCuO_4$. For $x = 0.06$, the SLR below T_c decreased exponentially down to 0.13 T_c, indicating unambiguously that the energy gaps were fully opened. The T variation of the SLR below T_c for other x was rendered nonexponential by impurity scattering.

The T-dependence of ^{121}Sb and ^{123}Sb NQR parameters including the SLR has been monitored from 4.2–395 K for Ag_5SbS_4, stephanite.[33] A phase transition at 140 K and internal motions with an activation energy of 0.29 eV were detected. Relaxation was deduced to operate *via* a quadrupolar mechanism. Internal rotation was due to the diffusion of silver ions over the structure of the crystal, and the nature of the phase transition was discussed. The T-dependence of Sb NQR parameters from 0–300 K has been followed for the filled skutterudite compounds RFe_4Sb_{12} (R = La or Ce), synthesised under high pressure (HP), and compared with those from samples prepared at ambient P.[34] The NQR spectra were much sharper for the HP samples, suggesting that the R-site filling fraction was close to unity and that the electronic states about Sb became uniform. In the La compound, the SLR as a function of T exhibited a Curie-Weiss dependence above 30 K, with the Weiss temperature $\theta = 12$ K, indicating the development of ferromagnetic fluctuations. The magnitude of θ was smaller for the HP sample, suggesting that the system became close to the vicinity of ferromagnetic instability as the La-site filling factor increased. In the Ce sample prepared at high P, the SLR followed an exponential decrease above 100 K, with a gap of 270 K, larger than that for the ambient P sample. The results suggested that c-f hybridisation was enhanced by increasing the Ce-site filling fraction, causing semiconducting behaviour at low T.

2.5 Group 16 (Oxygen-17)

Theoretical calculations of NQR parameters for this nucleus have been mentioned briefly in the sub-section on Group 1 (^{2}H).[7,8]

2.6 Group 17 (Chlorine-35 and -37, Bromine-79 and -81, and Iodine-127)

^{35}Cl NQR frequencies at 298 K have been recorded for three titanocene dichloride derivatives (3g samples).[35] The data were combined with mass spectrometric results to study the electronic structures of the compounds. The ^{35}Cl resonance frequency increased with methylation of Cp ligands, but showed a significant decrease for the *ansa* form of a titanocene. *Ab initio* calculations at the RHF/6-31G(d) and MP2/6-31G(d) levels have been applied to two possible structures of 4-(trichlorogermyl)butan-2-one, one with four-coordinate Ge and the second with five-coordinate Ge, formed by interaction of the carbonyl oxygen with the Ge atom, with total optimisation of geometry.[36] The pentacoordinate structure was found to be energetically more advantageous.

The ^{35}Cl NQR parameters for this structure were calculated, and the frequencies agreed satisfactorily with the experimental values at 77 K. Calculations were carried out for various Ge - - - O distances, and the charge distribution in the compound was discussed on the basis of the results. A similar treatment has been applied to 1-(1-trichlorogermyl)pyrrolidine-2-one, also with a pentacoordinate Ge atom from the X-ray structure, and the results were compared with experimental ^{35}Cl NQR data.[37] A four-coordinate form higher in energy by 2.7 $kcal \cdot mol^{-1}$ could also be stable.

Identical procedures have been described for 2- and 4-methoxyphenyl and 2,6-dimethoxyphenyl trichlorostannanes.[38] In the 4-MeO compound, the Sn atom was tetracoordinate, whereas in the 2-MeO and 2,6-diMeO derivatives it was pentacoordinate, due to an intramolecular interaction of the O atom with Sn. In the diMeO compound, only one oxygen was involved in this interaction. The charges on the atoms were calculated, and the variations in electron density due to the intramolecular O - - - Sn interaction were discussed, as well as the experimental ^{35}Cl NQR spectra at 77 K. This work has subsequently been extended to $2\text{-}EtOC_6H_4SnCl_3$ and to the 2-MeO compound with further substitution in the phenyl ring.[39] The results of the calculations were again compared with experimental ^{35}Cl NQR data. The Sn atom was pentacoordinate in all of the compounds studied. Replacing the Me group by Et in the alkoxy group (a better donor) increased the strength of the Sn←O coordinate bond. The same effect was seen on the introduction of an electron-releasing substituent into the aromatic ring. Similar calculations have been performed for 1 : 1 complexes $MCl_4\text{–}C_6H_5COCl$ (M = Si, Ge or Sn), and the corresponding 1 : 2 complexes.[40] The 1 : 1 complexes were energetically more advantageous, provided that the M - - - O distance was considerably less than the sum of the van der Waals radii (realised only for M = Sn). The M - - - O distances in the 1 : 2 complexes were close to the sum of the covalent radii, but solid complexes were not formed, even for Sn (not energetically favoured). From experimental ^{35}Cl NQR data, $SiCl_4$ and $GeCl_4$ did not form complexes with benzoyl chloride, either 1 : 1 or 1 : 2, whereas $SnCl_4$ yielded a 1 : 1 complex at 77 K.

^{79}Br NQR data, including the T-dependence of the SLR, together with high-field magnetisation and ^{125}Te NMR results, have been recorded for high quality single crystals of $Cu_2Te_2O_5Br_2$, which is a weakly coupled Cu^{2+} ($S=\frac{1}{2}$) tetrahedral system.[41] A strong divergence of the SLR at $T_0 = 13.5$ K was seen. In the paramagnetic state, the SLR revealed an effective singlet-triplet spin gap much larger than observed by static bulk measurements. The 3D magnetic order was deduced to emerge from a singlet background, as expected in a simple isolated spin-$\frac{1}{2}$ tetrahedral system. The data suggested that intertetrahedral coupling not only induced magnetic order, but also increased the effective spin gap by enhancing intratetrahedral spin-exchange processes. This unusual feature was attributed to frustrated intertetrahedral interactions, possibly accounting for discrepancies between experiment and theory. Bromine NQR (and NMR) investigations of $V_4S_9Br_4$, which has a tetragonal crystal structure including square-planar V_4 clusters, have been described.[42]

Magnetic measurements revealed a spontaneous phase separation to AF and spin-singlet (S) phases below $T_N = 18$ K at ambient *P*. The S phase appeared only when the crystal was sufficiently large, and was easily stabilised under hydrostatic *P*. The results suggested that, although the ground-state phase at ambient *P* was an AF phase, the AF and S phases were nearly degenerate. The spontaneous phase separation at ambient *P* probably arose from internal inhomogeneities in the crystal, together with possible distortion of the AF phase matrix below T_N.

^{127}I NQR measurements (and ^{69,71}Ga solid-state NMR) have been used to elucidate the composition of "GaI".[43] NQR spectroscopy clearly showed that GaI_3 was absent, and that GaI_2 was a significant component. (Three of the four expected ^{127}I signals from GaI_2 could be seen, but detection of the fourth site could not be attempted because of probe tuning limitations.) From the overall results, a tentative composition was proposed of two equivalents of Ga metal with two equivalents of GaI_2, the latter composed of Ga^+ and GaI_4^- ions, hence $[Ga^0]_2[Ga]^+[GaI_4]^-$. In an extension of previous work,[44] further ^{127}I NQR results at 77 K have been reported for $(PbI_2)_{1-x}(BI_3)_x$.[45] (Note that this is the reverse formulation of that used previously by these authors.[44]) For $0 \leq x \leq 0.02$, the material had the properties of an impure crystal of PbI_2, containing intralayer BI_3 clusters. As well as the known phase transition at $x \sim 0.80$, another was found at $x = 0.10$. In the range $0.1 \leq x \leq 0.8$, a new crystal $(PbI_2)_{1-x}(BiI_3)_x$ was formed, in which PbI_2 or BiI_3 units were intercalants, fully or partially ordered in the crystal system. Other ^{127}I results are described in the subsection on ^{187}Re below.

3 Transition metals and lanthanides

3.1 Cobalt-59

^{59}Co NQR including the *T*-dependence of the SLR and spin-spin relaxation has been performed to clarify the complex magnetic phases in noncentrosymmetric $CeCoGe_3$ approaching the quantum critical point (QCP).[46] At ambient *P*, successive transitions were found at $T_{N2} = 12$ K and $T_{N3} = 8$ K, after the ferrimagnetic-like order at $T_{N1} = 21$ K was confirmed. No critical slowing down of the relaxation rates was observed, suggesting first order transitions. On application of *P*, there were large spectral changes, even at 0.3 GPa. The Co NQR spectrum became simpler above *ca.* 0.7 GPa, with two Co sites in a 2 : 1 ratio. No successive transitions were observed at 1.52 GPa, indicating that they were confined to a relatively low pressure region. ^{59}Co NQR (and NMR) data have been reported for UCoGe.[47] The SLR and Knight shift provided clear evidence that both the static and dynamic susceptibilities were ferromagnetic, with strong Ising anisotropy along the *c* axis. The results suggested an intimate relationship between Ising magnetisation and anisotropic superconductivity in this compound. DFT calculations have been performed for twenty Co^{2+} and Ni^{2+} complexes, both four- and six-coordinate, to obtain various parameters including η and e^2Qq/h for NQR.[48] The nuclei concerned did not appear to be specified in this paper. The Laplace equation was verified from the data, but there were no experimental results for comparison.

3.2 Copper-63 and -65

^{63}Cu NQR spectra at low *T* (4.2 and 3.1 K), including the SLR, have been recorded for $CeCu_2Si_2$ as a function of *P* up to 5.4 GPa, using an opposed-anvil high pressure cell, and daphne oil 7474 as the pressure-transmitting medium.[49] Above 4 GPa, the NQR frequency suddenly decreased from the linear *P*-dependence observed up to 3.5 GPa. This decrease was associated with an increase of the Cu valence. Above 4.5 GPa, a linear dependence was again found, thought to be related to a sharp crossover to the high valence state. ^{63,65}Cu NQR spectra from 77 – 300 K have been recorded for several polycrystalline samples of Cu_xAlO_2.[50] Analysis of the asymmetric lineshapes led to an assumption regarding the defect structure of the crystals. A nutational NQR experiment on ^{63}Cu nuclei at 297 K supported axial symmetry for the efg tensor, *i.e.* $\eta = 0$. Similar studies have been carried out on $Cu_{1.1}Al_{0.9}O_{2-x}$ and $Cu_{0.9}Al_{1.1}O_{2-x}$, as well as the stoichiometric compound.[51] The resonances were significantly broader than that of the reference compound Cu_2O, possibly due to additional structural defects. Similar linewidths were detected for all three samples. Analysis of Cu NQR SLR (and NMR) data for copper oxide superconductors has provided evidence for s + d wave pairing, using a two-band model for superconductivity with coupled s + d wave superconducting gaps.[52] The data were consistent with 40% s-wave and 60% d-wave gaps.

The dimensionality of inhomogeneous spin modulation patterns has been examined theoretically for doped lanthanum cuprates (high-T_c superconductors), from the perspective of NQR and μSR experiments.[53] It was found that a spin vortex lattice led to a singularity in the distribution of local magnetic fields, quite similar to that produced by a stripe pattern. Hence no clear qualitative differences emerged, although noticeable quantitative discrepancies remained with the experimental results. Similar discrepancies were also found for 2D superposition of spin spirals. Hence a spin vortex checkerboard model could not be ruled out on the basis of the available experimental results. The local structure in high quality single crystals of $La_{1.93}Sr_{0.07}CuO_4$ has been investigated *via* NQR and NMR of ^{63}Cu and ^{139}La nuclei.[54] A sharper ^{139}La NQR signal at 40 K was obtained than from a previous literature report, indicating less chemical inhomogeneity in the present sample. The SLR for ^{139}La nuclei exhibited a sharp anomaly at T_s (387 K), consistent with an abrupt change of the spectrum and implying a first-order-like structural transition. There was also a dramatic change below ~ 70 K, with strong enhancement of the SLR, arising from collective glassy spin freezing. The *T*-dependence results suggested that the *T*-independent paramagnetic regime at high *T* gradually changed to a regime where scattering with doped holes dominated the relaxation. Other results for Cu nuclei have been described in the sub-section on Group 2.[9]

3.3 Molybdenum-97

The ^{97}Mo NQR spectrum for Mo_3Al_2C, a noncentrosymmetric superconductor, has been recorded at 25 K; the ^{95}Mo NMR spectrum of a powder sample was also obtained at 10 K.[55] The data enabled the efg at the Mo site to be determined, and compared with the results of a point charge calculation. The latter predicted a much larger value of the efg, ascribed to a contribution from the conduction electrons.

3.4 Ruthenium-101

In a further investigation of the 'hidden order' in URu_2Si_2, ^{101}Ru NQR measurements as a function of T have been carried out for a pure single crystal sample.[56] Some ^{73}Ge NMR results were also obtained. Precise measurement of the ^{101}Ru NQR frequency detected an anomaly just below T_0 (17.5 K). Its T-dependence showed a linear relationship with that of the in-plane lattice parameter, at least between 4 and 45 K, possibly giving a clue to a modification in the microscopic charge distribution at T_0.

3.5 Lanthanum-137

Results for this nucleus have been described in the sub-section on copper nuclei above.[54]

3.6 Rhenium-187

A new fast graphical program QUEST (Quadrupolar Exact Software) has been derived for the exact simulation of NMR and NQR spectra for quadrupolar nuclei.[57] It was applied to ^{187}Re NQR from $Re_2(CO)_{10}$ and the ^{127}I overtone NQR spectrum of SrI_2, as well as to NMR examples. Experimental results were obtained for comparison, and good agreement was obtained. This was believed to be the first direct observation of overtone NQR (for SrI_2).

References

1 J. Yarwood, R. Douthwaite and S. B. Duckett (eds), *Spectroscopic Properties of Inorganic and Organometallic Compounds*, The Royal Society of Chemistry, Cambridge, UK, 2013, vol. 44.
2 A. N. Egorochkin, O. V. Kuznetsova, N. M. Khamaletdinova and L. G. Domratcheva-Lvova, *Magn. Reson. in Chem.*, 2012, **50**, 40.
3 J. Yang and G.-Q. Zheng, *Modern Phys. Lett. B*, 2012, **26**, 1230008.
4 H.-J. Grafe, *Physica C*, 2012, **481**, 93.
5 J. D. Thompson and Z. Fisk, *J. Phys. Soc. Jpn.*, 2012, **81**, 011002.
6 J. Seliger and V. Žagar, *Appl. Magn. Reson.*, 2012, **43**, 469.
7 H. Aghabozorg, B. Sohrabi, S. Mashkouri and H. R. Aghabozorg, *J. Mol. Model*, 2012, **18**, 929.
8 Z. Shariatinia, C. O. Della Védova, M. F. Erben, V. Tavasolinasab and K. Gholivand, *J. Mol. Struct.*, 2012, **1023**, 18.
9 A. A. Gippius, N. E. Gervits, A. V. Tkachev, I. S. Maslova, O. S. Volkova, A. N. Vasiliev, N. Büttgen, W. Kraetschmer and A. S. Moskvin, *Phys. Rev. B*, 2012, **86**, 155114.
10 M. T. Baei, *Monatsh. Chem.*, 2012, **143**, 881.
11 M. T. Baei, A. A. Peyghan and M. Moghimi, *Monatsh. Chem.*, 2012, **143**, 1627.
12 M. T. Baei, S. Z. Sayyad-Alangi, A. V. Moradi and P. Torabi, *J. Mol. Model*, 2012, **18**, 881.
13 M. Matsumura, T. Inagaki, H. Kato, T. Nishioka, H. Tanida and M. Sera, *J. Phys.: Conf. Series*, 2012, **400**, 032052.

14 P. Khuntia, A. M. Strydom, L. S. Wu, M. C. Aronson, F. Steglich and M. Baenitz, *Phys. Rev. B*, 2012, **86**, 220401(R).
15 P. Khuntia, A. M. Strydom, L. S. Wu, M. C. Aronson, F. Steglich and M. Baenitz, *Phys. Rev. B*, 2013, **88**, 019902(E).
16 M. T. Baei, *Monatsh. Chem.*, 2012, **143**, 545.
17 M. T. Baei, A. A. Peyghan, K. Tavakoli, A. K. Babaheydari and M. Moghimi, *J. Mol. Model*, 2012, **18**, 4427.
18 V. Yu. Verchenko, M. S. Likhanov, M. A. Kirsanova, A. A. Gippius, A. V. Tkachev, N. E. Gervits, A. V. Galeeva, N. Büttgen, W. Krätschmer, C. S. Lue, K. S. Okhotnikov and A. V. Shevelkov, *J. Solid State Chem.*, 2012, **194**, 361.
19 Z. D. Kovalyuk, O. N. Sydor, G. I. Lastivka and A. G. Khandozhko, *Semiconductors*, 2012, **46**, 1145.
20 A. C. Shockley, A. P. Dioguardi, N. apRoberts-Warren, P. Klavins, C. Capan, Z. Fisk and N. J. Curro, *J. Supercond. Nov. Magn.*, 2012, **25**, 2141.
21 N. apRoberts-Warren, A. P. Dioguardi, A. C. Shockley, C. H. Lin, J. Crocker, P. Klavins and N. J. Curro, *J. Phys.: Conf. Series*, 2012, **344**, 012027.
22 H. Sakai, Y. Tokunaga, S. Kambe, H. Lee, V. A. Sidorov, P. H. Tobash, F. Ronning, E. D. Bauer and J. D. Thompson, *J. Phys.: Conf. Series*, 2012, **391**, 012057.
23 H. Fukazawa, R. Nagashima, S. Shimatani, Y. Kohori and D. Kaczorowski, *Phys. Rev. B*, 2012, **86**, 094508.
24 H.-J. Lee and T. Takimoto, *J. Phys. Soc. Jpn.*, 2012, **81**, 104704.
25 M. Yashima, N. Tagami, S. Taniguchi, T. Unemori, K. Uematsu, H. Mukuda, Y. Kitaoka, Y. Ōta, F. Honda, R. Settai and Y. Ōnuki, *Phys. Rev. Lett.*, 2012, **109**, 117001.
26 M. W. Malone and K. L. Sauer, *Appl. Magn. Reson.*, 2012, **43**, 541.
27 D. W. Prescott, M. W. Malone, S. P. Douglass and K. L. Sauer, *J. Chem. Phys.*, 2012, **137**, 214201.
28 J. A. DeBardelaben, J. K. Miller, W. L. Myrick, J. B. Miller, G. C. Gilbreath and B. Bajramaj, *Proc. of SPIE*, 2012, **8382**, 838206.
29 T. N. Rudakov, *Appl. Magn. Reson.*, 2012, **43**, 557.
30 S.-H. Baek, H.-J. Grafe, F. Hammerath, M. Fuchs, C. Rudisch, L. Harnagea, S. Aswartham, S. Wurmehl, J. van den Brink and B. Büchner, *Eur. Phys. J. B*, 2012, **85**, 159.
31 M. Hirano, Y. Yamada, T. Saito, R. Nagashima, T. Konishi, T. Toriyama, Y. Ohta, H. Fukazawa, Y. Kohori, Y. Furukawa, K. Kihou, C.-H. Lee, A. Iyo and H. Eisaki, *J. Phys. Soc. Jpn.*, 2012, **81**, 054704.
32 T. Oka, Z. Li, S. Kawasaki, G. F. Chen, N. L. Wang and G.-q. Zheng, *Phys. Rev. Lett.*, 2012, **108**, 047001.
33 A. Yu. Orlova, R. R. Gainov, A. V. Dooglav, I. N. Pen'kov and E. A. Korolev, *JETP Lett.*, 2012, **96**, 370.
34 K.-i. Magishi, H. Sugawara, M. Takahashi, T. (Takahito) Saito, K. Koyama, T. (Takashi) Saito, S. Tatsuoka, K. Tanaka and H. Sato, *J. Phys. Soc. Jpn.*, 2012, **81**, 124706.
35 J. Kubišta, M. Civiš, P. Španěl and S. Civiš, *Analyst*, 2012, **137**, 1338.
36 V. P. Feshin and E. V. Feshina, *J. Mol. Struct.*, 2012, **1011**, 149.
37 V. P. Feshin and E. V. Feshina, *Chem. Heterocyclic Cpds.*, 2012, **47**, 1378.
38 V. P. Feshin and E. V. Feshina, *Russ. J. Gen. Chem.*, 2012, **82**, 853.
39 V. P. Feshin and E. V. Feshina, *Russ. J. Gen. Chem.*, 2012, **82**, 1073.
40 V. P. Feshin and E. V. Feshina, *J. Coord. Chem.*, 2012, **65**, 2021.
41 S.-H. Baek, K.-Y. Choi, H. Berger, B. Büchner and H.-J. Grafe, *Phys. Rev. B*, 2012, **86**, 180405(R).
42 H. Nakamura, *AIP Conf. Proc.*, 2012, **1447**(Pt.1, Solid State Phys.), 40.

43 C. M. Widdifield, T. Jurca, D. S. Richeson and D. L. Bryce, *Polyhedron*, 2012, **35**, 96.
44 A. I. Barabash, I. G. Vertegel, E. D. Chesnokov, A. I. Ovcharenko and Yu. P. Gnatenko, *Ukr. J. Phys.*, 2011, **56**, 158.
45 O. I. Barabash, I. G. Vertegel, E. D. Chesnokov, O. I. Ovcharenko and L. S. Ivanova, *Functional Materials*, 2012, **19**, 330.
46 T. Inagaki, M. Matsumura, M. Mizoo, Y. Kawamura, H. Kato and T. Nishioka, *J. Phys.: Conf. Series*, 2012, **400**, 032026.
47 T. Hattori, Y. Ihara, K. Karube, Y. Nakai, K. Ishida, K. Deguchi, N. K. Sato and I. Satoh, *J. Phys.: Conf. Series*, 2012, **391**, 012017.
48 H. Singh, A. K. Bhardwaj, M. L. Sehgal and S. K. Mittal, *Int. J. Curr. Res. Rev.*, 2012, **4**, 12.
49 K. Fujiwara, M. Iwata, Y. Okazaki, Y. Ikeda, S. Araki, T. C. Kobayashi, K. Murata, C. Geibel and F. Steglich, *J. Phys.: Conf. Series*, 2012, **391**, 012012.
50 V. L. Matukhin, I. H. Khabibullin, D. A. Shulgin, S. V. Schmidt and E. I. Terukov, *Semiconductors*, 2012, **46**, 1102.
51 V. L. Matukhin, I. Kh. Khabibulin, D. A. Shul'gin and S. V. Smidt, *Russ. Phys. J.*, 2012, **55**, 186.
52 A. Bussmann-Holder, *J. Supercond. Nov. Magn.*, 2012, **25**, 155.
53 J. G. Brandenburg and B. V. Fine, arXiv.org, e-Print Archive, *Cond. Mat.*, 2012, 1-6, arXiv: 1209. 1934v1 [cond-mat. str-el].
54 S.-H. Baek, A. Erb, B. Büchner and H.-J. Grafe, *Phys. Rev. B*, 2012, **85**, 184508.
55 Y. Ihara, Y. Kimura, K. Kumagai, E. Bauer, G. Rogl and P. Rogl, *J. Phys.: Conf. Series*, 2012, **400**, 022033.
56 T. Mito, M. Hattori, G. Motoyama, Y. Sakai, T. Koyama, K. Ueda, T. Kohara, M. Yokoyama and H. Amitsuka, *J. Phys.: Conf. Series*, 2012, **391**, 012044.
57 F. A. Perras, C. M. Widdifield and D. L. Bryce, *Solid State Nucl. Magn. Reson.*, 2012, **45–46**, 36.

Variable-pressure luminescence and Raman spectroscopy of molecular transition metal complexes: spectroscopic effects originating from small, reversible structural variations

Christian Reber,*[a] Camille Sonneville,[a] Stéphanie Poirier,[a] Nicolas Bélanger-Desmarais,[a] William B. Connick,[b] Sayandev Chatterjee,[b] Patrick Franz[c] and Silvio Decurtins[c]
DOI: 10.1039/9781782621485-00260

The past ten years have seen a significantly increasing number of published crystal structures for molecular transition metal complexes at variable pressure, providing quantitative information on structural variations. Spectroscopic measurements at variable pressure have been reported over the past 60 years for a variety of complexes, but luminescence measurements were mostly limited to intense signals until early in this century. The combination of variable-pressure structure variations with spectroscopic trends can lead to detailed new insight on a variety of aspects of electronic structure. This approach holds promise for the in-depth study of many categories of functional materials.

1 Introduction

Vibrational and electronic spectroscopic techniques have been traditionally used as experimental probes for specific functional groups or symmetry aspects, *i.e.* as indirect techniques to determine partial structures. This is well illustrated by the title and a historical perspective in the introduction of the recent new edition of the graduate-level textbook "Structural Methods in Molecular Inorganic Chemistry" by Rankin, Mitzel and Morrison,[1] discussing many spectroscopic techniques as well as diffraction methods. In the following, we present a number of examples where the spectroscopic study goes beyond this traditional perspective toward an approach based on both structural data and spectroscopic information, leading to possibilities to correlate quantitative variations of spectroscopic properties with structure. External pressure applied to solids leads to structure modifications,[2,3] and often to very significant spectroscopic effects.[4–7] Examples presented in the following overview are limited to reversible phenomena showing the spectroscopic effects resulting from structure changes at pressures below 10 GPa, a range easily accessible through standard diamond anvil cells.

[a]*Département de chimie, Université de Montréal, Montréal QC H3C 3J7, Canada. E-mail: christian.reber@umontreal.ca*
[b]*Department of Chemistry, University of Cincinnati, Cincinnati OH 45221-0172, USA*
[c]*Departement für Chemie und Biochemie, Universität Bern, CH-3012 Bern, Switzerland*

Spectroscopic measurements under variable conditions have a long history, with variable-temperature studies[8,9] being more abundant than variable-pressure spectra. Figure 1 shows an example of variable-temperature effects in the polarized absorption spectra of $[Nb^{IV}\{(\mu\text{-}CN)_4Mn^{II}(H_2O)_2\}_2 \cdot 4H_2O]_n$, a high-symmetry three-dimensional network with interesting magnetic properties that have been studied both at variable temperature and pressure.[10] As temperature increases, a broad, completely polarized band appears between 16000 and 16500 cm^{-1}. This "hot band" becomes intense at temperatures above 100 K, whereas the magnetic properties change in a much lower temperature range. Its integrated intensity is given in the inset to Fig. 1 and shows a behavior typical for a thermal activation barrier of approximately 150 cm^{-1}, the order of magnitude of low-frequency metal-ligand vibrational modes. This is a typical signature for a vibronic intensity mechanism and the observed new "hot band" is not due to structural changes as temperature increases, but arises from varying populations of vibrational levels. Such effects make the analysis of spectroscopic changes due to temperature-dependent structure variations highly challenging if not impossible, an intrinsic problem of variable-temperature spectroscopic measurements.

Variable-pressure measurements, often carried out at room temperature, do not have this problem. In general, structural changes induced by pressure do not lead to drastic changes of vibrational frequencies, and therefore thermal populations are constant and spectroscopic features are more easily analyzed in terms of structure variations. Until very recently, only a handful of variable-pressure structural studies had been published. As an example, a search in the Cambridge Structural Database published in 2006 lists 7 structures. Our literature search has revealed more than 30 structures published since 2006, as summarized in Table 1. This increase has been made possible by advances in X-ray diffraction

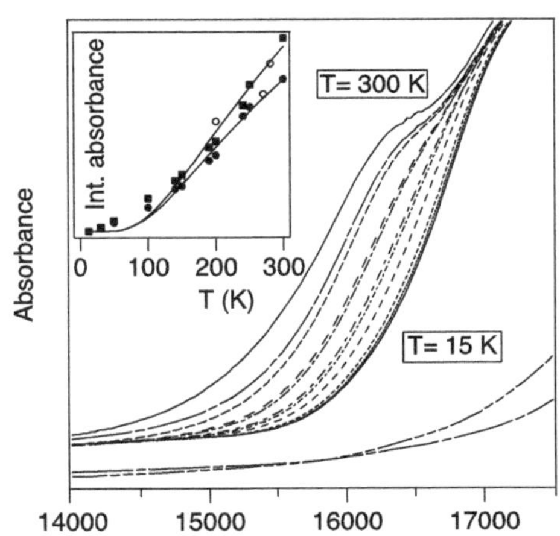

Fig. 1 Temperature-dependent, polarized absorption spectra of $[Nb^{IV}\{(\mu\text{-}CN)_4Mn^{II}(H_2O)_2\}_2 \cdot 4H_2O]_n$. Inset: integrated absorption intensity between 15000 cm^{-1} and 16700 cm^{-1} with solid traces calculated for activation energies of 150 cm^{-1} and 200 cm^{-1}.

Table 1 Overview of published variable-pressure structural studies.

Metal	Com ref	Max. Pressure (GPa) (# pressures)	Techniques (Year)
Li	$LiB(Im)_4$ (ZIF)[11]	1.69 (5)	XRD, Nano-Indentation (2010)
Gd(III)	$[Gd(PhCOO)_3(DMF)]_n$[12]	5.01 (8)	XRD (2010)
Mn(II), Nb(IV)	$\{[Mn(pydz)(H_2O)_2][Mn(H_2O)_2][Nb\ (CN)_8]_3H_2O\}_n$[13]	1.8 (1)	XRD, Magnetism (2012)
Mn(III)	$[Mn(pyrol)_3tren]$[14]	1 (2)	XRD (2005)
Mn(II)	$[(CH_3)_4N][MnCl_3]$[15]	1.7 (5)	XRD, Magnetism (2006)
Mn(III)	$[Mn_6O_2(Et\text{-}sao)_6(O_2CPh(Me)_2)_2\text{-}(EtOH)_6$[16]	1.5 (3)	XRD, Magnetism (2008)
Mn(III)	$[Mn_6O_2(Et\text{-}sao)_6(O_2C\text{-}naphth)_2(EtOH)_4(H_2O)_2]$[17]	1.5 (3)	XRD, Magnetism (2009)
Mn(III)/Mn(II)	$[Mn_3(Hcht)_2(bpy)_4](ClO_4)_3Et_2O_2MeCN$[17]	1.25 (3)	XRD, Magnetism (2009)
Mn(III)/Mn(IV)	$[Mn_2O_2(bpy)_4](PF_6)_3 \cdot 2CH_3CN \cdot 1H_2O$[18]	4.55 (5)	XRD, Magnetism (2010)
Mn(III)/Mn(IV)	$[Mn_2O_2(bpy)_4](ClO_4)_3 \cdot 3CH_3CN$[18]	2 (3)	XRD, Magnetism (2010)
Mn(III)	$[Mn_{12}O_{12}(O_2CCH_2tBu)_{16}(H_2O)_4]CH_2C_{l2}MeNO_2$[18]	2.5 (2)	XRD, Magnetism (2010)
Fe(II)	$Fe(1,10\text{-}Phen)_2(NCS)_2$[19]	1 (1)	XRD (1993)
Fe(II)	$Fe(Btz)_2(NCS)_2$[19]	0.95 (1)	XRD (1993)
Fe(II)	$[Fe(bapbpy)(NCS)2]$[20]	1.09 (2)	XRD, Raman (2011)
Fe(II)	$[Fe(dpp)_2(NCS)_2] \cdot pyridine$[21]	2.48 (7)	XRD, Magnetism, Raman (2012)
Ru(0)	$Ru_3(CO)_{12}$[22]	8.14 (7)	XRD, Raman, IR (2004)
Co(0)	$[Co_2(CO)_6(PPh_3)_2]$[23]	4.4 (6)	XRD (2005)
Co(II)	$(4\text{-Chloropyridinium})_2[CoCl_4]$[24]	3.73 (9)	XRD (2008)
Co (II)	$(4\text{-Chloropyridinium})_2[CoBr_4]$[24]	2.92 (9)	XRD (2008)
Co(0)	$[Co_2(CO)_6(AsPh_3)_2]$[25]	4.1 (8)	XRD (2009)

Ni(II)	bis(3-fluoro-salicylaldoximato)Ni[26]	5.4 (7)	XRD, Absorption (2012)
Ni(II)	bis(3-methoxy-salicylaldoximato)Ni[26]	5.56 (7)	XRD, Absorption (2012)
Pd(II)	*cis*-$[PdCl_2(1,4,7\text{-trithiacyclononane})]$[27]	5.35 (6)	XRD (2006)
Pd(II)	$Cs_2[PdI_4]I_2$[28]	3.7 (12)	XRD (2006)
Pd(II)	$Cs_2[PdBr_4]I_2$[28]	2.18 (9)	XRD (2006)
Pd (II)	$Cs_2[PdCl_4]I_2$[28]	4.1 (16)	XRD (2006)
Cu(II)	$[Cu_2(OH)_2(H_2O)_2$ (tetramethylethylenediamine)$_2](ClO_4)_2$[29]	2.5 (4)	XRD, Magnetism (2009)
Cu(II)	$[Cu_2(OH)_2$(di-tbutylethylenediamine)$_2](ClO_4)_2$[29]	0.9 (2)	XRD, Magnetism (2009)
Cu(II)	$[Cu_2(OH)_2$(2,2′-bipyridine)$_2](BF_4)_2$[29]}	4.7 (9)	XRD, Magnetism (2009)
Cu(II)	$[Cu_2(OH)$(citrate)(Guanidine)$_2]$[30]	4.23 (4)	XRD (2009)
Cu(II)	$[GuH]_4[Cu_2(cit)_2]_2H_2O$[31]	2.2 (3)	XRD (10), Absorption
Cu(II)	$[Cu$(L-Aspartate)$(H_2O)_2]$(CuAspartate)[32]	7.9 (8)	XRD (2012)
Cu(II)	$[Cu(CO_3)_2](CH_6N_3)_2$[33]	3.95 (3)	XRD (2012)
Cu(II)	$[CuF_2(H_2O)_2(pyz)]$[34]	3.3 (8)	XRD, Magnetism (2012)
Ag(I)	Ag 2-methylimidazolate[35]	6.4 (8)	XRD, Piezoelectric response (2012)
Au	[Au(trimethylene tetrathiafulvalenedithiolate)$_2$][36]	10.7 (6)	XRD (2009)
Au(I)	$[(C_6F_5Au)_2$(m-1,4-diisocyanobenzene)][37]	4.39 (7)	XRD, Luminescence (2013)
Zn(II)	$[Zn_2(C_3H_3N_2)_4]$n (ZIF: $ZnIm_2$)[38]	0.54 (4)	XRD (2009)
Zn(II)	Porous Zn(2-methylimidazolate)$_2$ (ZIF-8)[38]	1.47 (6)	XRD (2009)
Zn(II)	$[Zn(Im)_2]$ (ZIF-4)[39]	4.39 (6)	XRD (2011)
Zn(II)	$[NH_4][Zn(HCOO)_3]$[40]	0.94 (7)	XRD (2012), Nano-indentation

equipment, from intense synchrotron sources to highly sensitive detectors and advanced computational tools. This very significant increase in available data is likely to continue, and the combination of structural and spectroscopic data promises to lead to significant new knowledge relevant to many areas of chemistry, in particular to inorganic functional materials, where both the electronic structure of molecular units and their intermolecular interactions are of key importance to materials characteristics.

One example of structural variations induced by pressure is shown in Fig. 2 for a square-planar nickel(II) complex, one of many d^8 compounds with this coordination geometry. The unit cell of the crystal decreases by more than 25% over the pressure range studied, which is a very significant decrease. In contrast, metal-ligand bond lengths show a much less obvious variation with pressure, also illustrated in Fig. 2. Similar small changes are observed for all intramolecular distances. Despite their small changes, spectroscopic effects are easily measured, as shown in the following. The pressure-induced volume decrease mainly causes intermolecular distances to decrease, as shown in Fig. 2, illustrating the potential of variable-pressure spectroscopy to probe intermolecular effects.

Figure 3 schematically illustrates luminescence spectroscopic effects for square-planar transition metal compounds, a vast class of compounds similar to the example for which structural changes are shown in Fig. 2. Intramolecular effects on luminescence spectra can be qualitatively predicted from the bonding characteristics of the HOMO and LUMO levels: the σ^* LUMO is destabilized very strongly by even the smallest bond length decrease, an effect that dominates the destabilization of the π^* HOMO levels in Fig. 3a. The result is an increase of the luminescence energy, leading to a shift of the maximum to higher energy or shorter wavelength, often denoted a blue shift in the literature. d-d transitions as illustrated in Fig. 3a have been investigated for many compounds, and shifts on the order of $+10\,cm^{-1}$/kbar to $+30\,cm^{-1}$/kbar

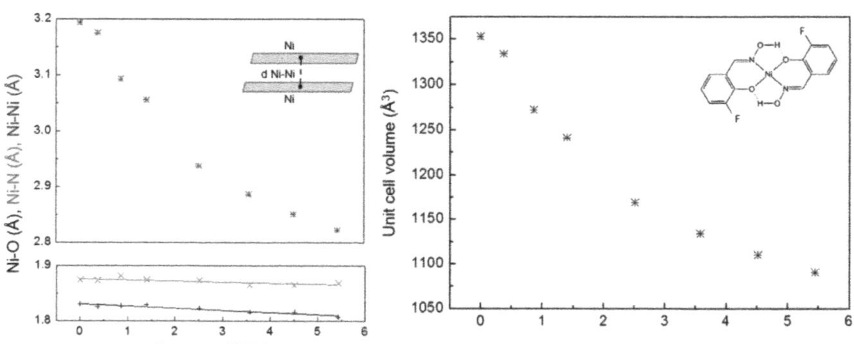

Fig. 2 Variable-pressure unit cell volume of bis(3-fluorosalicylaldoximato)nickel(II) (right) and variable-pressure inter- and intramolecular distances: Ni-O (+), Ni-N (×) and $d_{Ni\text{-}Ni}$ (*): Intramolecular Ni-N and Ni-O distances vary by −0.002 Å/GPa and −0.004 Å/GPa, respectively. The intermolecular Ni–Ni distance varies by −0.07 Å/GPa, greater by at least an order of magnitude.

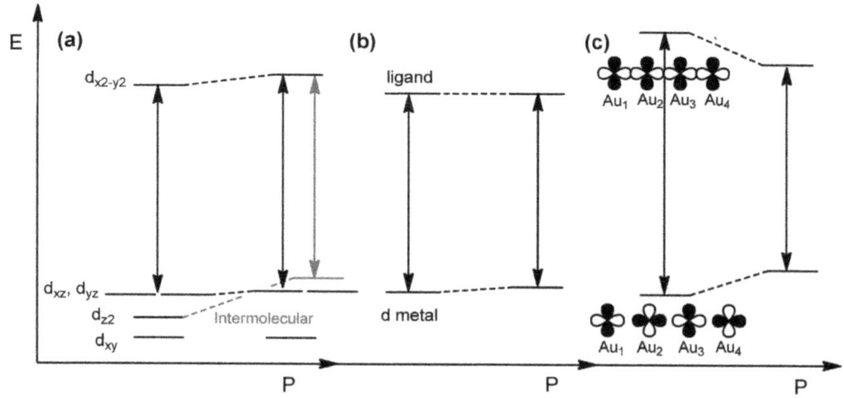

Fig. 3 Schematic representation of pressure effects on HOMO and LUMO energies for typical d-d transitions in square-planar complexes (**a**), MLCT transition (**b**) and for a gold(I) dithiocarbamate polymer chain (**c**).

are typical (shifts are given in cm^{-1}/kbar units throughout this report in order to be easily compared to the majority of published values, multiply by 10 to obtain cm^{-1}/GPa).[6,7] Intermolecular effects change the expected trend, also shown in Fig. 3a. They lead to an energy increase of the d_{z2} level and therefore to a decrease of the HOMO-LUMO energy difference, a variation in the opposite direction of the trend derived from the intramolecular changes in Fig. 3a. From experimental spectra it is therefore straightforward to determine which effects dominate, but obviously the spectra do not give access to quantitative structural changes – the combination with diffraction data is essential. Figure 3b represents other types of transitions, such as charge-transfer bands, with different variable-pressure properties than those in Fig. 3a. Figure 3c shows an intermediate case, involving gold(I)-gold(I) interactions with distances that are easily shortened through the application of pressure, leading to very characteristic spectroscopic effects such as the strong decrease of the luminescence energy schematically illustrated in Fig. 3c. Experimental examples for these categories of phenomena will be presented in the following sections.

2 Structural changes and variable-pressure Raman spectroscopy

There is a rich literature on variable-pressure vibrational spectroscopy.[41,42] Changes in molecular vibrational frequencies are small in the pressure range of interest here, and not obvious to quantitatively correlate with small structural changes, such as those shown in Fig. 2. In contrast, structural phase transitions are revealed through characteristic discontinuous variations of frequencies, which makes this a very sensitive technique for detection of such changes. An illustrative class of compounds with intriguing changes of structures and properties in response to external stimuli are spin-crossover materials. A recent perspective on these materials emphasizes the importance of combining an

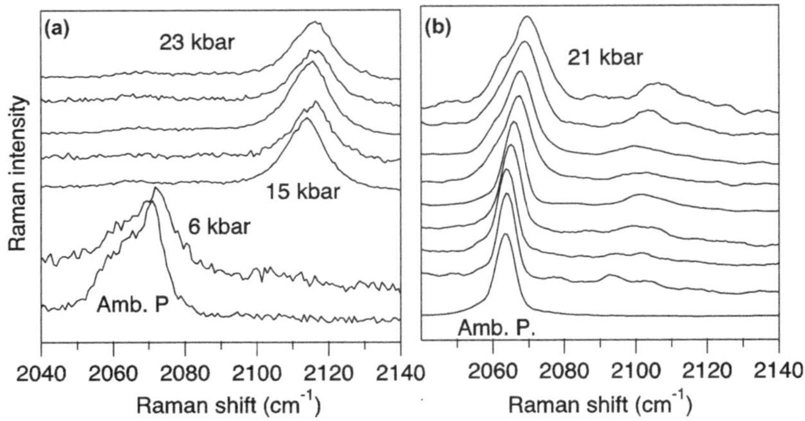

Fig. 4 Variable-pressure Raman spectra in the region of the $\nu_{N\text{-}C}$ stretching vibration for *cis*-Fe(1,10-phenanthroline)$_2$(NCS)$_2$, (a) and *trans*-Fe(pyridine)$_4$(NCS)$_2$ (b).

array of techniques and conditions such as variable pressure and temperature. Very significant shifts of vibrational frequencies occur as a consequence of the spin transition and its changes in metal-ligand bonding.

Figure 4 shows partial Raman spectra of two such compounds, *cis*-Fe(1,10-phenanthroline)$_2$(NCS)$_2$, a thoroughly studied complex with a low-spin form observed at low temperature and high pressure, and *trans*-Fe(pyridine)$_4$(NCS)$_2$, a complex that does not undergo a spin transition at low temperature but for which Raman spectra show evidence for the presence of a minority low-spin form at high pressure.[43] The $\nu_{N\text{-}C}$ stretching mode of the NCS^- ligands is a sensitive probe for this electronic structure change. Its frequency is 2052 cm^{-1} for uncoordinated NCS^- ions, similar to the frequencies of 2063 cm^{-1} and 2069 cm^{-1} for high-spin *trans*-Fe(pyridine)$_4$(NCS)$_2$ and *cis*-Fe(1,10-phenanthroline)$_2$-(NCS)$_2$, respectively, at ambient temperature and pressure. The Raman spectra in Fig. 4 show a small frequency increase by approximately +0.2 cm^{-1}/kbar for both compounds as pressure increases, a typical behavior for stretching frequencies. At pressures above 10 kbar, a new peak at 2116 cm^{-1} dominates in Fig. 4a, assigned to the $\nu_{N\text{-}C}$ mode of the low-spin form. This change is sudden in the Raman spectra in Fig. 4a, but very gradual in Fig. 4b, which shows the corresponding spectra for *trans*-Fe(pyridine)$_4$(NCS)$_2$, where the peak corresponding to the high-spin frequency persists at all pressures. The $\nu_{N\text{-}C}$ stretching frequency is compared to the N-C bond length in a variety of iron(II) complexes in Fig. 5.

The comparison shows that bond length changes are very small and no quantitative trend can be established for the variation of the frequency as a function of bond length change over this small range, indicating that metal-ligand bonds and other interactions also have a significant effect on the $\nu_{N\text{-}C}$ vibrational frequencies compared in Fig. 5. This overview illustrates that a combination of Raman spectroscopic and structural techniques is useful to understand and guide research on phenomena and properties based on both electronic and structural effects.

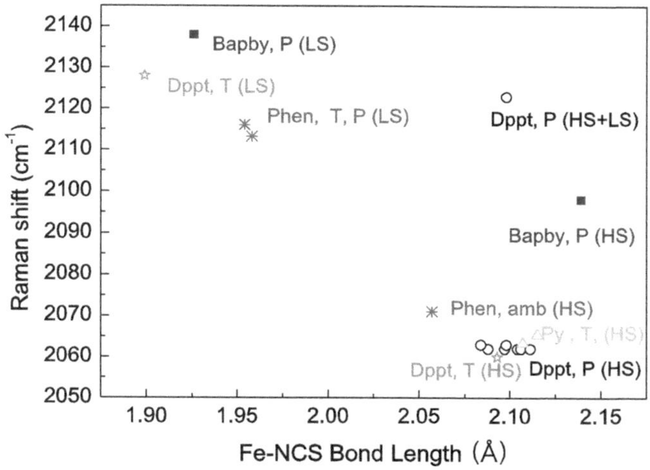

Fig. 5 Comparison of the $\nu_{N\text{-}C}$ stretching frequency of $(NCS)^-$ and the Fe-N distance for different iron(II) complexes: Bapby ($[Fe(bapbpy)(NCS)_2]$[22]), Dppt ($[Fe(dpp)_2(NCS)_2]\cdot$pyridine[23]), Phen ($Fe(1,10\text{-}Phen)_2(NCS)_2$)[47] and Py (trans-Fe(pyridine)4$(NCS)_2$).[47] Complexes are high spin (HS) or low spin (LS) depending on the temperature (T) and pressure (P) conditions.

3 Luminescence of square-planar d^8 complexes

Square-planar complexes are particularly attractive for variable-pressure luminescence and structural studies, as both intra- and intermolecular effects can dominate the observed properties. Overviews of the variable-pressure behavior of d-d transitions have been published.[6,7] The typical behavior is illustrated in Fig. 6a for a bis-dithiocarbamato complex of palladium(II).[44] It shows a broad luminescence band characteristic for a d-d transition with a maximum shifting to higher energy by $+12\,cm^{-1}$/kbar, in the expected range from $+10\,cm^{-1}$/kbar to $+30\,cm^{-1}$/kbar. The shift of the maximum is given in the inset at the center of the figure. The comparison with variable-temperature spectra in Fig. 6b is revealing: as temperature increases, the band maximum shifts to lower energy, a trend also shown in the inset. This shift is a consequence of the band broadening on its low-energy side, and indicates an increase of excited-state distortions at higher temperatures. High pressure also prevents the band from broadening on the low-energy side, acting qualitatively similar to low temperature. The comparison in Fig. 6 shows characteristic differences between variable-pressure and variable temperature luminescence due to intramolecular effects including emitting-state properties.

An attractive class of complexes for exploring significant pressure-induced structural changes are square-planar complexes with a dangling nucleophile above the coordination plane. Variable-pressure structures for one such compound, Pd(1,4,7-trithiacyclononane)Cl_2 have been published[27] and show a significant pressure-induced decrease of the distance between the dangling nucleophile and the metal center, a structural change comparable in magnitude to the change in intermolecular distance in Fig. 2. Variable-pressure d-d luminescence spectra of such systems show distinct variations for structurally similar

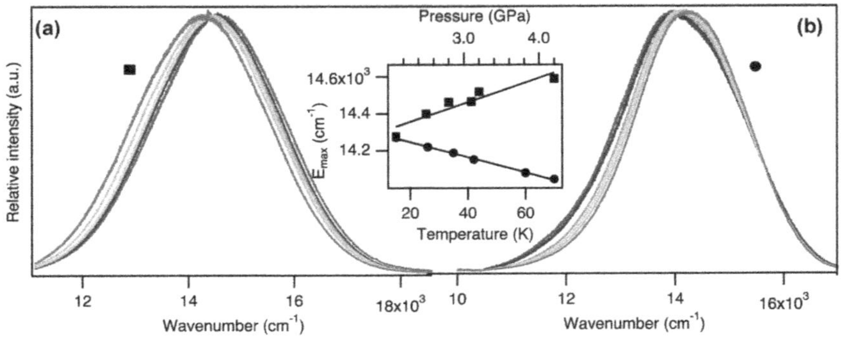

Fig. 6 (a) Pressure-dependent luminescence spectra of $Pd(PDTC)_2$ crystals shown at pressures between 2.1 and 4.2 GPa (left to right). (b) Temperature-dependent luminescence spectra of $Pd(PDTC)_2$ crystals shown at temperatures ranging from 15 to 70 K (right to left). Inset: variable-pressure luminescence maxima pressure (squares) and variable-temperature luminescence maxima (circles).

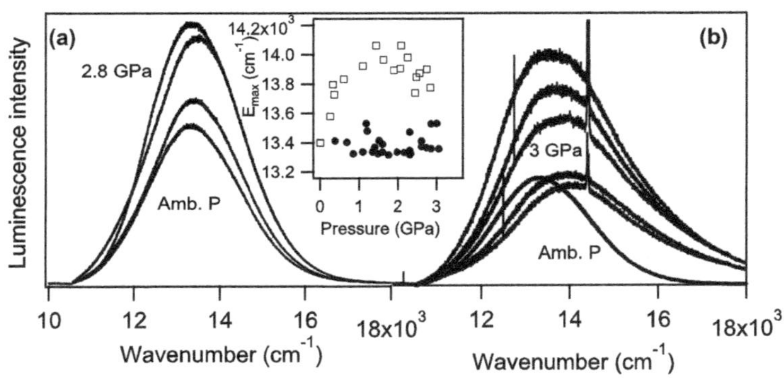

Fig. 7 Pressure-dependent charge-transfer luminescence spectra of Pd(1,4,7-trithiacyclononane)(phenylpyridine)$(PF_6)\cdot CH_3CN$ (a), maxima given by solid circles in the inset, and Pt(1,4,7-trithiacyclononane)(5-NO_2-phenanthroline)$(PF)_6$ (b)., maxima given as open squares in the inset.

palladium(II) and platinum(II) complexes with identical ligands:[45] a shift to higher energy by $+6\,cm^{-1}$/kbar for Pd(1,4,7-trithiacyclononane)Cl_2, but to lower energy by $-19\,cm^{-1}$/kbar for the platinum(II) analog. This trend is qualitatively rationalized by the destabilization of the d_{z2} level in Fig. 3a, labeled intermolecular in the Figure. These shifts are both lower than the typical range given above for d-d transitions, providing experimental evidence for the significant influence of the approaching dangling nucleophile and illustrating that variable-pressure spectra show the trend resulting from the changes to electronic states from all structural modifications. Shifts of band maxima with different signs generally occur because of different electronic effects arising from the small, pressure-induced structural changes. The variable-pressure luminescence spectra therefore show that the effect of the approaching nucleophile is stronger for platinum(II) complexes than for their palladium(II) analogs.

Charge-transfer luminescence transitions for complexes in this class are an appealing next category to explore. Examples are shown in

Fig. 7 for representative palladium(II) and platinum(II) complexes with 1,4,7-trithiacyclononane ligands. The observed trends are clearly different from those described above for d-d emissions: no significant shift occurs for the palladium(II) complex and only a small shift to higher energy for the platinum(II) complex. Both bands broaden significantly as pressure increases, indicating strong inhomogeneous broadening. The spectroscopic results suggest that charge-transfer transitions do not react as strongly as d-d bands to structural changes induced by pressure. One possible reason for this difference is that the charge-transfer excited state is ligand-centered and therefore less sensitive to pressure, as strong bonds show only small pressure-induced changes, illustrated by the $\nu_{N\text{-}C}$ stretching frequency and N-C distance in Figs. 4 and 5, and only the metal-centered ground state is affected by pressure. In the palladium(II) compound, there appears to be an almost perfect balancing of the effects induced by pressure, leading to no net change of the band maximum, as shown in the inset to Fig. 7. The energy of the charge-transfer luminescence band maximum of the platinum(II) complex increases with pressure, again shown in the inset to Fig. 7, a trend in the opposite direction from the d-d luminescence for platinum(II) complexes with 1,4,7-trithiacyclononane ligands. This difference indicates that the effect of the approaching nucleophile on the d_{z2} level, leading to an energy increase of the ground state and to a decrease of the d-d luminescence energy, is not the dominant effect for this complex. An energy increase of the excited state at higher pressure, likely to occur for the typically π^* ligand orbital accepting the metal-centered electron after the excitation, is a probable physical origin of the observed shift. These examples indicate that the electronic structure of the states involved in charge-transfer and d-d transitions do not adapt in the same way to the pressure-induced change of the molecular structure, resulting in different pressure-induced variations.

4 Aurophilic interactions

Variable-pressure luminescence spectra of dicyanoaurates have been studied in detail as they show dramatic variations of luminescence properties[46] as a consequence of small structural changes, in particular the metal-metal distances. Structures and spectra at variable pressure have been reported, but a correlation is not obvious, as a variety of similar Au(I)-Au(I) distances have to be considered, a challenging situation when several comparable but not identical distances occur.[47] Luminescence spectra show very significant shifts to lower energy on the order of $-100\ cm^{-1}$/kbar to $-300\ cm^{-1}$/kbar due to the effect of shorter metal-metal distances on the luminescence energy. The schematic view in Fig. 3c qualitatively illustrates the origin of these shifts to lower energy, as the energy of the ground state increases and the emitting-state energy decreases at higher pressure. Dithiocarbamate ligands lead to relatively simple structures, one-dimensional chains with alternating ligand-bridged and unbridged gold(I) centers, a situation closely resembling the schematic view in Fig. 3c.[48] Different substituents of the ligands lead to

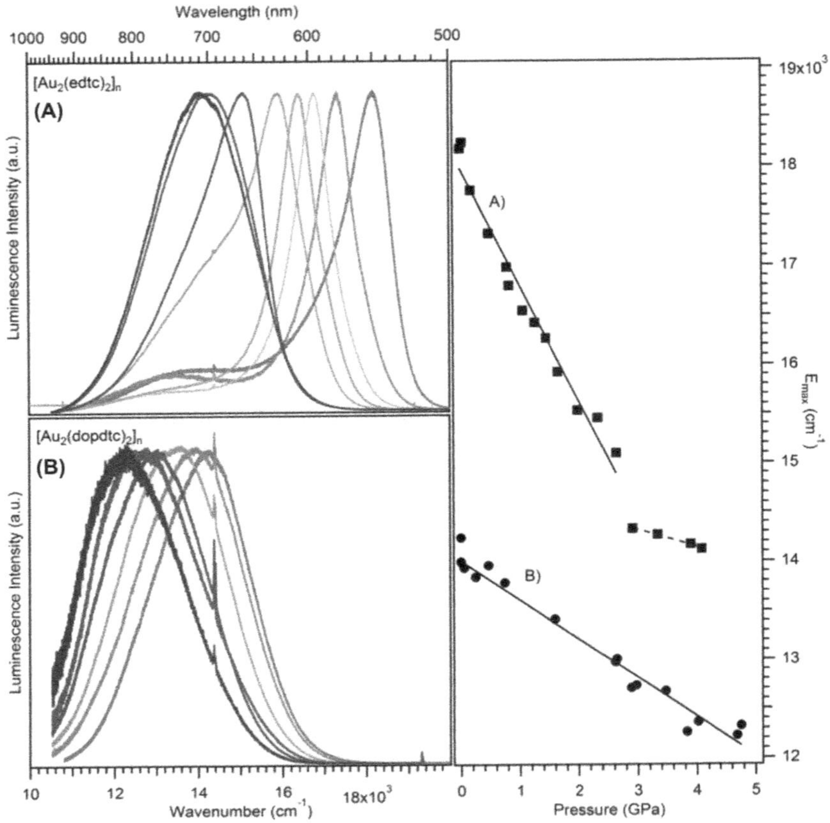

Fig. 8 Variable-pressure luminescence spectra and shift of luminescence maxima for two gold(I) dithiocarbamates between ambient pressure and 4.3 GPa.

polymers with almost identical Au(I)-Au(I) distances, but very distinct luminescence properties. Figure 8 shows variable-pressure luminescence spectra for two such systems, with obvious differences in luminescence band shape and maximum energy, and widely different pressure-induced shifts of the luminescence maxima by $-120\,cm^{-1}$/kbar and $-40\,cm^{-1}$/kbar. The key reason for this difference by a factor of three is the angular geometry, as schematically indicated by the linear and kinked chains on the right-hand side of Fig. 8, significantly changing the electronic structure of the states involved in the transitions despite the very similar metal-metal distances.[48,49]

At pressures above 2 GPa the linear-chain polymer shows broad-band luminescence with a smaller shift, a spectroscopic signature similar to characteristics of the kinked chain and a possible indication of pressure-induced structural defects. This comparison illustrates that pressure-induced structural changes other than distance decreases can have a very important influence on luminescence properties, a very attractive starting point to explore the variable-pressure structures and luminescence spectra of other molecular gold(I) systems in view of applications such as luminescent sensors.

Conclusion

The preceding overview aims to illustrate the potential of a combined structural-spectroscopic approach based on high-quality data obtained with new-generation instrumentation. Quantitative trends indicate very characteristic differences. This approach is particularly promising for guiding the chemistry of inorganic functional materials, where both the electronic structure of molecular units and their intermolecular interactions have to be understood and controlled in order to optimize materials characteristics.

Acknowledgements

The authors thank the Natural Sciences and Engineering Research Council (Canada) for research funding. CR and WBC also thank the Hans and Marlies Zimmer visiting scholar program at the University of Cincinnati for financial support.

References

1 D. W. H. Rankin, N. Mitzel and C. Morrison, *Structural Methods in Molecular Inorganic Chemistry*, Wiley, 2013.
2 W. Grochala, R. Hoffmann, J. Feng and N. Ashcroft, *Angew. Chem. Int. Ed.*, 2007, **46**, 3620.
3 R. J. Hemley, *Ann. Rev. Phys. Chem.*, 2000, **51**, 763.
4 H. G. Drickamer, *Annu. Rev. Mater. Sci.*, 1990, **20**, 1.
5 H. G. Drickamer and K. L. Bray, *Acc. Chem. Res.*, 1990, **23**, 55.
6 C. Reber, J. K. Grey, E. Lanthier and K. A. Frantzen, *Comments on Inorg. Chem.*, 2005, **26**, 233.
7 K. L. Bray, *Top. Curr. Chem.*, 2001, **213**, 1.
8 P. Day, *Angew. Chem. Int. Ed.*, 1980, **19**, 290.
9 P. Day, *Acc. Chem. Res.*, 1988, **21**, 250.
10 J. M. Herrera, P. Franz, R. Podgajny, M. Pilkington, M. Biner, S. Decurtins, H. Stoeckli-Evans, A. Neels, R. Garde, Y. Dromzée, M. Julve, B. Sieclucka, K. Hashimoto, S. Okhoshi and M. Verdaguer, *C. R. Chimie*, 2008, **11**, 1192.
11 T. D. Bennett, J. C. Tan, S. A. Moggach, R. Galvelis, C. Mellot-Draznieks, B. A. Reisner, A. Thirumurugan, D. R. Allan and A. K. Cheetham, *Chem.-Eur. J.*, 2010, **16**, 10684.
12 P. Parois, S. A. Moggach, A. R. Lennie, J. E. Warren, E. K. Brechin, M. Murrie and S. Parsons, *Dalton Transactions*, 2010, **39**, 7004.
13 D. Pinkowicz, K. Kurpiewska, K. Lewinski, M. Balanda, M. Mihalik, M. Zentkova and B. Sieklucka, *Crystengcomm*, 2012, **14**, 5224.
14 P. Guionneau, M. Marchivie, Y. Garcia, J. A. K. Howard and D. Chasseau, *Phys. Rev. B*, 2005, **72**, 8.
15 S. Tancharakorn, F. P. A. Fabbiani, D. R. Allan, K. V. Kamenev and N. Robertson, *J. Am. Chem. Soc.*, 2006, **128**, 9205.
16 A. Prescimone, C. J. Milios, S. Moggach, J. E. Warren, A. R. Lennie, J. Sanchez-Benitez, K. Kamenev, R. Bircher, M. Murrie, S. Parsons and E. K. Brechin, *Angew. Chem. Int. Ed.*, 2008, **47**, 2828.
17 A. Prescimone, C. J. Milios, J. Sanchez-Benitez, K. V. Kamenev, C. Loose, J. Kortus, S. Moggach, M. Murrie, J. E. Warren, A. R. Lennie, S. Parsons and E. K. Brechin, *Dalton Trans.*, 2009, 4858.

18 A. Prescimone, J. Sanchez-Benitez, K. V. Kamenev, J. E. Warren, A. R. Lennie, M. Murrie, S. Parsons and E. K. Brechin, *Z.Naturforsch.(B)*, 2010, **65**, 221.

19 T. Granier, B. Gallois, J. Gaultier, J. A. Real and J. Zarembowitch, *Inorg. Chem.*, 1993, **32**, 5305.

20 H. J. Shepherd, S. Bonnet, P. Guionneau, S. Bedoui, G. Garbarino, W. Nicolazzi, A. Bousseksou and G. Molnar, *Phys. Rev. B*, 2011, **84**, 144107.

21 H. J. Shepherd, T. Palamarciuc, P. Rosa, P. Guionneau, G. Molnar, J. F. Letard and A. Bousseksou, *Angew. Chem. Int. Ed.*, 2012, **51**, 3910.

22 C. Slebodnick, J. Zhao, R. Angel, B. E. Hanson, Y. Song, Z. X. Liu and R. J. Hemley, *Inorg. Chem.*, 2004, **43**, 5245.

23 N. Casati, P. Macchi and A. Sironi, *Angew. Chem. Int. Ed.*, 2005, **44**, 7736.

24 G. M. Espallargas, L. Brammer, D. R. Allan, C. R. Pulham, N. Robertson and J. E. Warren, *J. Amer. Chem. Soc.*, 2008, **130**, 9058.

25 N. Casati, P. Macchi and A. Sironi, *Chemistry Eur. J.*, 2009, **15**, 4446.

26 P. J. Byrne, P. J. Richardson, J. Chang, A. F. Kusmartseva, D. R. Allan, A. C. Jones, K. V. Kamenev, P. A. Tasker and S. Parsons, *Chemistry Eur. J.*, 2012, **18**, 7738.

27 D. R. Allan, A. J. Blake, D. G. Huang, T. J. Prior and M. Schröder, *Chem. Commun.*, 2006, 4081.

28 P. Heines, H. L. Keller, M. Armbrüster, U. Schwarz and J. Tse, *Inorg. Chem.*, 2006, **45**, 9818–9825.

29 A. Prescimone, J. Sanchez-Benitez, K. K. Kamenev, S. A. Moggach, J. E. Warren, A. R. Lennie, M. Murrie, S. Parsons and E. K. Brechin, *Dalton Transactions*, 2009, **39**, 113.

30 S. A. Moggach, K. W. Galloway, A. R. Lennie, P. Parois, N. Rowantree, E. K. Brechin, J. E. Warren, M. Murrie and S. Parsons, *Crystengcomm*, 2009, **11**, 2601.

31 K. W. Galloway, S. A. Moggach, P. Parois, A. R. Lennie, J. E. Warren, E. K. Brechin, R. D. Peacock, R. Valiente, J. Gonzalez, F. Rodriguez, S. Parsons and M. Murrie, *Crystengcomm*, 2010, **12**, 2516.

32 J. A. Gould, M. J. Rosseinskya and S. A. Moggach, *Dalton Trans.*, 2012, **41**, 5464.

33 E. C. Spencer, N. L. Ross and R. J. Angel, *J. Mat. Chem.*, 2012, **22**, 2074.

34 A. Prescimone, C. Morien, D. Allan, J. A. Schlueter, S. W. Tozer, J. L. Manson, S. Parsons, E. K. Brechin and S. Hill, *Angew. Chem. Int. Ed.*, 2012, **51**, 7490.

35 J. M. Ogborn, I. E. Collings, S. A. Moggach, A. L. Thompson and A. L. Goodwin, *Chem. Sci.*, 2012, **3**, 3011.

36 Y. Okano, B. Zhou, H. Tanaka, T. Adachi, Y. Ohishi, M. Takata, S. Aoyagi, E. Nishibori, M. Sakata, A. Kobayashi and H. Kobayashi, *J. Am. Chem. Soc.*, 2009, **131**, 7169.

37 C. H. Woodall, C. M. Beavers, J. Christensen, L. E. Hatcher, M. Intissar, A. Parlett, S. J. Teat, C. Reber and P. R. Raithby, *Angew. Chem. Int. Ed.*, 2013, **52**, 9691.

38 S. A. Moggach, T. D. Bennett and A. K. Cheetham, *Angew. Chem.-Int. Edit.*, 2009, **48**, 7087.

39 T. D. Bennett, P. Simoncic, S. A. Moggach, F. Gozzo, P. Macchi, D. A. Keen, J. C. Tan and A. K. Cheetham, *Chem. Commun.*, 2011, **47**, 7983.

40 W. Li, M. R. Probert, M. Kosa, T. D. Bennett, A. Thirumurugan, R. P. Burwood, M. Parinello, J. A. K. Howard and A. K. Cheetham, *J. Am. Chem. Soc.*, 2012, **134**, 11940.

41 Y. Suffren, F.-G. Rollet and C. Reber, *Comm. Inorg. Chem.*, 2011, **32**, 246.

42 C. M. Edwards and I. S. Butler, *Coord. Chem. Rev.*, 2000, **199**, 1.

43 Y. Suffren, F.-G. Rollet, O. Levasseur-Grenon and C. Reber, *Polyhedron*, 2013, **52**, 1081.
44 C. Genre, G. Levasseur-Theriault and C. Reber, *Can. J. Chem.*, 2009, **87**, 1625.
45 E. Pierce, E. Lanthier, C. Genre, Y. Chumakov, D. Luneau and C. Reber, *Inorg. Chem.*, 2010, **49**, 4901.
46 H. Yersin and U. Riedl, *Inorg. Chem.*, 1995, **34**, 1642.
47 P. Fischer, J. Mesot, B. Lucas, A. Ludi, H. H. Patterson and A. Hewat, *Inorg. Chem.*, 1997, **36**, 2791.
48 F. Baril-Robert, M. A. Radtke and C. Reber, *J. Phys. Chem. C*, 2012, **116**, 2192.
49 R. J. Roberts, N. Bélanger-Desmarais, C. Reber and D. B. Leznoff, *Chem. Commun.*, 2014, **50**, 3148.

Near-infrared spectroscopic studies of nanostructured materials

Christian W. Huck
DOI: 10.1039/9781782621485-00274

The measurement of the chemical and physical properties of nanostructured materials, including *e.g.* silicate, carbon nanotubes and fullerenes, is time-consuming, requires expensive instrumental equipment and a lot of user experience. Infrared spectroscopy, especially near-infrared spectroscopy (NIR), coupled with multivariate data analysis (MVA) can be used as a non-destructive, fast, reliable and robust technique for the characterization and classification of nanostructured materials. In this contribution the most relevant milestones reached so far in NIR spectroscopic characterization of nanostructured materials are reviewed, summarized and discussed.

1 Introduction

The word "nano" is derived from the Greek term for dwarf (nánnos) and is nowadays used as a prefix for a measuring unit. One nanometer corresponds to 10^{-9} meters. The human hair for example has an average diameter of approximately 80.000 nm, bacteria of 2.000 nm and the DNA of 2 nm. "Nano" is, in our daily routine, a collective name for investigations and work being carried out with all different kinds of matter between 0.1 and some 100 nm. With matter in this size range specific physical,[1,2] chemical,[3] biological, medical, electrical, mechanical properties can be achieved, offering new application fields in the macroscopic world.

For the production of nano-materials two main approaches exist: The "bottom-up" approach, where materials are built up, atom by atom, and the "top-down" approach, where materials are synthesized, or larger entities are brought to nano-level. Additionally, nano-materials can cause potent health risks. Due to the increasing importance of matter at the nano-scale in many fields of application it's physico-chemical and morphological characterization is an urgent and demanding challenge.

Particle size is conventionally determined by electron microscopy (ELMI), light scattering or the Coulter-Counter method. For the determination of the specific surface area, nitrogen adsorption experiments according to the Brunauer-Emmett-Teller (BET) theory; for pore size and volume, mercury intrusion porosimetry (MIP) or size-exclusion chromatography (SEC) are applied, respectively. Chemical materials' properties include, after the identification of the material itself, qualitative analysis of chemical derivatization and also surface coverage. For this purpose, nuclear magnetic resonance (NMR) spectroscopy and/or elementary analysis are usually applied. The disadvantages of all these traditional

Institute of Analytical Chemistry and Radiochemistry, CCB-Center for Chemistry and Biomedicine, Innrain 80/82, 6020 Innsbruck, Austria.
E-mail: christian.w.huck@uibk.ac.at

techniques are that they are time-consuming; require expensive analytical instrumentation, and a lot of experience from the operating staff. The investigation is invasive and finally, also expensive.

Nanostructured materials and/or nanostructured properties of carrier materials in analytical chemistry play a permanently increasing role. They are employed as a stationary phase in liquid-chromatography (LC)[4] and also for selective enrichment techniques such as solid-phase extraction (SPE).[5] In 2006, nanostructured materials were introduced for the biomarker profiling of prostate cancer related human serum samples, and the technique was given the new term "material enhanced laser desorption ionization" (MELDI) mass spectrometry (MS).[6] It was demonstrated that this novel method clearly benefited from the increased loading capacity, due to the high ratio of surface-to-size.[3]

In pharmacy and pharmaceutical technology nanostructured (delivery) systems become more and more important.[7] Studies have illustrated that in some, particularly promising ways, nano-materials as drug or vaccine carriers can assist in navigating barriers (plasma membrane, endo-lysosomes, nuclear membrane, multiple drug resistance mechanisms), with a particular focus on administration by injection.

The pharmaceutical industry has realized the special advantages of NIR spectroscopy to investigate the physical and morphological properties of liquid, solid and semisolid products. Blend uniformity analysis, raw material identification, process analysis and finished product quality monitoring must be carried out non-invasively and fast. Parameters like size and composition (physical and chemical, "physicochemical") must be analyzed not only in parallel, but simultaneously.[8]

It is the aim of the present contribution to firstly introduce the principle of NIR and then to explain the measurement technique. Secondly, the efficiency of NIR to monitor the physic-chemical properties of nanostructured materials is demonstrated, discussed and evaluated. In the third part the application of nanostructured materials as a selective carrier for distinct qualitative/quantitative analyses of single molecules is described.

2 Fundamentals of near-infrared spectroscopy and applied chemometrics

The region of the electromagnetic spectrum between the visible (VIS) and the microwave wavelength is called infrared radiation (IR).[9] The near-infrared (NIR) spectroscopic excitation of molecules is accomplished in a wavelength range between 750 and 2500 nm, corresponding to a wavenumber range between 4.000 and 13.000 cm^{-1}.[10] Samples of all different aggregation states can absorb parts of the incoming IR radiation at specific wavelengths resulting in a fingerprint or a characteristic spectrum.[11] Molecules containing C–H, C–O, C=O, N–H and O–H functional groups are excited to perform stretching-, deformation- and scissor-vibrations. In the mid-infrared (MIR) region only fundamental vibrations ("signatures") can be observed. In the NIR region overtones and combination vibrations can be found ,containing a manifold of information compared to MIR,[12] which often results in a crowded spectrum with overlapping peaks. Today, there are highly sensitive spectrometers available in order to overcome the 10–1000 times lower NIR-intensities. The light recorded by the detector contains compositional information which can be unravelled by special software to report multiple analyses almost instantaneously. NIR spectroscopy is able to provide simultaneous, rapid and non-destructive qualitative and quantitative analysis of compounds even in complex matrices.[13]

2.1 Model of the harmonic and an-harmonic oscillator

The physical principle behind the observed effects both in the MIR as well as in the NIR region is the model of the harmonic and anharmonic oscillator. Thereby, the reduced mass μ performs vibrations with the frequency ν_{osc}. In the MIR this vibration follows more or less the equation for the harmonic oscillator, whereas in the NIR the equation for the anharmonic oscillator is valid, describing the excitation into higher energy states. In the following chemometric approaches based, in most cases, on multivariate statistical analyses are performed.

2.2 Instrumentation

An NIR spectrometer consists of a light source (*e.g.* tungsten halogen lamp), sample presentation accessories, monochromator, detector and different optical components including lenses, collimators, beam splitters, integrating spheres and optical fibers.[14]

In NIR spectroscopy there is only little, and in most cases no, sample preparation required prior to analysis. Transparent materials are analysed in transmission and/or transflection, solid materials in diffusive reflection and/or interaction mode (using light fiber optics), respectively. In reflection mode, the light source and detector are mounted at a specific angle, in order to avoid specular reflection. In transmission mode, the light source is positioned opposite to the detector. In interaction mode, the light source and detector are positioned in parallel to each other.[14]

Basically speaking, spectrometers can be divided into dispersive and non-dispersive instruments. Dispersive filter instruments often contain a monochromator which is a wheel holding a number of absorption of interference filters. In scanning monochromator instruments a grating or a prims is used for the separation of the individual frequencies. Non-dispersive, Fourier transform spectrophotometers take advantage of interferometers using modulated light and time domain signals, which must be converted into a spectrum *via* Fourier transformation.[15] In most instruments of this type a Michelson or polarisation interferometer is used.

Photodiode array (PDA) spectrophotometers consist of a fixed grating that focuses the dispersed radiation onto a detector array of silicon (Si, 350–1100 nm) or indium gallium arsenide (InGaAs, 1100–2500 nm) offering high acquisition speed. As an alternative, laser based systems, diffraction-based optical band-pass filters and/or liquid tune-able filter (LCTF) instruments can be used.[16]

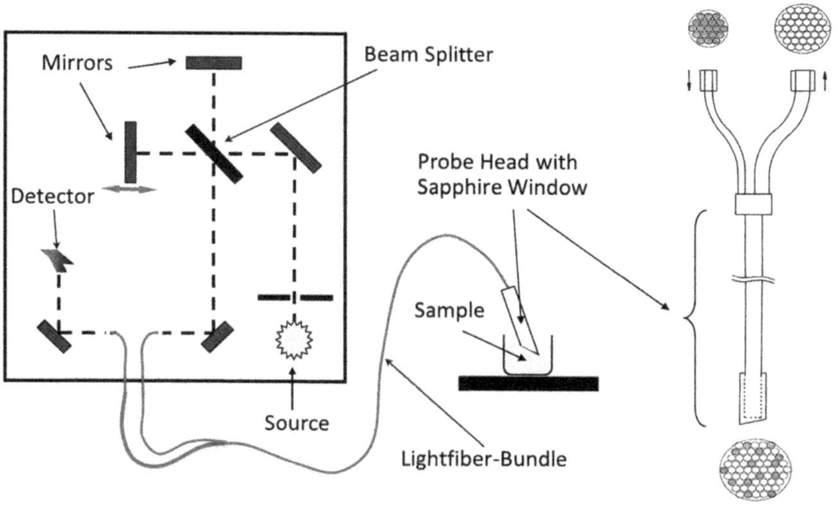

Fig. 1 Sample presentation for non-destructive diffuse reflection NIR spectroscopy.

Hyper-spectral NIR spectroscopic imaging is becoming more and more popular.[17] Thereby, a multi-wavelength hyper-spectral cube is recorded consisting of spectra recorded at every defined 2-D spatial position. The implementation of sensitive detectors (MCT, mercury cadmium telluride) and attenuated total reflection (ATR) units enables highly efficient analysis down to a spatial resolution of 1.2 μm.[18]

For quantitative analysis the amount of absorbed radiation is dependent upon the Lambert-Beer's law in transmission, and upon the Kubelka-Munk equation in diffuse reflection mode.

The basic principle for non-destructive diffuse reflection NIR spectroscopy can be deduced from Fig. 1.

2.3 Chemometrics and multivariate data analysis

The NIR spectrum consists of a huge number of partially overlapping overtones and combination vibrations. In addition, scattering effects, instrumental noise and/or sample in-homogeneities might occur.[9] Therefore, for many applications it is impossible to correctly assign the corresponding vibration bands. NIR in combination with multivariate statistical analysis (MVA) has been demonstrated to form a powerful synergistic combination enabling the extraction of the required information from the spectrum.[11] The most appropriate and frequently used chemometric procedure is principal component analysis (PCA) for reducing the number of variables, facilitating both qualitative and quantitative analysis. Data pre-treatment is applied to minimize inhomogeneities which might originate from the recording of the spectra and allows the elimination of baseline shifts. Normalization algorithms can be applied to eliminate different sample positioning. Diffusion and/or unexpected scattering caused by particle size effects can be compensated by a multiplicative scatter correction (MSC). Spectral noise can be reduced by performing the first or second derivative of the original spectrum. Calibration development is mathematically achieved by the co-variation between defined variables or by a mathematical function (regression model). In a regression model the values of the dependent variables are calculated from values of the measured variables (Siebert). It is important to select an optimum number of variables or components. The system can become over-fitted if too many variables are used and then poor prediction results will result. Using too few components will cause under-fitting and the model is too small to capture the variability in the data. This fitting effect is strongly dependent on the number of samples and, in general, more samples give rise to more accurate predictions.[20]

The calibration procedure of the NIR spectrometer can be summarised in five steps: 1) Choice of a representative sample set; 2) Recording of the NIR spectra; 3) Measurement of the reference values; 4) Multivariate modelling in order to generate a relationship between the recorded spectra and the reference values; 5) Validation of the system. The most frequently regression methods comprise principal component regression (PCR) and partial least square regression (PLSR), discriminant analysis (DA) and artificial neural networks (ANN).[18]

The choice of the highest suitable regression model[19] is based on the calculation of the following values:

1. BIAS, *i.e.*, the average deviation between the predicted values (y_n) and the actual values (x_n), in the calibration-set, should be close to zero.

$$\text{Bias} = \frac{1}{N}\sum(x_n - y_n) \tag{1}$$

2. PRESS, Predicted Residual Error Sum of Squares is the sum of the squares of the deviations between predicted and reference values. The PRESS value of the validation set should be as small as possible and similar to that of the calibration set.

$$\text{PRESS} = \sum(x_n - y_n)^2 \tag{2}$$

3. Standard error of estimation (SEE), *i.e.*, the standard deviation of the differences between reference values and NIRS-results in the calibration set.

$$\text{SEE} = \sqrt{\frac{1}{N}\sum(x_n - y_n - \text{Bias})^2} \tag{3}$$

4. Standard error of prediction (SEP), *i.e.*, the counterpart for the test-set samples. SEE and SEP should be as small as possible.

$$\text{SEP} = \sqrt{\frac{1}{N}\sum(x_n - y_n - \text{Bias})^2} \tag{4}$$

5. The correlation coefficient (R^2) should approach 1.

3 Properties of nanostructured materials and their characterization by near-infrared spectroscopy

In the first section of this chapter, materials properties are described. In the second section tailored NIR based technologies are introduced and reviewed.

3.1 Silicate materials

In industry, silica gel is used as an adsorbent for vapors from *e.g.* alcohol, ether, benzene and from air *e.g.* in explosives factories. It is applied as an efficient desiccant used for drying liquids, gases and solids, for purification and de-colorization of liquids and lipids. In analytical chemistry silica gel plays an important role as important stationary phase and carrier material in chromatographic separation science, including the thin-layer chromatography (TLC) and the high pressure liquid chromatography (HPLC) in the fields of phytomics, proteomics and metabolomics.[4] Physical properties, including pore size, particle size and surface area, play an important role in attaining the highest separation efficiencies, because merely well optimized and characterized stationary phases can be used in challenging fields such as bioanalysis.[3] The most popular method of HPLC is the reversed-phase high pressure liquid chromatography (RP-PLC). Stationary phases for RP-LC can be purchased with different carbon chain length, for example, C4, C8, C18, C30, *etc.* Other derivatisations comprise functionalisation with *e.g.* anion- or cation-exchange or affininty groups.

3.2 Carbon nanomaterials

Carbon nanomaterials such as carbon nanotubes (CNTs) and fullerenes can be applied in different scientific fields like medicine, drug delivery, optics, electronics and analytical chemistry because of their unique biological, physical and chemical properties.[21]

Carbon nanotubes can be approximated as rolled-up sheets of graphite and are formed in two principal types: one type consists of a single tube of graphite, the so called single-walled carbon nanotubes, and the second type, which consists of several concentric tubes of graphite, fitted one inside the other; the so called multi-walled carbon nanotubes. The length of the tubes is usually in the micrometer range, the diameter of the tubes depends on the type. Whereas the diameter of the single-walled carbon nanotubes is just a few nanometers, the one of the multi-walled several ten of nanometers. Carbon nanotubes have been applied as a matrix for matrix-assisted laser desorption/ionization (MALDI) time-of-flight mass-spectrometry and for material-enhanced laser desorption/ionization (MELDI) time-of-flight mass-spectrometry.[22]

Fullerenes: In 1966 the idea came up, that a graphite-layer can be formed to a hollow ball, but to achieve that the addition of some pentagons between the hexagons is required. In 1970 Eiji Osawa[23] predicted the existence of the C60-molecule, but only in 1985 was the first fullerene produced by laser evaporation of graphite.[24] The name of this new class of molecules, which denotes all closed carbon cages, including twelve pentagons and more than one hexagon which forms a sp^2 network, was 'Buckminster Fuller', who invented the geodesic domes.[25] C60, C70, C76, C78 and C84 are commonly available fullerene-molecules.

To obtain a material for solid phase extraction, silica was derivatized with fullerenes. The silica sorbents were derivatized with an amino propyl linker and then covalently bound to C60-epoxyfullerenes or C60-fullerenoacetic acid.[1]

4 Characterisation of nanomaterials

4.1 Silicate materials

For the NIR spectroscopic characterization of scientific materials, including silicate material, carbon nanotubes and fullerenes, a method is described and protected by patent US007715002 B2 (Method for classifying scientific materials such as silicate materials, polymer materials and/or nanomaterials).[26] The method offers operation at reduced costs. it is non-invasive and fast. It enables improved sample handling and the possibility to measure physical and chemical properties simultaneously. NIR spectroscopy is coupled with statistical software to perform principle component analysis (PCA), regression (PCR) or partial least squares regression (PLSR).

The described procedure provides a method for characterization of silicate material in the following steps:

1. A number of material classes is provided, each with a plurality of calibration and material specimens per class with different class-characteristics. By irradiating a measuring light of the NIR-VIS wavelength range (400 to 250 nm, respectively 4000 to 14186 cm^{-1}) into the material specimen and record reflected and/or reemitted measuring light the calibration model is established.
2. Afterwards a ratio, depending on the wavelength of irradiated to detected measuring light (spectrum), for each material of a class is determined. In the following numerical-mathematical processing of the spectral data of each material specimen is carried out.
3. A database is generated by the correlation of the spectra of a majority of material specimens with a predetermined material class.
4. With a measuring light to numerical-mathematical processing of the spectral data are repeating at least once with at least one material sample with partially unknown characteristics to validate the model established.

To improve the accuracy and/or minimized the noise by avoiding or minimizing interactions of the measuring radiation and/or the reflected and/or reemitted radiation, two different strategies can be applied: One possibility is the analysis of the sample in a vacuum, the other possibility, which is preferred and easier to use, is that the measuring radiation is conducted to the material *via* a light conductor and/or is absorbed by the material. For high quality spectra, the thickness of the specimen layer and the temperature of the sample are of importance. The optical thickness of the material specimens is in the range of approximately 0.2 to 5 mm. Thinner layers can be measured as long as there is still enough material, so that any effects can be measured, and thicker layers as long as still a measurable signal can be obtained. To save energy, reduce the preparation time for the samples and to ensure a good comparability and reproducibility of the measured data, the maintenance of a constant temperature, approximately 23 °C, is important. In order to reduce file size or to eliminate noise, it is convenient to process the obtained raw data, by at least one of; normalization, smoothing, 1st derivation, 2nd derivation, multiplicative scatter correction, reciprocal value, square, mean centering, Kubelka Munk transformation, absorption, baseline correction, addition of a constant, shift negative to zero, PCA, PCR, or partial least square regression (PLS). The characteristics of the individual material classes determined by the numerical-mathematical processing are at least one of particle surface, particle size, pore diameter, pore volume, pore porosity and derivatization. The correlation of the numerically-mathematically processed spectral data is at least one multivariate method selected from PCA and/or a smoothing, and/or at least one series expansion and/or artificial neuronal network algorithm, dynamic learning vector quantization (DLVQ-algorithm), radial basis functions (RBF), wherein the correlation is caused by the formation of clusters and the tolerance areas of individual clusters are adjustable during calibration. Approximately 3 to 20 spectral-scans/material specimens are recorded, 30% of all material specimens measured by class are used for validating the procedure, 70% are used for calibration and the classification results can be shown as a 2D/3D-plot. In Fig. 2 the corresponding calibration and validation curves for the quantitative determination of particle size and specific surface area are depicted.

4.2 Pharmaceutical nanomaterials

A NIR spectroscopic method allowing the characterization of pharmaceutical nanomaterials is protected by patent WO2009/137855 A1 (Characterization of physicochemical properties of a solid) respectively AT506681A1 (Charakterisierung von physiko-chemischen Eigenschaften eines Feststoffes).[7] The method described enables the determination of several characteristics of a solid, such as particle size, porosity and/or specific surface area with one measurement simultaneously. The characterization of physico-chemical properties of a solid is an important concern, for example, to guarantee a consistent quality in the production of the solid. A solid, in particular a pharmaceutical composition, is measured with NIR spectroscopy and the measured data are compared with values from calibration tables, which were created prior to the determination for the given characterization of the solid composition by an appropriate reference method. The measuring light of the NIR wavelength range (4000 to 12800 cm^{-1}) is irradiated onto the sample and the transmittance, diffuse reflectance or transflectance of the light, notable diffuse reflectance is recorded. With the aid of mathematic, statistic and multivariate methods and chemometric software the calibration tables can be established. Basis of known and measured values, qualitative and quantitative calibration tables are created, corresponding to the different characteristics of the solid.

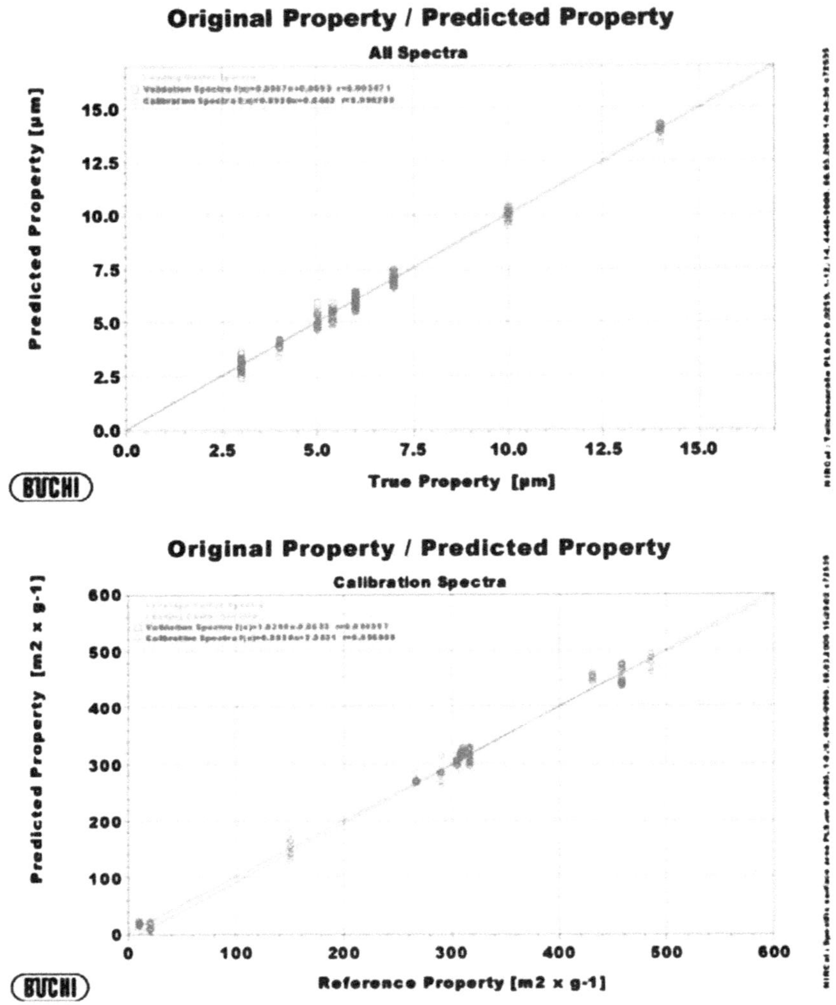

Fig. 2 Calibration and validation curves for the quantitative determination of particle size and surface area.

These calibration models can serve as a basis for the determination of unknown samples by comparing the respective measured values. This method enables the determination with a non-invasive measurement (NIR spectroscopy) coupled with multivariate data analysis, simultaneously, several characteristics of a solid, including particle size and surface area[7] (Fig. 3).

4.3 C60-fullerene-silica

C60-fullerene-derivatised silica is used as carrier in material laser desorption/ionization mass spectrometry (LDI-MS) for the analysis of molecules having a molecular weight of <1500 Da, such as *e.g.* sugars, peptides, amino acids and lipids.[1] A clear identification of small molecules through intensive signals without matrix-related background disturbances is the result benefitting from this progress.

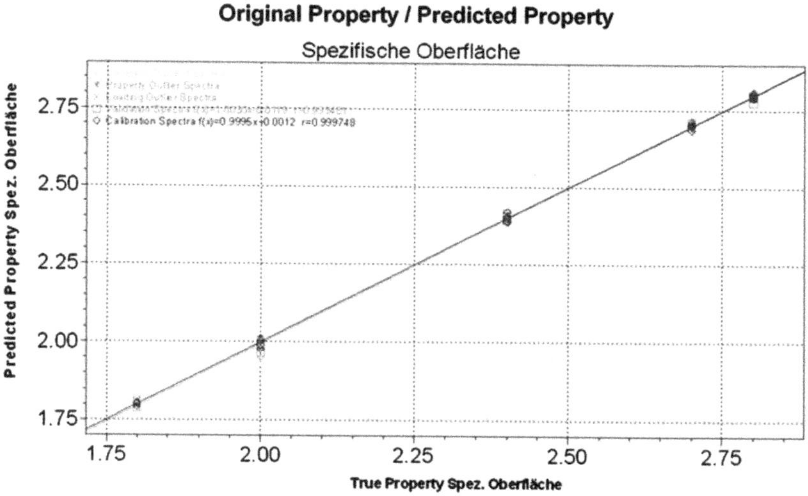

Fig. 3 Determination of specific surface area (m^2/g) for amoxicillin trihydrate.

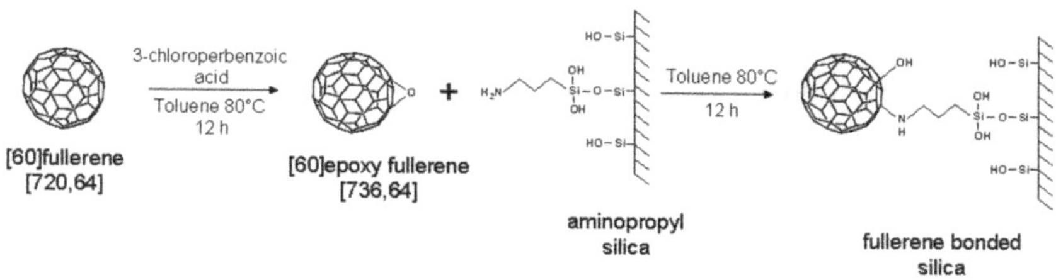

Fig. 4 Synthesis of C60-fullerene bound silica.

Even though the acceptance of MALDI-MS for biomolecules was growing, its use in low-molecular weight analysis and tandem MS experiments was limited. This is mainly due to issues with uniform energy transfer from matrix to analyte upon UV laser excitation, and/or interference from matrix molecules. An obstacle to rapid analysis of a diverse set of samples is that known matrices are specific for certain molecules. For this reason there is a need of materials, which can measure molecules having a molecular weight of <1500 Da, such as C60-fullerene-derivatised silica. The C60-fullerene-derivatised silica is fullerene-bounded silica, which is a result from the reaction of aminopropyl silica and C60-fullerenoacetyl chloride or of aminopropyl silica and C60-epoxy fullerene[1] (Fig. 4).

The fullerene-derivatized silica has a pore size in the range of 0 (*i.e.* non-porous) to 100 nm; preferably 30 nm and an increased specific surface area relative to silica alone. To prepare the matrix material, the fullerene-derivatized silica is suspended in a solvent, preferably selected from the group consisting of methanol, acetonitrile, acetone and a mixture of acetonitrile and water, applying to a MALDI target and drying the material. Surface area and pore size are both an essential factor in the desorption/ionization process. High enough surfaces are needed to allow the laser energy to be forwarded from the fullerenes to the analytes, and from larger pores the analytes are capable of easier desorption.

5 Application of nanostructured materials for single molecule near-infrared spectroscopic investigations

Water is a strong absorber in the near infrared (NIR) region. The most efficient way to analyze single molecules in complex fluids is to bind the analyte of interest with high efficiency and selectivity to a well-designed carrier material. This strategy enables two fundamental analytical improvements: On one side the direct measurement can be achieved and on the other side an increase in sensitivity due to the appearing pre-concentrating effects can be observed. A strategy for the analysis of low and high density lipoprotein (LDL and HDL) in human serum applying NIR spectroscopy and multivariate calibration techniques was described.[27] During method development it was useful to evaluate the feasibility of NIRS for classifying and identifying different analytes by establishing a qualitative principal component analysis (PCA) based cluster model. In case of LDL and HDL analysis, titanium oxide (TiO_2) nano-beads offer an efficient material for selective immobilization, including incubation and a washing step. For quantification, a principal component regression (PCR) model of standards in a range from 500–3000 ppm (clinical value is 1500 ppm) and a partial least squares regression (PLSR) model of HDL standards in a range from 100–1000 ppm (clinical value is 400 ppm) could be demonstrated as being highly efficient. Wavenumber region selection allowed spectral information to be gained between 4000 and 7240 cm^{-1}. For the analysis of real samples it is necessary to analyze HDL and LDL in chronological order by employing precipitation. It could be demonstrated that this NIRS method is a highly useful potential alternative or even supplementary clinical method for the fast determination of single molecules in complex biological fluids.

6 Conclusion

The described methods demonstrate the use of NIR spectroscopy coupled with multivariate data analysis for the characterization of nano-structured silicate materials, carbon nano-tubes and C60-fullerene-derivatised silica and pharmaceutical material. NIRS is used a non-destructive method with high precision, speed and ease of use, without laborious sample pre-treatment. The quality analysis can quickly identify the kind of material and control derivatization steps, the quantitative regression models can be used to determine physico-chemical properties simultaneously with high precision and non-invasively. The application of titanium oxide nano-beads enables the selective enrichment of single biological molecules and the efficient analysis by NIRS.

References

1 R. M. Vallant, Z. Szabo, S. Bachmann, R. Bakry, M. Najam-ul-Haq, M. Rainer, N. Heigl, C. Petter, C. W. Huck and G. K. Bonn, *Anal. Chem.*, 2007, **79**, 8144–53.

2 F. R. Kloss, M. Najam-Ul-Haq, M. Rainer, R. Gassner, G. Lepperdinger, C. W. Huck, G. Bonn, F. Klauser, X. Liu, N. Memmel, E. Bertel, J. A. Garrido, D. Steinmüller-Nethl and Doris, *J. Nanosci. Nanotechnology*, 2007, **7**, 4581–4587.
3 M. Najam-Ul-Haq, M. Rainer, N. Heigl, Z. Szabo, R. Vallant, C. W. Huck, H. Engelhardt, K.-D. Bischoff and G. K. Bonn, *Amino Acids*, 2008, **34**, 279–86.
4 G. Stecher, C. W. Huck, W. M. Stöggl and G. K. Bonn, *TrAC Trends Anal. Chem.*, 2003, **22**, 1–14.
5 C. W. Huck and G. K. Bonn, *J. Chromatogr. A*, 2000, **885**, 51–72.
6 R. M. Vallant, Z. Szabo, L. Trojer, M. Najam-ul-Haq, M. Rainer, C. W. Huck, R. Bakry and G. K. Bonn, *J. Proteome Res.*, 2007, **6**, 44–53.
7 L. K. H. Bittner, N. Heigl, C. H. Petter, M. F. Noisternig, U. J. Griesser, G. K. Bonn and C. W. Huck, *J. Pharm. Biomed. Anal.*, 2011, **137**, 1584–1595.
8 J. Märk, M. Karner, M. Andre, J. Rueland and C. W. Huck, *Anal. Chem.*, 2010, **82**, 4209–15.
9 W. F. Mc Clure, *J. near infrared Spectrosc.*, **11**, 487–518.
10 W. Herschel, *Philos. Trans. R. Soc. London*, 1800, **90**, 284–292.
11 M. Blanco and I. Villarroya, *TrAC Trends Anal. Chem.*, 2002, **21**, 240–250.
12 F. Barton, *Spectrosc. Eur.*, 2002, **14**, 12–18.
13 B. G. Osborne, T. Fearn, and P. H. Hindle, *Practical NIR spectroscopy with applications in food and beverage analysis.*, Longman Scientific and Technical, 1993.
14 P. Williams and K. Norris, *Near-infrared technology in the agricultural and food industries*, American Association of Cereal Chemists, Inc., 1987.
15 O. Faix, *Methods in Lignin Chemistry*, Springer Berlin Heidelberg, Berlin, Heidelberg, 1992.
16 B. M. Nicolaï, K. Beullens, E. Bobelyn, A. Peirs, W. Saeys, K. I. Theron and J. Lammertyn, *Postharvest Biol. Technol.*, 2007, **46**, 99–118.
17 R. Salzer and H. W. Siesler, *Infrared and Raman Spectroscopic Imaging*, Wiley-VCH, 2009.
18 J. D. Pallua, W. Recheis, R. Pöder, K. Pfaller, C. Pezzei, H. Hahn, V. Huck-Pezzei, L. K. Bittner, G. Schaefer, E. Steiner, G. Andre, S. Hutwimmer, S. Felber, A. K. Pallua, A. F. Pallua, G. K. Bonn and C. W. Huck, *Analyst*, 2012, **137**, 1584–1595.
19 K. J. Siebert, *J. Am. Soc. Brew. Chem.*, 2001, **59**, 147–156.
20 D. Cozzolino, *Planta Med.*, 2009, **75**, 746–56.
21 R. Bakry, R. M. Vallant, M. Najam-ul-Haq, M. Rainer, Z. Szabo, C. W. Huck and G. K. Bonn, *Int. J. Nanomedicine*, 2007, **2**, 639–49.
22 M. Najam-ul-Haq, M. Rainer, T. Schwarzenauer, C. W. Huck and G. K. Bonn, *Anal. Chim. Acta*, 2006, **561**, 32–39.
23 E. Osawa, *Kagaku*, 1970, **25**, 854–863.
24 H. Kroto, J. Heath, S. O'Brien, R. Curl and R. 318, 162–163 (1985) Smalley, *Nature*, 318, 162–163.
25 H. W. Kroto, A. W. Allaf and S. P. Balm, *Chem. Rev.*, 1991, **91**, 1213–1235.
26 C. W. Huck, C. Pezzei, V. A. Huck-Pezzei, J. D. Pallua, S. A. Schoenbichler, L. K. Bittner and G. K. Bonn, *Recent Pat. Nanotechnol.*, 2012, **6**, 135–41.
27 C. H. Petter, N. Heigl, R. Bakry, G. K. Bonn, A. Ritsch and C. W. Huck, *Curr. Med. Chem.*, 2009, **16**, 8.